"A kind of Rosetta Ston[e] ... Age of Discovery."
—Laurence Bergreen, author of *Marco Polo*
and *Over the Edge of the World*

"One of the most readable and satisfying books of the year." —*The Cleveland Plain Dealer*

"An intellectual detective story. Lester has penned a provocative, disarming testament to human ambition and ingenuity." —*The Boston Globe*

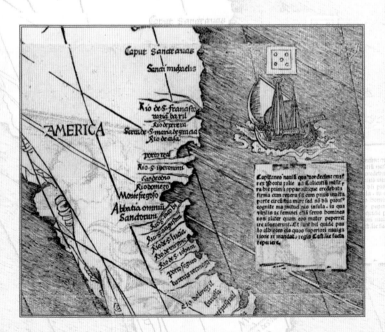

Praise for *The Fourth Part of the World*

"Marvelously imaginative, exhaustively researched.... Guiding the reader Virgil-like through the Age of Discovery, Lester introduces a chronologically and conceptually vast array of Great Men (Columbus, Vespucci, Polo, Copernicus, et al.), competing theories, monastic sages, forgotten poets, opportunistic merchants, unfortunate slaves, and more. That he relates it all so cleanly and cogently—via elegant prose, relaxed erudition, measured pacing, and purposeful architecture—is a feat. That he proffers plentiful visual delights, including detailed views of the legendary document, is a gift. This map, Lester writes, 'draws you in, reveals itself in stages, and doesn't let go.' Nor does this splendid volume."

—*The Atlantic*

"Lester pulls on the threads of Waldseemüller's map and finds an extraordinary braid of influences. [He] builds a cumulative tale of rich, diverse influences that he juggles with gathering speed and showmanship until the whir of detail coalesces into an inspired, imaginative piece of mapmaking."

—*San Francisco Chronicle*

"One of the most readable and satisfying books of the year. [A] gracefully concise, richly illustrated, wonderfully detailed compression of dozens of stories. An offbeat, hybrid, labor-of-love books that charms readers with its eclecticism and sheer love of knowledge. [It is] a history of mapmaking; of the reclamation of Greek and Roman geographical knowledge in the Renaissance; of Europe's conception of and exploration of the rest of the world; and finally, of the life, travels and literary career of one Amerigo Vespucci and of the group of landlocked German cosmographers who gave us his name."

—*The Cleveland Plain Dealer*

"Compelling...allows us to see how a group of European Renaissance scholars 'managed to arrive at a new understanding of the world as a whole.' Mr. Lester bravely ventures where few have gone before."

—*The New York Times*

"The story of Waldseemüller's map is impassioning: as a source of insight into the history of our knowledge of our world; as an object lesson in the gropings and failings of Renaissance humanism; as a detective story in which a vital document mysteriously disappears to be startlingly rediscovered; as an instance of the role of chance and error in making history; as a cautionary tale of the overlap of obscurity and influence, notoriety and fame; and as a case study of stunning historical *supercherie*. Lester's deftness in narrating a long and complex tale is impressive: fluent, clear, well informed, and perfectly paced. In short, he is an example of a phenomenon increasingly embarrassing to professional historians: a journalist who writes history better than we can. Lester makes a formidable contribution."

—Felipe Fernandez-Armesto, *The Wilson Quarterly*

"In [this] page-turner, Mr. Lester chronicles how a dreamy German youth yearning for the glories of ancient Greece and Rome assembled probably the most influential cartographic document ever drawn. The Waldseemüller map in 1507 gave the West its first view of its known world. Make this book compulsory reading in high schools."

—*The Washington Times*

"Europe's discovery of the rest of the world during the Renaissance is combined with a history of mapmaking in one of this year's most captivating and richly detailed histories."

—*Kansas City Star*

"Fascinating. Without Toby Lester's fine book, the Waldseemüller Map might remain an interesting historical footnote. Instead, one now understands the creation of the map as a world-changing moment, 'a birth certificate for the world that came into being in 1492—and a death warrant for the one that was there before.'"

—*Minneapolis Star-Tribune*

"Lester captures the passion, curiosity and, at times, the hubris behind the European explorations. His real interest lies in the evolution of Europeans' perception of the world, as reflected by their maps, an approach that works splendidly. To mid-millennial Europeans, there was nothing over the western sea but mystery and legends about islands, monsters and mythical beings. It took courage to sail off into that unknown, and Lester's book offers a clear survey of how people came to understand the world in which they lived."

—*The Washington Post*

"Maps—intricate, absurd, fantastical, ridiculous—fill this beautiful book, reinforcing Lester's thesis that they tell us as much about their makers as our surroundings. The heretofore unknown fourth part of the world was an enormous, unspoiled continent whose natural resources could be exploited and whose natives could be converted, sold into slavery, or exterminated. Like any train wreck, the controversies of this historical moment fascinate."

—*The Christian Science Monitor*

"With the excitement and exhilaration of an explorer, *Atlantic* contributor Lester sets off on his own journey of discovery across the seas of cartography and history. . . . Lester traces the map's journey to America over the next century in a majestic tribute to a historic work."

—(Starred) *Publishers Weekly*

"An engrossing adventure for both general and informed lay readers. . . . Highly recommended for anyone with an interest in cartography, the Age of Exploration, or European intellectual history."

—*Library Journal*

"A swift, sweeping primer on the Age of Discovery and the legacy of mapmaking. Lester begins with the amazing story of an obscure German cartographer, Martin Waldseemüller. . . . As the Age of Discovery progressed, with the likes of Columbus, Cabot, Vespucci, and the Portuguese navigators testing new margins, the race of the scientists and cartographers to keep up—separating self-promotion from fact—becomes a fascinating saga, ably captured in Lester's hands."

—*Kirkus Reviews*

"The complex artistry of the beautiful German map that first identified 'America' five centuries ago provides, for a truly imaginative writer, the opportunity to tell a wonderful and exciting story. Toby Lester, seizing this opportunity, has risen to the occasion brilliantly, creating a masterpiece of cartographic literature that will be of lasting importance."

—Simon Winchester, author of *The Map That Changed the World*

"The right technology at the right time can change the world. Toby Lester has written a page-turning story of the creation of what amounts to a sixteenth century Google Earth, a revolutionary way to see the world. It inspired generations of explorers then and will inspire readers now."

—Chris Anderson, author of *The Long Tail*, editor in chief of *WIRED*

"Sherlock Holmes once claimed he could deduce the Atlantic Ocean from a drop of water. Toby Lester performs a similar feat. He sets out to tell the story of a single ancient map, but into this yarn he sneaks the whole saga of planetary exploration. Intellectual ingenuity meets swashbuckling audacity, until at last a picture emerges of the earth as we know it today. *The Fourth Part of the World* reminds us that our maps aren't just about where we are—they're also a record of every place we've ever been."

—Cullen Murphy, editor at large, *Vanity Fair*

"A big-picture history done in the finest, most-engaging style. Toby Lester's easy control of cultural, technological, and diplomatic history allows him to connect themes in new and revealing ways, all of it driven with vivid narrative. This is a wonderfully entertaining and instructive book."

—James Fallows, *The Atlantic*, author of *Postcards from Tomorrow Square*

"Toby Lester's stupendously well-researched adventure story treats maps as cultural documents with stories to tell of the way the Old World cartographers visualized the New World: America. His book is a wonderful addition to the history of the imagination."

—Vincent Virga, author (with the Library of Congress) of *Cartographia: Mapping Civilizations*

"Brilliantly conceived and painstakingly researched, an original take on the European discovery of America."

—Robert D. Kaplan, author of *Balkan Ghosts* and *Eastward to Tartary*

"What distinguishes civilized people from barbarians? It's the map of the world they have in their minds. A barbarian's map marks the spot of just a few things: herds of sheep to steal, convenience stores to rob, political opponents to condemn on talk radio or the internet. A civilized person tries to see the world as a whole. Toby Lester's brilliant work explains how Western Europeans ceased to be a horde of pillaging bloggers and blowhards (intellectually speaking) and became upstanding citizens (intellectually speaking) of Western Civilization."

—P. J. O'Rourke

*f***P**

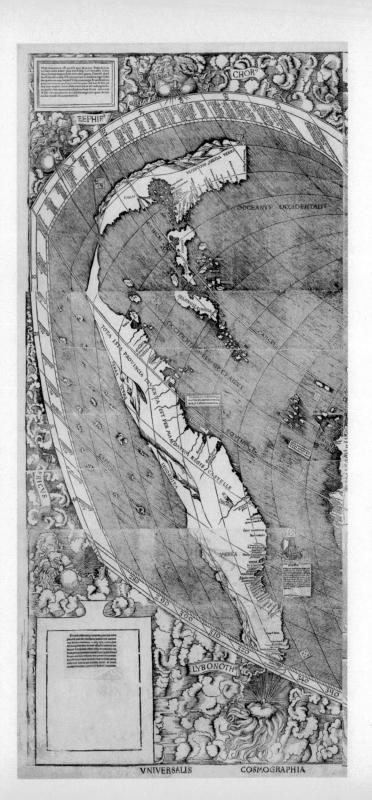

The
Fourth Part
of the
World

*An Astonishing Epic of Global
Discovery, Imperial Ambition,
and the Birth of America*

TOBY LESTER

Free Press
New York London Toronto Sydney

FREE PRESS
A Division of Simon & Schuster, Inc.
1230 Avenue of the Americas
New York, NY 10020

First Free Press trade paperback edition July 2010

FREE PRESS and colophon are trademarks of Simon & Schuster, Inc.
For information about special discounts for bulk purchases, please contact
Simon & Schuster Special Sales at 1-866-506-1949
or business@simonandschuster.com

The Simon & Schuster Speakers Bureau can bring authors to your live event. For
more information or to book an event contact the Simon & Schuster Speakers Bureau
at 1-866-248-3049 or visit our website at www.simonspeakers.com.

Book design by Ellen R. Sasahara

Manufactured in the United States of America

1   3   5   7   9   10   8   6   4   2

Library of Congress Cataloging-in-Publication Data
Lester, Toby.
The fourth part of the world : an astonishing epic of global discovery, imperial ambition, and the birth of America / Toby Lester.
p. cm.
Includes bibliographical references and index.
1. America—Maps—History. 2. America—Name. 3. Waldseemüller,
Martin, 1470-1521? 4. World maps—History. 5. Cartography—History.
6. Voyages and travels—History—To 1500. 7. Discoveries in
geography—History—To 1500. 8. Paris, Matthew, 1200–1259. 9. Polo,
Marco, 1254-1323? 10. Travelers' writings, European. I. Title.
E18.75.L47 2009
912.73—dc22   2009001230

ISBN 978-1-4165-3531-7
ISBN 978-1-4165-3534-8 (pbk)
ISBN 978-1-4391-6042-8 (ebook)

To the four parts of my little world:
Catherine, Emma, Kate, and Sage

Contents

Timeline

A select chronology of people, events, and maps discussed in this book

- Fourth century B.C.: **Aristotle** explains the makeup of the spherical cosmos in *On the Heavens*.

- First century B.C.: **Cicero** describes the Earth's five climate zones in *The Dream of Scipio*.

- First century B.C.: **Virgil** predicts in the *Aeneid* that Rome will rule a great southern land across the Ocean.

- First century A.D.: **Strabo**, **Pomponius Mela**, and **Pliny the Elder** write influential geographical works.

- Second century: **Claudius Ptolemy** compiles the *Geography*.

- Fourth century: **Solinus** catalogs the monstrous races in his *Gallery of Wonderful Things*.

- Fourth century: **Constantine the Great** makes Christianity the official religion of Rome.

- Fifth century: **St. Augustine** dismisses the possibility of human habitation on the other side of the globe.

- Fifth century: **Macrobius** includes a map of the world's climate zones in his *Commentary on the Dream of Scipio*.

- Sixth century: **Cassiodorus** refers his readers to Ptolemy's *Geography*, the last Latin reference to the text for more than 800 years.

- Seventh century: **Isidore of Seville** describes the known world in his *Etymologies* and includes a T-O map.

- Eighth century: **Arab scholars** begin translating and studying classical treatises of the ancient Greeks.

- Twelfth century: Latin translations of works by **Al-Battani** and **Ibn al-Saffar** include mentions of Ptolemy's *Geography*.

- Thirteenth century: **Sacrobosco** writes *The Sphere*, and **Matthew Paris** and **Roger Bacon** make their maps.

- Thirteenth century: **The Mongols** invade Europe, and **John of Plano Carpini** and **William of Rubruck** trek into Central Asia in search of the Great Khan.

- Mid- to late thirteenth century: Production of the **Psalter map** and other elaborate Christian *mappaemundi*.

- Late thirteenth century: **Marco Polo** spends years in the Far East and writes *The Description of the World*.

- *c.* 1275: Production of the **Carte Pisane**, the earliest surviving European marine chart.

- 1291: **The Vivaldi brothers** sail off into the Atlantic in search of a sea passage to India.

- *c.* 1300: **Maximos Planudes** rediscovers Ptolemy's *Geography* in Byzantium.

- 1321: **Petrus Vesconte** combines features of a *mappamundi* and marine charts on a new kind of hybrid world map.

- Early fourteenth century: **Sir John Mandeville** describes the full circuit of the earth.

- Early fourteenth century: **Lanzarotto Malocello** discovers two of the Canary Islands.

- 1368: China expels the **Mongols**.

- Mid-fourteenth century: **Petrarch** and **Boccaccio** pioneer the study of classical geography.

- 1375: The **Catalan Atlas** portrays the Far East as described by Marco Polo.

- 1397: **Manuel Chrysoloras** carries Ptolemy's *Geography* to Florence.

- *c.* 1406–1409: **Jacopo Angeli** completes the first Latin translation of the *Geography*.

- 1413: The **Viladestes chart** shows the extent of the West African coast known to Europeans.

- 1414–1418: **Poggio Bracciolini** and other humanists at the **Council of Constance** exchange rediscovered classical texts and discuss ancient and modern geographical ideas.

- 1415: **Prince Henry** and the Portuguese take Ceuta and begin to explore North Africa's Atlantic coast.

- *c.* 1415–1420: The Italian merchant **Niccolò Conti** begins twenty-five years of eastern travel.

- 1424: **Claudius Clavus** makes the first Ptolemaic map of lands not included in the *Geography* .

- 1434: **Gil Eanes**, a Portuguese squire, rounds Cape Bojador, on the West Africa coast.

- 1439–1443: **Italian humanists** meet with Greek, African, and Asian delegates at the **Council of Florence** and press them for geographical information.

- 1453: **Constantinople** falls to the Turks.

- 1453–1454: **Johannes Gutenberg**, inventor of the European printing press, produces the Gutenberg Bible.

- *c.* 1459: **Fra Mauro**, relying on Islamic sources, makes a *mappamundi* for the Portuguese that shows a sea route around southern Africa to the Far East.

- Mid- to late-fourteenth century: **The Portuguese** develop the **caravel** and the **mariner's astrolabe**, and pioneer new techniques of **celestial navigation**.

- 1474: **Paolo Toscanelli** proposes to the Portuguese that the quickest route to the Far East is to sail west across the Atlantic.

- 1477: First printed edition of **Ptolemy's** *Geography*

- 1476–1485: **Christopher Columbus** arrives in Lisbon and sails with the Portuguese up and down the Atlantic seaboard, to Iceland, Ireland, and West Africa.

- 1487: **Bartolomeu Dias** sails the length of Africa's west coast and reaches the Cape of Good Hope.

- *c.* 1489–1490: **Henricus Martellus** produces Ptolemaic world maps that include Portugal's recent African discoveries and the Far East as described by Marco Polo.

- 1492: **Martin Behaim** makes his famous globe, showing the world as Europeans knew it on the eve of Columbus's first voyage.

- 1492: Spain **defeats the Moors** on the Iberian Peninsula.

- 1492–1506: **Columbus**, sailing for Spain, makes four voyages of discovery across the Atlantic, explores the Caribbean and parts of South America, and dies believing he has reached the vicinity of Japan, China, and the Earthly Paradise.

- 1493: Unaware of Columbus's successful first voyage, **Hieronymous Muntzer** proposes to Portugal's king that Martin Behaim set out on a similar expedition across the Atlantic in search of the Far East.

- 1494: Spain and Portugal divide the Atlantic and the New World between themselves in the **Treaty of Tordesillas**.

- 1497–1500: **John Cabot**, the **Corte-Real brothers**, and **Pedro Álvarez Cabral** sail across the Atlantic and explore separate mainland portions of the New World.

- 1499–1502: **Amerigo Vespucci** makes at least two voyages to the New World and sends letters to Florence describing a coastline that extends thousands of miles into the southern hemisphere.

- 1500–1505: Production of the earliest surviving **New World marine charts**, among them the **La Cosa chart** (1500), the **Cantino chart** (1502), and the **Caverio chart** (1500–1505).

- 1503: Florentine printers publish Vespucci's *Mundus Novus* letter and (1504) his *Letter to Soderini*.

- 1505: **Matthias Ringmann** publishes the *Mundus Novus* letter in Strassburg, under the title *Concerning the Southern Shore*.

- *c.* 1505–1506: **Ringman** and **Martin Waldseemüller** form the **Gymnasium Vosagense** in Saint-Dié, Lorraine, and begin work on a modern edition of Ptolemy's *Geography*.

- 1507: **Waldseemüller** and **Ringman**, in possession of New World marine charts, and letters by Vespucci, coin the name America in Vespucci's honor and print a giant wall map, a tiny globe, and the *Introduction to Cosmography*.

- *c.* 1507–1514: **Nicholas Copernicus** describes America as depicted on the Waldseemüller map and uses it as geographical evidence to support his theory that the earth revolves around the sun.

- 1510: **Henricus Glareanus**, a young Swiss humanist studying at the University of Cologne, acquires the *Introduction to Cosmography* and draws the earliest known copy of the Waldseemüller map.

- 1538: **Gerardus Mercator**, the Flemish master cartographer, draws up a map on which he places the name America for the first time on the northern and southern portions of the New World.

- 1544: **Sebastian Münster** publishes the first edition of his *Cosmography*, one of the best-selling books of the sixteenth century. He borrows from Waldseemüller and Ringmann's *Introduction to Cosmography* and uses the name America on maps, helping to ensure the acceptance of the name.

Cast of Characters

Ailly, Pierre d' (*c.* 1351–1420), French cardinal and theologian, whose *Image of the World*, which summed up medieval geographical thought, greatly influenced Christopher Columbus, who filled it with annotations.

Bacon, Roger (1214–1294), English polymath and Franciscan friar, who urged Christians to study geography in new ways and who suggested, borrowing from ancient authors, that the ocean between Europe and Asia was "of no great extent."

Boccaccio, Giovanni (1313–1375), Florentine poet and scholar, who recorded the earliest known account of a European expedition to the Canary Islands and who compiled a practical, one-volume guide to the geography of antiquity.

Bracciolini, Poggio (1380–1459), leading Italian humanist, who rediscovered lost classical manuscripts, interviewed the merchant Niccolò Conti about his travels in the Far East, and helped write papal bulls governing the Portuguese exploration of Africa.

Brendan, Saint (*c.* 484—*c.* 577), Irish monk, whose legendary seven-year Atlantic wanderings in a tiny sailboat culminated in the discovery of a great new western island, and the story of whose voyage circulated widely in medieval Europe, influencing geographers and explorers alike.

Cabot, John (*c.* 1450—*c.* 1508), Venetian explorer, who, sailing for the English in 1497, reached the shores of North America and returned announcing that he had reached "the country of the Great Khan."

Cantino, Alberto (dates uncertain), secret agent, sent by the Duke of Ferrara to Portugal to uncover information about new Atlantic discoveries, and who in 1502 managed to smuggle back to Italy one of the earliest surviving sailor's charts of the emerging contours of the New World.

Cão, Diogo (*c.* 1450—*c.* 1487), Portuguese sea captain, who in the 1480s made voyages along almost the full length of Africa's west coast, which convinced King João that sailing around Africa was the best way to reach the Far East, not sailing west, as proposed to him by Columbus.

Celtis, Conrad (1459–1508), itinerant German humanist, who urged his countrymen to study the classics and classical geography, in order to challenge the Italians as heirs to the legacy of Greece and Rome, and whose message would influence the makers of the Waldseemüller map.

Chrysoloras, Manuel (*c.* 1355–1415), Byzantine scholar and diplomat invited to Florence in 1397 to teach Greek to the city's humanists—and who brought with him a manuscript of Ptolemy's *Geography*, the first to appear in Europe for almost 1000 years.

Columbus, Christopher (*c.* 1451–1506), Genoese sailor, whose epochal first voyage across the Atlantic in 1492, in the service of Spain, was followed by three others, each of which he believed took him to the Far East—and, on the third voyage, to the out-skirts of the Earthy Paradise.

Conti, Niccolò (1395–1469), Venetian merchant who in 1414 embarked on what would become a 25-year journey to Far East, and whose reports about his travels upon his return thrilled humanist geographers in Florence and confirmed the fantastic stories told by Marco Polo.

Copernicus, Nicholas (1473–1543), Polish astronomer, who in *On the Revolutions of the Heavenly Spheres* proposed the revolutionary idea that the earth revolves around the sun—an insight he supported by describing America as it appeared on the Waldseemüller map.

Covilhã, Pêro da (*c.* 1460—*c.* 1526), Arabic-speaking Portuguese squire, who in the late 1480s explored the Middle East and the west coast of India, and reported news back to Portugal that sailing from Europe around Africa and into the Indian Ocean was indeed possible.

Eanes, Gil (dates uncertain), sea captain in the service of Prince Henry of Portugal, who in 1434 sailed south along Africa's northwest coast and successfully rounded the dreaded Cape Bojador, where legend had it that "no ship having once passed . . . will ever be able to return."

Da Gama, Vasco (*c.* 1460/1469–1524), Portuguese sea captain, who in 1497 became the first European to sail right around Africa and across the Indian Ocean to India.

Dias, Bartolomeu (*c.* 1451–1500), Portuguese sea captain, who in 1489 became the first European known to have reached the Cape of Good Hope, at the southern tip of Africa.

Fillastre, Guillaume (1348–1428), French cardinal and humanist, one of the first outside of France to study Ptolemy's *Geography* and to try to synthesize the different geographical ideas circulating in the Middle Ages.

Fischer, Joseph (1858–1944), German Jesuit priest and professor of geography, who, in 1901, while visiting Wolfegg Castle, in southern Germany, accidentally rediscovered the sole surviving copy of the Waldseemüller map.

Henry, Prince of Portugal (1394–1460), Portuguese royal, also known as Henry the Navigator, who sponsored a series of voyages south along the coast of West Africa, seeking gold, slaves, Christians, and the legendary Prester John.

Humboldt, Alexander von (1769–1859), prolific German scholar and explorer and naturalist, who in 1839 announced that after years of study he had discovered "the name and the literary connections of the mysterious character who first proposed the name *America*."

John, Prester, legendary Christian priest-king from the East, who was believed for centuries in medieval Europe to be amassing wealth and armies in preparation for a campaign—and the search for whom fueled much of the early European exploration of Africa and Asia.

Julian, Friar (dates uncertain), Dominican missionary, who in 1237 traveled to the Ural Mountains, where he learned that Mongol forces had amassed to the east and intended to "devastate all the lands which they could subjugate."

Khan, Chingis (*c.* 1162–1227), also known as Genghis Khan, unifier of the nomadic tribes of Central Asia and founder of the Mongol empire, whose advance into eastern Europe in the early 1200s prompted distressed Western Christians to begin expanding their knowledge of the east.

Khan, Kubilai (1215–1294), ruler of the Mongol empire, based in China, where in the late decades of his life he employed the young Italian merchant Marco Polo as an adviser.

Lud, Walter (*c.* 1448—*c.* 1527), secretary to René II, Duke of Lorraine, who helped found and support the Gymnasium Vosagense, the small group of printers and scholars who in 1507 would publish the Waldseemüller map.

Mandeville, Sir John (dates uncertain), purported world traveler and possibly fictional figure who, in 1356 published *Mandeville's Travels*, an account of his journeys that would become hugely popular in Europe and would be studied carefully by Columbus and other explorers.

Medici, Lorenzo di Pierfrancesco de' (1463–1503), Florentine banker and politician, for whom the young Amerigo Vespucci worked as a commercial agent, and to whom later in life Vespucci sent letters describing his visits to the New World.

Paris, Matthew (*c.* 1200–1259), Benedictine monk based at St. Albans, in England, whose historical chronicle, the *Chronica majora*, records the arrival of the Mongols in Europe and contains an abundance of maps representative of medieval geographical thought.

Petrarch, Francesco (1304–1374), Italian poet and scholar, often called the "father of humanism," who pioneered the study of ancient geography and helped launch the quintessentially humanist effort to reconcile ancient and modern conceptions of the world.

Piccolomini, Aeneas Sylvius (1405–1464), Italian humanist and future pope, who in the 1440s wrote the *History of Matters Conducted Everywhere*, which would become one of Columbus's favorite books, because of its discussion of geography.

Plano di Carpini, Friar John (*c.* 1180–1252), Franciscan missionary, who, overweight and in his sixties, traveled overland through eastern Europe and Russia deep into Central Asia in search of the Mongols, whom he found, met, observed, and then returned home to describe to the pope.

Planudes, Maximos (*c.* 1260–1330), Byzantine scholar and poet, who during the final years of the 1200s obsessively hunted for and rediscovered the lost *Geography* of Ptolemy—and who, using the data provided in the text, became the first to reconstruct Ptolemy's map of the world.

Plethon, George Gemisthos (*c.* 1355–1452/4), Byzantine sage, who as an octogenarian discussed ancient and modern geographical ideas with Latin humanists at the

Council of Florence, and introduced them to the important work of the ancient geographer Strabo.

Polo, Marco (*c.* 1254–1324), Venetian merchant, who spent 24 years in the Far East and whose book describing his travels exerted a profound influence on European geographers and explorers, among them Christopher Columbus, who heavily marked up his copy.

Ptolemy, Claudius (*c.* 90—*c.* 168), Greek astronomer and geographer, who, in the *Geography*—a work lost for centuries but brought back to Europe in about 1400—laid out the principles of latitude and longitude and map projection, and used them to map the ancient world.

René II, Duke of Lorraine (1451–1508), French patron of the Gymnasium Vosagense, the group of scholars and printers who created the Waldseemüller map.

Ringmann, Matthias (1482–1511), German classics scholar and member of the Gymnasium Vosagense, who collaborated with Martin Waldseemüller to create the Waldseemüller map, and probably himself coined the name America.

Rubruck, Friar William of (*c.* 1220–1293), Flemish Franciscan missionary, who from 1253 to 1255, like John of Plano Carpini, made an improbable journey far out into Central Asia to find the Mongols—whom he, too, met, observed, and returned home to describe to the pope.

Salutati, Coluccio (1331–1406), chancellor of Florence, who helped launch the study of ancient Greek in Italy by inviting Manuel Chrysoloras to teach in Florence, thus indirectly helping bring Ptolemy's *Geography* back to Europe.

Schöner, Johannes (1477–1547), German mathematician and astronomer who between 1515 and 1517 acquired a copy of the Waldseemüller map and had it bound into the beech-wood folio in which Joseph Fischer would discover it almost 400 years later.

Toscanelli, Paolo dal Pozzo (1397–1482), Florentine physician, who in 1474 wrote a letter to the Portuguese suggesting that they sail west to the Indies rather than around Africa and to the east, and who at the end of his life may have corresponded with Columbus about this very idea.

Vespucci, Amerigo (1454–1512), Florentine merchant who in the late 1400s and early 1500 made voyages to the New World, and whose letters describing what he had found inspired Martin Waldseemüller and Matthias Ringmann to depict the New World as a new continent called America.

Martin Waldseemüller (*c.* 1470–1520), German humanist and cartographer who, as part of the Gymnasium Vosagense, designed, compiled, and oversaw the production of the Waldseemüller map of 1507—the map that gave America its name.

Preface

OLD MAPS LEAD you to strange and unexpected places, and none does so more ineluctably than the subject of this book: the giant, beguiling Waldseemüller world map of 1507. The map draws you in, reveals itself in stages, and doesn't let go. When I first imagined telling the story of the map, several years ago, I thought I knew where I was going. But then the map took over. For two years I followed its lead, and this unconventional book is the result.

I first heard about the map in 2003, when I came across a press release announcing that the Library of Congress had just bought the sole surviving copy of the map for the staggering sum of $10 million. This was the highest price ever paid publicly for a historical document, almost $2 million more than the previous record holder, an original copy of the Declaration of Independence. I was intrigued. What was this map, why was it considered so valuable, and how was it that I hadn't heard about it before?

The map has a memorable claim to fame: it gave America its name. Proudly describing the map as "America's birth certificate," the Library of Congress's press release laid out the basic story. The map was the creation of a small group of scholars and printers based in the mountains of Lorraine, in eastern France—among them a German cartographer named Martin Waldseemüller, whose name is now attached to the map. Improbably, Waldseemüller and his colleagues, almost entirely forgotten today, had decided, when they set about mapping the New World some fifteen years after its initial discovery by Columbus, that they would give it a new name. But what they called America had nothing to do with what's colloquially called America today: the United States. Waldseemüller's America was South America: a vast part of the New World that Waldseemüller and his colleagues believed had first been explored and described only a few years earlier, by the Italian merchant Amerigo Vespucci, in whose honor they coined their new name.

The map did more than just introduce the name America to the world. For the first time—breaking with the prevailing notion that Columbus and Vespucci had reached some unknown part of Asia—it confidently depicted the New World as surrounded by water. Theirs was the first map, in other words, to depict the western hemisphere roughly as we know it today. And, mysteriously, for reasons that have yet to be satisfactorily explained, it did so years before Europeans are supposed to have first learned of the existence of the Pacific Ocean, in 1513.

A thousand copies of the map were printed, and within a decade, aided by the spread of the newly invented printing press, the name America caught on around Europe. But the map itself soon went out of date, replaced and discarded in favor of maps that provided increasingly full pictures of the New World. By the end of the sixteenth century the map had faded entirely from view. It would stay out of sight for centuries, forgotten or considered lost, until 1901, when Father Joseph Fischer, a middle-aged Jesuit priest visiting a minor castle in southern Germany, stumbled across a copy of the map in an out-of-the-way corner of the castle's library. Fischer, an expert in the history of cartography, recognized what he had found and soon announced to the world that he had discovered the mother of all modern maps: the map that had introduced the New World to Europe and given America its name.

It was a captivating little story, and I decided that I would try to write a book about it. But the more I looked at the map itself, the more I saw. The view quickly became kaleidoscopic: a constantly shifting mosaic of geography and history, people and places, stories and ideas, truth and fiction. Before long I realized that the map offers a window on something far vaster, stranger, and more interesting than just the story of how America got its name. It provides a novel way of understanding how, over the course of several centuries, Europeans gradually shook off long-held ideas about the world, rapidly expanded their geographical and intellectual horizons, and eventually—in a collective enterprise that culminated in the making of the map—managed to arrive at a new understanding of the world as a whole.

This book tells the story of the Waldseemüller map in two distinct ways: as microhistory that focuses on the little-known and fascinating story of the making of the map itself, in the years leading up to 1507; and as a macrohistory that traces the convergence of ideas, discoveries, and social forces that together made the map possible—a series of overlapping voyages, some geographical and some intellectual, some famous and some forgotten, that made it possible to depict the world as we know it today.

The book is divided into four parts. The prologue, titled "Awakening," introduces the map, tells the story of the naming of America, and concludes with the dramatic rediscovery of the map in 1901. The first two parts of the book, titled "Old World" and "New World," introduce a multitudinous and diverse cast of characters, all of whom, broadly speaking, and in their own idiosyncratic ways, helped make the map possible. Amerigo Vespucci plays a prominent role in the story, as does Christopher Columbus, but so do legions of others: Europe's medieval monks and sages, who preserved ancient teachings about the nature of the world and the cosmos; the early missionaries and traders who journeyed east in search of the Mongols, the Great Khan, and a mythical Christian king known as Prester John; Marco Polo and other travelers and writers who introduced medieval Europe to the wonders of the Far East; the scholars and poets of the Renaissance, who began reviving ancient works of geography as a literary and patriotic exercise, only to find themselves suddenly able to see the world in fundamentally new ways; the Church officials who dreamed of Christendom as a global power and pursued the study of geography as a way of making it a reality; the mariners and merchants who sailed farther and farther away from home and returned with new charts of what they had found—and, of course, in the midst of it all, the dutiful geographers and mapmakers who, year after year, century after century, tried to make sense of it all.

What binds the book together is the Waldseemüller map itself. The story, like the map, is Eurocentric. Each chapter opens with a close-up detail of the map and then brings to life some of the many stories and ideas, familiar and unfamiliar, that can be found embedded there. Starting in the early 1200s, in England, at the very western edge of the known world, the chapters gradually sweep across the map, in a narrative progression that is both chronological and geographical. The book first moves east across the map, as Europeans begin to explore and change their ideas about Asia; then it travels south, as they do the same for Africa and the Indian Ocean; and finally it heads west, across the Atlantic, as they first set eyes on the New World and try to make sense of what they have found. Only after all of this groundwork has been laid does the Waldseemüller map itself come back into view, in the book's third and final section, "The Whole World," which describes the making of the map and then concludes with a story that's almost never told: the story of how, not long after its publication, in 1507, the map made its way to Poland and helped the young Nicholas Copernicus develop the startling idea that earth was not fixed at the center of the cosmos but instead revolved around the sun.

It's a fitting ending to the story. The Waldseemüller map appeared on the scene at a time of convulsive social and intellectual change, just as Europeans were famously rethinking their place in the cosmos. Serendipitously, implausibly, indelibly, the map captured a new worldview as it was coming into being—and that worldview, of course, is our own.

The earth is placed in the central region of the cosmos, standing fast in the center, equidistant from all other parts of the sky. . . . It is divided into three parts, one of which is called Asia, the second Europe, the third Africa. . . . Apart from these three parts of the world there exists a fourth part, beyond the ocean, which is unknown to us.

—Isidore of Seville, *Etymologies* (circa A.D. 600)

The

Fourth Part

of the

World

PROLOGUE

AWAKENING

But where is this Waldseemüller map?... Somewhere, in some dark corner of a monastic library, folded away in some oak-bound volume, a copy may be sleeping.

—John Boyd Thacher (1896)

IT WAS A curious little book. When copies began resurfacing, in the eighteenth and nineteenth centuries, nobody quite knew what to make of them. One-hundred-and-three pages long and written entirely in Latin, the book laid out its contents on its title page (*Figure 1*).

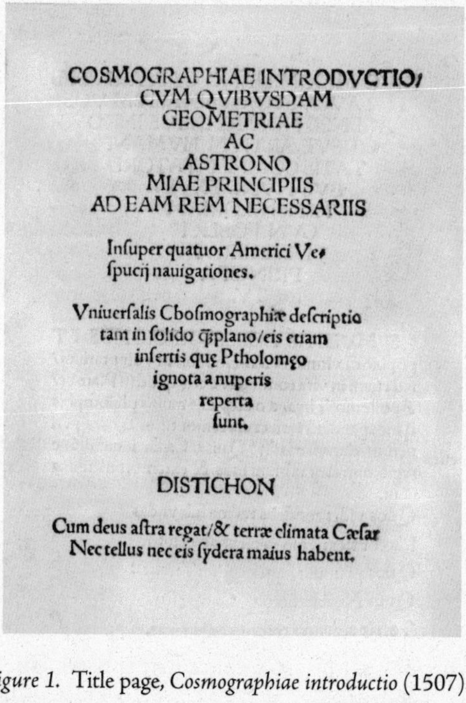

COSMOGRAPHIAE INTRODVCTIO/
CVM QVIBVSDAM
GEOMETRIAE
AC
ASTRONO
MIAE PRINCIPIIS
AD EAM REM NECESSARIIS

Inſuper quatuor Americi Ve-
ſpucij nauigationes.

Vniuerſalis Cboſmographiæ deſcriptio
tam in ſolido q̄plano/eis etiam
inſertis quę Ptholomęo
ignota a nuperis
reperta
ſunt.

DISTICHON

Cum deus aſtra regat/& terræ climata Cæſar
Nec tellus nec eis ſydera maius habent.

Figure 1. Title page, *Cosmographiae introductio* (1507).

INTRODUCTION TO COSMOGRAPHY
WITH CERTAIN PRINCIPLES OF GEOMETRY AND
ASTRONOMY NECESSARY FOR THIS MATTER

ADDITIONALLY, THE FOUR VOYAGES OF AMERIGO VESPUCCI

A DESCRIPTION OF THE WHOLE WORLD ON BOTH A GLOBE
AND A FLAT SURFACE WITH THE INSERTION OF
THOSE LANDS UNKNOWN TO PTOLEMY
DISCOVERED BY RECENT MEN

DISTICH
SINCE GOD RULES THE HEAVENS, AND CAESAR THE LANDS,
NEITHER THE EARTH NOR THE STARS HAVE GREATER THAN THESE.

The book—known today as the *Cosmographiae introductio*, or *Introduction to Cosmography*—listed no author. But it did provide a few scattered clues about its origins. A printer's mark at the back of the book recorded that it had been published "seven years after the Sesquimillennium," in "the Vosgian Mountains," in a town named after Saint Deodatus. This was easy enough to decipher: the date of publication was 1507, and the town was Saint-Dié, a little town in eastern France, located some sixty miles to the southwest of Strasbourg, in the Vosges Mountains of Lorraine. More clues appeared in the book's two opening dedications, both of which paid tribute to a certain Maximilian Caesar Augustus. Only one figure at this time in European history carried this title: Maximilian I, a German monarch known as King of the Romans, who would soon become Europe's Holy Roman Emperor. The authors of the two dedications were less easy to identify. Although they seemed to have played roles in the preparation and printing of the book, they identified themselves only as Philesius Vogesigena and Martinus Ilacomilus—pen names that had long ago been forgotten.

The word *cosmography* isn't used much today, but educated readers in 1507 knew what it meant: the study of the known world and its place in the cosmos. At the time, as had been the case for centuries, scholars believed the cosmos to consist of a set of giant concentric spheres. The moon, the sun, and the planets each had their own sphere, and beyond them was a single sphere containing all of the stars: the firmament. The earth sat motionless at the center of these spheres, each of which wheeled grandly around the globe day and night, in a never-ending procession. It was the job of the cosmographer—who was at once a geographer, an astronomer, a mathematician, and even a philosopher—to describe the visible makeup of the cosmos and to explain how its various parts fit together.

The *Introduction to Cosmography* devoted its opening thirty-five pages to this task. Methodically and authoritatively, the work laid out a traditional model of the cosmos. First came the abstract definitions: a *circle*, a *line*, an *angle*, a *solid*, a *sphere*, an *axis*, a *pole*, a *horizon*. Then came descriptions of the order of the cosmos, supported with practical diagrams and quotations from ancient authorities. Finally came the earth itself: its place at the center of the cosmos; its shape, which everybody knew was spherical; its various climate zones; its winds; and its many habitable and uninhabitable regions.

This was familiar material. Countless other textbooks and treatises had covered the very same ground. That was precisely the point: the world and the cosmos were what they always had been, and the *Introduction to Cosmography* taught its readers how to study them. But near the end of the book, in the chapter devoted to the makeup of the earth, the author did something extremely unusual: he elbowed his way onto the page to make an oddly personal announcement. It came just after he had introduced readers to Asia, Africa, and Europe—the three parts of the world that been known since antiquity. "These parts," he wrote,

> have in fact now been more widely explored, and a fourth part has been discovered by Amerigo Vespucci (as will be heard in what follows). Since both Asia and Africa received their names from women, I do not see why anyone should rightly prevent this [new part] from being called Amerigen—the land of Amerigo, as it were—or America, after its discoverer, Americus, a man of perceptive character.

It was a profoundly strange moment. With no fanfare, near the end of a minor Latin treatise on cosmography published in the mountains of eastern France, a nameless sixteenth-century author had stepped briefly out of obscurity to give America its name.

And then he had disappeared again.

*   *   *

ONLY THE FIRST THIRD of the *Introduction to Cosmography* actually provided an introduction to cosmography. The rest of the work consisted of a long letter written by Amerigo Vespucci and addressed to one René II, the Duke of Lorraine. In colorful and often lurid detail, Vespucci, a Florentine merchant, described four voyages that he had made—or at least claimed to have made—to the New World between 1497 and 1504. It was the search for this letter, in fact, that had brought copies of the *Introduction to Cosmography* back to light in the eighteenth century. At issue was a question that historians of the Age of Discovery had been rancorously debating for centuries. Who had reached the mainland of the New World first: Christopher Columbus or Amerigo Vespucci?

Nobody doubted that Columbus had sailed west across the Atlantic before Vespucci. As was well-known, Columbus had made four voyages of western discovery, starting in 1492—but during the first two voyages, which took place between 1492 and 1496, he had only visited the islands of the Caribbean, which he called the Indies, believing them to lie on the outskirts of Asia. Not until 1498, during his third voyage, did he finally set foot on continental soil, on the coast of present-day Venezuela. This was the moment, Columbus partisans felt, that represented the true European discovery of the New World. One of the first to put this view into writing, sometime before the middle of the sixteenth century, was the Spanish friar Bartolomé de Las Casas. "It is manifest," he thundered in the first book of his *General History of the Indies*, "that the Admiral Don Cristóbal Colón was the first by whom Divine Providence ordained that this our great continent should be discovered," adding, "No one can presume to usurp the credit, nor to give it to himself or to another, without wrong, injustice, and injury committed against the Admiral, and consequently without offence against God."

What provoked this outburst was the little *Introduction to Cosmography*. Las Casas had studied the book carefully and was deeply offended by its Vespucci letter, which suggested that Vespucci had visited the coast of South America in 1497—a year before Columbus. Columbus had been the first to reach the Indies, the book suggested, but Vespucci was the one who had discovered the New World. No evidence in the historical record corroborated this claim, however, and Las Casas decided it was a wrong he had to right. Writing with the *Introduction to Cosmography* explicitly in mind, he dismissed Vespucci as a fraud and set in motion a debate about the discovery of the New World that would rage internationally for centuries.

It is well here to consider the injury and injustice which that Americo Vespucio appears to have done to the Admiral, or that those who published his *Four Voyages* have done, in attributing the discovery of this continent to himself, without mentioning anyone but himself. Owing to this all the foreigners who write of these Indies in Latin, or in their own mother tongue, or who make charts or maps, call the continent *America*, as having been first discovered by Americo . . . Certainly these *Voyages* unjustly usurp from the Admiral the honor and privilege of having been the first who, by his labors, industry, and the sweat of his brow, gave to Spain and the world the knowledge of this continent.

The *Introduction to Cosmography* faded quickly from memory after Las Casas's death, in 1566. But the name America stuck—to the consternation of many Spaniards, who considered it an affront not only to Columbus but also to their national honor. Others carried Las Casas's anti-Vespucci campaign forward, and by the nineteenth century even Americans themselves had joined in. "Everybody knows the crafty wiles of these losel Florentines," Washington Irving wrote in 1809, "by which they filched away the laurels from the arms of the immortal Colon (vulgarly called Columbus) and bestowed them on their officious townsman Amerigo Vespucci." Ralph Waldo Emerson weighed in later in the century, famously reflecting on how strange it was that "broad America must wear the name of a thief . . . [who] managed in this lying world to supplant Columbus and baptize half the Earth with his own dishonest name."

The people of Florence naturally had a different take on the matter. The Florentine astronomer Stanislao Canovai summed it up in a public eulogy he delivered for Vespucci on October 15, 1788.

The universe, astonished at his deeds, regarded him as the confidant of the stars, as the father of cosmography, as the wonder of navigation, and, having by the unanimous suffrages of all nations abolished that primitive denomination the *New World*, willed that the continent should derive its name from Americus alone, and with sublime gratitude and justice secured that reward to him, and an eternity of fame. But will you believe it? [Some countries] still nourish hearts so ungrateful and minds so narrow that they have not only dishonored with satire the incomparable deeds of Vespucius but, expostulating loudly against the unanimous decree of the nations, have made it criminal in Americus that his name has thus been adopted, and have depicted him in the black colors of an ambitious usurper.

All things considered, it was an epic fuss—quite something for an obscure little treatise on cosmography to have stirred up.

* * *

THE FIRST WIDELY read author to focus popular attention on the *Introduction to Cosmography* was Washington Irving. While working on a biography of Christopher Columbus in the 1820s, Irving decided that he should try to settle the Columbus-Vespucci debate—in Columbus's favor. Combing through the scholarly literature on Vespucci in an attempt to confirm that Vespucci's 1497 voyage had never taken place, Irving discovered the work of an Italian abbot named Francesco Cancellieri, who some decades earlier had chanced across a copy of the *Introduction to Cosmography* in the Vatican Library and had summarized its contents—briefly but in enough detail to make it clear that Vespucci had had nothing to do with the naming of America. When he published his Columbus biography, in 1828, Irving therefore set the record straight about Vespucci. "His name," he wrote, "was given to that part of the continent by others." The deed, he explained in a footnote, had been done "in St. Diey, in Lorraine, in 1507." On the question of who those "others" might have been, however, he had nothing to say.

The person who finally worked out the answer was Baron Alexander von Humboldt, one of the great scholars and explorers of the nineteenth century. Intrigued by what Irving had written about the *Introduction to Cosmography*, Humboldt developed an obsession with the book, and in 1839, after much investigation, announced that he had unearthed its secret.

"This extremely rare book has taken up much of my time in recent years," he wrote in his monumental *Critical Examination of the History of the Geography of the New Continent*, adding, "I have had the pleasure of recently discovering the name and the literary connections of the mysterious character who first proposed the name *America* to designate the New Continent, and who hid himself under the Hellenized name *Hylacomylus*."

The story that Humboldt had pieced together went something like this. Hylacomylus—Martinus Ilacomylus, that is, the author of one of the *Introduction to Cosmography*'s dedications—was Martin Waldseemüller, a German cartographer from near Freiburg; Philesius Vogesigena, the author of the other dedication, was Matthias Ringmann, a young poet and classicist from a town in the Vosges Mountains. The two men were friends, and in the very early years of the sixteenth century they had made their way to Saint-Dié. There they became part of a tiny group of scholars and printers who, with the patronage of Duke René, established a scholarly press and began calling it the Gymnasium Vosagense. After the duke received his famous Vespucci letter and shared it with the members of the Gymnasium, they decided it was time to put Vespucci's discoveries into theoretical—that is, a cosmographical—context. And so it was, in 1507, that America got its name, and the *Introduction to Cosmography* came into being.

Humboldt's reconstruction was a tour de force of literary-historical detective work: meticulously researched, closely argued, and elegantly written. His announcement inspired others to seek out the *Introduction to Cosmography*—but now instead of flipping directly to the naming-of-America passage and the Vespucci letter they began to read the whole work, and those who studied it carefully realized that Humboldt had sidestepped one important aspect of the book's story. The American politician and statesman Charles Sumner acquired a partial copy of the book in the middle of the nineteenth century, and on its front flyleaf he penned a thought that was occurring to more and more people. "Whoever he was," Sumner wrote, "the author of this cosmography has done more than give America the name it bears at present. For it may be seen in different passages of his work that already at the commencement of 1507 he had prepared maps of the world confirming all that was known of the New World."

The maps! All that fuss about Vespucci and the naming of America had made readers overlook the last item that had been announced on the *Introduction to Cosmography*'s title page: the map and the globe. Students of the book now recognized that these maps were the main reason the book had been written. The book's author himself made this point unambiguously— but only in an easy-to-miss paragraph printed on the back of a foldout diagram. "The purpose of this little book," he declared, "is to write a sort of introduction to the whole world that we have depicted on a globe and on a flat surface." The rest of the paragraph went on to describe the map in considerable detail.

The globe, certainly, I have limited in size. But the map is larger. As farmers usually mark off and divide their farms by boundary lines, so it has been our endeavor to mark the chief countries of the world by the emblems of their rulers. And (to begin with our own continent) in the middle of Europe we have placed the eagles of the Holy Roman Empire (which rule the kings of Europe), and with the key (which is the symbol of the Holy Father) we have enclosed almost the whole of Europe, which acknowledges the Roman Church. The great part of Africa and a part of Asia we have distinguished by crescents, which are the emblems of the supreme Sultan of Babylonia, the lord of all Egypt, and a part of Asia. The part of Asia called Asia Minor we have surrounded with a saffron-colored cross joined to a branding iron, which is the symbol of the Sultan of the Turks, who rules Scythia this side of the Imaus, the highest of the mountains of Asia and Samratian Scythia. Asiatic Scythia we have marked by anchors, which are the emblems of the great Tartar Khan. A red cross symbolizes Prester John (who rules both eastern and southern India, and who resides in Biberith); and finally, on the fourth part of the world, discovered by the kings of Castile and Portugal, we have placed the emblems of those Sovereigns. What is not to be ignored is that we have marked with crosses shallow places in the sea where shipwreck may be feared. With this we end.

Other information about the map appeared in Waldseemüller's dedication to Maximilian I. "To the best of my ability and with the help of others," Waldseemüller wrote, "I have studied the books of Ptolemy, from a Greek copy, and, having added the information from the four voyages of Amerigo Vespucci, I have prepared a map of the whole world. . . . This work I have determined to dedicate to your most sacred Majesty, since you are the lord of the world." At the end of the work itself, the author also mentioned making this map in "sheets," which suggested a map of considerable size, assembled from a number of separately printed pages. The author noted that in preparing his map he consulted not only the work of Ptolemy, an ancient Greek geographer, but also charts made by modern sailors—and immediately after the naming-of-America paragraph he dropped a bombshell. This new part of the world, he wrote, "is found to be surrounded on all sides by the ocean."

This was an astonishing statement to make in 1507. According to the history books, it was only in 1513—after Vasco Núñez de Balboa had first caught sight of the Pacific by looking west from a mountain peak in Panama—that Europeans began to conceive of the New World as a separate continent. Columbus had died in 1506 believing he had reached the vicinity of Japan and China, and Vespucci himself had explicitly referred to the region he had explored as an "endless Asian land." Only after 1520—once Magellan had rounded the bottom of the New World and sailed into the Pacific—were Europeans supposed to have confirmed the continental nature of the New World.

So why did the *Introducion to Cosmography* announce the New World to be surrounded by water? Was this just an inspired guess, made in the same poetic spirit of invention that had summoned the name America into being? Or was it an indication that Waldseemüller and his colleagues had had access to information about a forgotten voyage of discovery that had reached the west coast of South America before 1507—a voyage that perhaps had even involved Amerigo Vespucci himself? Nobody could say. All that was certain was that this map presented a revolutionary new geographical vision. Prepared with care by a team of scholars, printed on several sheets, and showing the world's political divisions, it synthesized the learning of the ancients and the discoveries of the moderns, and brought them together to create a new picture of the world—one that expanded it dramatically to include a new continent bearing the name America.

Unfortunately, however, the *Introduction to Cosmography* contained no such map.

* * *

THE MAP TOOK on an increasingly legendary aspect as the nineteenth century progressed. Scholars searched for it in libraries and map collections, and published extensive new studies concerning the activities of Waldseemüller, Ringmann, and the Gymnasium Vosagense. As the four hundredth anniversary of Columbus's first voyage approached, the idea of finding the map became something of a quest for the cartographical Holy Grail. "Ever since Humboldt first called attention to the 'Cosmographiae Introductio,'" the *Geographical Journal* declared at the turn of the century, "no lost maps have ever been sought for so diligently as these of Waldseemüller. It is not too much to say that the honour of being their lucky discoverer has long been considered as the highest possible prize to be obtained amongst students in the field of ancient cartography." Yet no copy of the map turned up—and as the final decade of the century approached, some specialists began to dismiss the search for the map as a waste of time. "Much of what is written about the gymnasium of St. Dié and Waldseemüller," wrote the towering cartographical authority Baron Adolf Erik Nordenskjold in 1889, "has been a useless exhibition of learning."

Less than a decade later, in 1897, the baron had to eat his words. That year the *Geographical Journal* announced the discovery in Germany of several small maps drawn by hand in 1510—documents "of special interest," the *Journal* reported, "as helping us in some measure to reconstruct the lost map of Waldseemüller." The maps were the work of a Swiss humanist named Henricus Glareanus. On one of his maps Glareanus had noted that in drawing it he had followed "the Deodatian or, preferably, Vosgean geographer"—a clear reference to Waldseemüller. On another of his maps—one of two discovered in Glareanus's own copy of the *Introduction to Cosmography*—Glareanus had provided even more information. The original Waldseemüller map was far too big to fit into the *Introduction to Cosmography*, he wrote, and so, in order to make his copy of the book complete, he had decided to make a reduced copy. "I have depicted it proportionally on a small scale," he wrote: "the three parts of the world and the recently discovered fourth American land." Oddly, the two maps found in Glareanus's copy of the *Introduction to Cosmography* seemed to be copies of two different maps; one showed the western hemisphere and the other showed

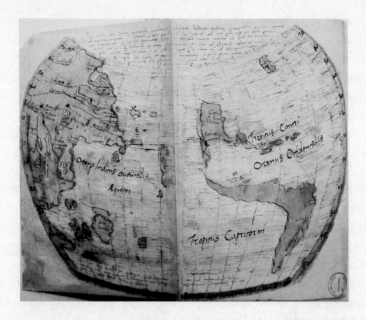

Figures 2 and 3. *Top:* map of the western hemisphere, Henricus Glareanus (1510). North and South America are on the right; Japan is in the middle; China and the Far East on the left. *Bottom:* map of the world, Henricus Glareanus (1510).

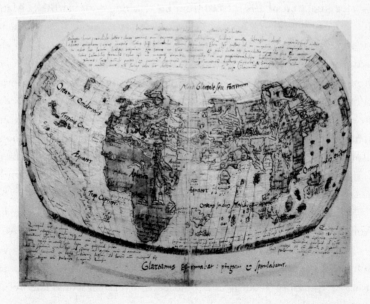

the world as a whole (*Figures 2 and 3*). This was hard to explain, given that the two maps were supposedly copies of the same original, but despite their differences they were clearly related. Both placed the words "American land" (*Terra America*) on the southern portion of the New World that Amerigo Vespucci claimed to have visited—and both depicted it as a new continent, completely surrounded by water.

It was a thrilling discovery. Scholars could now reconstruct with some certainty what the lost map had actually looked like. Waldseemüller, it seemed, had created the ur-document of modern world cartography: the first map to present a vision of the world's continents and oceans that corresponds roughly to the one we know today.

The search for the map continued in the 1890s, but it bore no fruit. Writing in 1896, in *The Continent of America: Its Discovery and Its Baptism*, the historian of discovery John Boyd Thacher admirably summed up everything known about the Waldseemüller story, but when it came time for him to address the question of the missing map, he simply threw up his hands. "The mystery of the map," he wrote, "is a mystery still."

* * *

IN THE SUMMER of 1901, freed from his teaching duties at Stella Matutina, a Jesuit boarding school in Feldkirch, Austria, Father Joseph Fischer packed his things and set out for Germany. Balding, bespectacled, and forty-four years old, Fischer was a professor of history and geography. For the previous seven years he had been haunting the public and private libraries of Europe in his spare time, hoping to find maps that showed evidence of the early Atlantic voyages of the Norsemen. This current trip was no exception. Earlier in the year, in correspondence with Father Hermann Hafner, a Jesuit in charge of the impressive collection of books and maps at Wolfegg Castle, in southern Germany, Fischer had received word of a rare fifteenth-century world map that depicted Greenland in an unusual way.

Fischer only had to travel some fifty miles from Stella Matutina to reach Wolfegg, a tiny town situated in the rolling countryside just north of Austria and Switzerland, not far from Lake Constance. He reached the town on July 15, and upon his arrival at the castle he was met by its owner, Prince Franz zu Waldburg-Wolfegg und Waldsee, who, Fischer would later recall, immediately offered him "a most friendly welcome and all the assistance that could be desired."

The map of Greenland turned out to be everything Fischer had hoped it would be. As was his custom on research trips, after studying the map Fischer began a systematic search of the prince's entire collection, in the hopes of finding other items of interest. First he made his way through the large inventory of maps and prints, which, to Fischer's delight, did indeed contain a number of treasures. He then spent hours immersed in the castle's library, inspecting many of its rare books. And then, on July 17, his third day at the castle, he walked over to the castle's south tower. On the second floor of the tower, he had been told, he would find a small garret containing what little he hadn't yet seen of the castle's collection.

The garret is a simple room. It's designed for storage, not show. Bookshelves line three of its walls, from floor to ceiling, and two windows let in a cheery amount of sunlight. The floors, uncarpeted, are lined with broad, unfinished planks. Wandering about the room and peering at the spines of the books on the shelves, Fischer soon came across something that piqued his curiosity: a large folio with red beech-wood covers, bound together with finely tooled pigskin. Two Gothic brass clasps held the folio shut, and Fischer pried them open gently. On the inside cover he found a small bookplate, bearing the date 1515 and a small inscription, which gave him the name of the folio's original owner: Johannes Schöner, a well-known sixteenth-century German mathematician and geographer. "Posterity," the inscription began, "Schöner gives this to you as an offering."

Fischer started leafing carefully through the folio and discovered, to his amazement, that it contained not only a rare 1515 star chart engraved by Albrecht Dürer, one of the most famous and sought-after of all Renaissance artists, but also two giant and unfamiliar world maps. Fischer had never seen anything quite like them. In pristine condition, printed from intricately carved woodblocks, each map was made up of twelve separate sheets that if removed from the folio and assembled would create maps approximately four and a half by eight feet in size.

Fischer began examining the first map in the folio. Its title, running in block letters across the bottom of the map, read, THE WHOLE WORLD ACCORDING TO THE TRADITION OF PTOLEMY AND THE VOYAGES OF AMERICUS VESPUCIUS AND OTHERS. This was language that immediately brought to mind the *Introduction to Cosmography*—as did what Fischer saw at the top of the map. On the folio's second page, on one of the map's top four sheets, Fischer found a finely executed portrait of Ptolemy, shown gazing east out over a small inset hemisphere that contained Europe, Africa, and Asia. The meaning of the image was clear: this was the Old World, as it had been known since ancient times and as Ptolemy himself had mapped it.

Another portrait, designed to join and complement the first, appeared on the following page. This one showed Amerigo Vespucci, gazing west out over the other half of the world, which contained what we know today as North and South America, the Pacific, and the Far East. This was the New World, as it had been explored and mapped in recent centuries—and it bore a startlingly close resemblance to Glareanus's map of the Western Hemisphere (*Figure 2*).

Fischer's pulse quickened. Could he be looking at . . . *the* map? He hardly dared think so, but the little hemisphere's rare depiction of the New World, along with the rare pairing of Ptolemy and Vespucci at the top of the map, was irresistibly suggestive.

Fischer began to study the map sheet by sheet. He quickly determined that the map's two center sheets—which showed Europe, northern Africa, the Middle East, and western Asia—derived straight from the work of Ptolemy, whose ancient maps had been revived and reproduced widely during the fifteenth century. Farther to the east, beyond the limits of the world as Ptolemy had known it, the map borrowed a depiction of the Far East that Fischer had seen on other maps of the period: a highly conjectural vision of the region based largely on the writings of Marco Polo. Fischer immediately noticed, however, that the map's depiction of southern Africa derived from a third source: sailors' charts that showed the sea route around the continent and into the Indian Ocean that the Portuguese had pioneered at the end of the fifteenth century.

It was an unusual mix of styles and sources—and precisely the sort of synthesis, that the *Introduction to Cosmography* had promised. There was more. Strewn across the map's pages was the very set of political emblems mentioned in the *Introduction to Cosmography*: the imperial eagle, the papal key, the branding iron, the anchor, and the various kinds of crosses. But

Fischer got his biggest thrill when he turned to the map's three western sheets. What he saw there, rising out of the water and stretching some three feet from the top of the map to its bottom, was a giant depiction of the New World surrounded by water, one that bore an uncanny resemblance to the New World as it appeared on the *other* map found in Glareanus's *Introduction to Cosmography* (*Figure 3*).

It was a breathtaking sight. A legend at the bottom of the page told readers what they were looking at—and the language corresponded verbatim to a paragraph that appeared in the *Introduction to Cosmography.* "A general description," the legend read, "of the various lands and islands, including some of which the ancients make no mention, discovered lately between 1497 and 1504 in four voyages over the seas, two by Fernando of Castile, and two by Manuel of Portugal, most serene monarchs, with Amerigo Vespucci as one of the navigators and officers of the fleet; and especially a delineation of many places hitherto unknown."

North America appeared on the top sheet, a runt version of its modern self, and at its western edge it bore the words DISTANT UNKNOWN LAND. Just to the south lay a number of Caribbean islands, among them two big ones identified as Spagnolla and Isabella, accompanied by a small legend that read, "These islands were discovered by Columbus, an admiral of Genoa, at the command of the King of Spain." What dominated this side of this map, however, filling two entire sheets, was a vast southern landmass, long and thin, that stretched from above the equator right down to the very bottom of the map. The words DISTANT UNKNOWN LAND appeared on the northwest coast of this landmass, too, and just below the equator another legend read THIS WHOLE REGION WAS DISCOVERED BY THE ORDER OF THE KING OF CASTILE. But what stunned Fischer was what he saw on the bottom sheet. Printed on what we know today as Brazil, and placed alongside the stretch of coastline described by Amerigo Vespucci in his letters, was a single word: AMERICA (*Plate 11*).

Flush with excitement, and now allowing himself to believe that he had indeed found the missing map, Fischer scoured it for a date or an author. But he found nothing anywhere—until he turned to the second map in the folio.

Dated 1516, this map called itself a *carta marina,* Latin for *marine chart.*

The map was gorgeous, illustrated and annotated in almost bewildering detail, but except for its size and twelve-sheet format it bore little resemblance to the first map. It made no nod to Ptolemy, it didn't show the New World as surrounded by water, and it used only the words NEW LAND to identify South America, which itself was depicted far more tentatively than on the first map. But at the bottom of the map Fischer found a note that must have made him sit up straight. "Arranged and completed by Martin Waldseemüller, Ilacomilus," it read, "in the town of Saint-Dié." Even better, he discovered the longest legend on the map to be a rambling address to the reader written by Waldseemüller himself. ("A greeting from Waldseemüller!" Fischer would later recall thinking.) And in that address Waldseemüller made a brief allusion to an earlier map of the world that he had designed—one that sounded a lot like the anonymous map Fischer had just been looking at. "In the past," Waldseemüller wrote, "we published an image of the whole world in 1,000 copies, which was completed in a few years, not without hard work, and based on the tradition of Ptolemy, whose works are known to few because of his excessive antiquity."

Fischer no longer had any doubts. The prominent juxtaposition of Ptolemy and Vespucci, the use of the word *America*, the echoes of the *Introduction to Cosmography*, the resemblance to the Glareanus maps, the New World shown surrounded by water, the reliance on sailors' charts, the statements of Waldseemüller himself—the evidence was overwhelming. Alone in the little garret in the tower of Wolfegg Castle, Father Fischer realized that he had discovered Waldseemüller's long-lost 1507 map—the most sought-after map of all time.

* * *

FISCHER TOOK THE news of his discovery straight to his mentor and former professor, the renowned historian of cartography Franz Ritter von Wieser—who, as it happened, was the one who had discovered the Glareanus maps in the *Introduction to Cosmography*. In the fall of 1901, after a period of intense study, Fischer and von Wieser went public with news of Fischer's discovery, and the reception was ecstatic. "Geographical students in all parts of the world have awaited with the deepest interest details of this most important discovery," the *Geographical Journal* declared, breaking the news in a February 1902 essay, "but no one was probably prepared for the gigantic cartographical monster which Prof. Fischer has now awakened from so many centuries of peaceful slumber." The *New York Times* followed suit on March 2. Under the headline LONG SOUGHT MAP DISCOVERED: EARLIEST KNOWN RECORD OF THE WORD AMERICA FINALLY BROUGHT TO LIGHT, the paper reported, "There has lately been made in Europe one of the most remarkable discoveries in the history of cartography, and one which must hereafter be referred to in all works pretending to give in detailed form the story of the New World."

More acclaim came the following year, when Fischer and von Wieser published *The Oldest Map with the Name America, and the Carta Marina of the Year 1516*. The book was a critical study of the Waldseemüller map of 1507 and the Carta Marina, but it also announced a new Waldseemüller discovery. In researching the book, Fischer and von Wieser wrote, they had managed to identify a tiny little printed world map—discovered in the late nineteenth century and initially believed to date from well after 1507—as an original copy of the globe mentioned on the title page of the *Introduction to Cosmography*. The map consisted of a set of interconnected oval strips, known as gores, that were designed to be cut out and pasted onto a ball—and it, too, identified the New World as America and showed it as surrounded by water. Here was yet another Waldseemüller first: history's first mass-produced globe (*Figure 4*).

The Oldest Map brought together the big Waldseemüller map, little globe, and the *Introduction to Cosmography* for the first time in centuries. It was a large and luxurious book designed to attract attention—and it got it. President Theodore Roosevelt acquired a copy not long after the book was published and was so pleased with it that he sent Fischer a handwritten note congratulating him on his discoveries. The book also found its way to Pope Pius X. During a private meeting with Fischer at the Vatican, the pope confessed that he been gratified to learn that the New World had been first named and put on the map by Waldseemüller, a Catholic cleric who had lived and worked "before the so-called Reformation."

Interest in the map continued to build, climaxing in 1907, when the London-based bookseller Henry Newton Stevens Jr., a leading dealer in Americana, secured the rights to put the 1507 map and the Carta Marina up

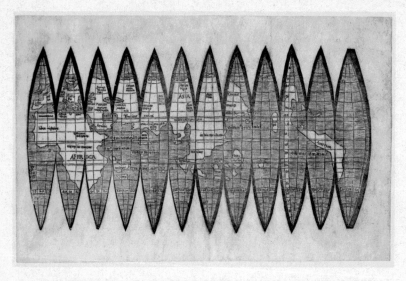

Figure 4. Globe gores, Martin Waldseemüller (1507). Africa and Europe are on the left; Asia is in the middle; North and South America are on the right.

for sale for $300,000—about $7 million in today's currency. Stevens knew that his asking price put the maps beyond the reach of private collectors, so in his auction catalog, after calling attention to the fact that 1907 was the four hundredth anniversary of the naming of America, he aimed his sales pitch elsewhere.

> What more suitable or patriotic commemoration of that auspicious event could possibly be conceived than the securing of this unique contemporary memento for some representative American Library, where it could be seen and studied by geographical students, admired by all lovers of art, and reverenced for all time by historians and antiquarians as the veritable fountain head from whence, in conjunction with the book *Cosmographiae Introductio,* America received its baptismal appellation!

Stevens talked a good game, but he found no takers. His asking price was too high. The four hundredth anniversary passed, two world wars and the Cold War engulfed Europe for much of the twentieth century, and the Waldseemüller map, left alone in its tower garret, went to sleep once more.

But today, at last, the map is awake again—this time for good. On April 30, 2007, almost exactly five hundred years after its making, the map was officially transferred to the United States by Germany's chancellor, Angela Merkel. In December of the same year the map was put on permanent public display at the Library of Congress—encased in the most sophisticated and expensive picture frame ever made—as the centerpiece of a permanent exhibit titled "Exploring the Early Americas."

It's a small exhibit. As you move through it you pass a variety of priceless cultural artifacts made in the pre-Columbian Americas, along with a choice selection of original texts and maps dating from the period of first contact between the New World and the Old. Finally you arrive at an inner sanctum. There, reunited with the Carta Marina, the Schöner folio, and the *Introduction to Cosmography*, is the Waldseemüller map itself. The room is quiet, the lighting dim. To see the map you have to move close and peer at it carefully—and when you do, it begins to tell its stories (*Figure 5*).

UNIVERSALIS   COSMOGRAPHIA   SECVNDVM PTHOLOMAEI TRA   DITIONEM

Figure 5. The Waldseemüller world map of 1507.

PART ONE

OLD WORLD

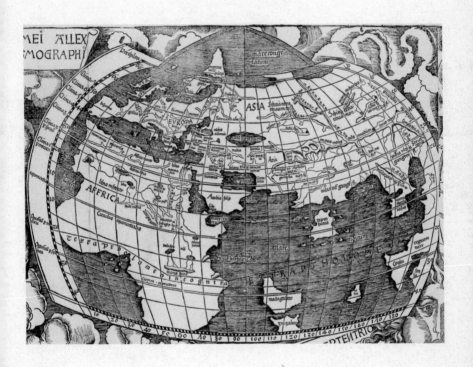

+ British Isles and northern Europe +

CHAPTER ONE

MATTHEW'S MAPS

It is the vocation of a monk to seek not the earthly but the heavenly Jerusalem, and he will do this not by setting out on his feet but by progressing with his feelings.

—Saint Bernard of Clairvaux (circa 1150)

IN THE EARLY 1200s the Benedictine monastery of St. Albans hummed with activity. Situated just a day's ride north of London, the monastery was one of the largest and most important in England, home to as many as two hundred monks. In the parlance of the times they were Latins: members of the greater community of Roman Catholics in Europe who submitted to the authority of the pope. But St. Albans wasn't just a religious retreat. It was a busy center of economic, political, and intellectual life, and even had played an important role as a meeting place in the Magna Carta crisis. It also ran a popular guesthouse—the first stopping point on one of the major routes north from London—and operated stables that could accommodate some three hundred horses at a time. By the beginning of the thirteenth century, St. Albans was feeding and lodging a steady stream of visitors on their way to and from London: Oxford professors, royal councilors, powerful bishops, papal emissaries and monks from elsewhere in Europe, a traveling delegation of Armenians, and even the king of England himself. It was a worldly place.

25

After a day of traveling, guests would unwind in the monastery's dormitories and refectory. Inevitably the talk would turn to where they had come from, and to what news and information they had picked up along the way. Again and again the same subjects came up: the weather, local crimes and misdemeanors, politics, the antics of the royals, the utterances of the pope, and the ill-conceived and apparently interminable series of wars being waged in the Middle East—the Crusades. One monk in particular had a special interest in stories from beyond the monastery's walls. A down-to-earth, willfully opinionated, and generally likable crank, he was Brother Matthew Paris: the greatest and most colorful of all medieval church chroniclers.

Born in about 1200, Matthew joined the Benedictine order at St. Albans in 1217, became its official chronicler in 1237, and died in 1259. The work for which Matthew is most famous is the *Chronica majora*, or *Great Chronicle*, a vast history of the world that, in typical medieval fashion, extends from the time of the creation right up to Matthew's own time. The first half or so of the *Chronicle* amounts to little more than Matthew's copying and fiddling with the chronicle of his predecessor, but from 1235 forward the entries are his own—and in one commonly consulted English translation they fill three five-hundred-page volumes. Yet despite its size the *Chronicle* is a wonderfully good read.

Matthew wrote and wrote and wrote. Keeping him properly supplied with writing materials alone was a tall order. In the thirteenth century the production of a book—that is, a manuscript scratched out with goose quill and ink, on page after page of parchment—amounted to a significant investment of a monastery's capital. A single book might well consume the skins from a whole flock of sheep. But Matthew's output justified this investment; it brought St. Albans great renown, even during Matthew's own lifetime.

Matthew was more than just a writer. He was also a gifted artist who illustrated his work with everything from tiny doodlings to lavishly executed portraits. Biblical figures, ancient emperors, popes, European kings, saints, monks, martyrs, battles, shipwrecks, eclipses, exotic animals—they all come to life on Matthew's pages, and not just as frivolous additions to his text. They were an integral part of his chronicle. "I desire and wish," he wrote, "that what the ear hears the eyes may see."

That brief reference to hearing, rather than reading, serves as a useful reminder: in thirteenth-century England reading was primarily an oral act, not a silent one. Monks in monastery libraries read aloud to themselves, and the din they created would have exasperated modern library patrons. Matthew read to himself, to his fellow monks, and to special guests visiting the monastery, and what he offered his readers and listeners was a captivating mix of words and pictures. "Turning the pages of Matthew's *Chronica majora*," one modern historian has written, "is like opening the door of a great abbey cupboard, from which spills forth a rich succession of disparate images and objects, each conjuring up its own compelling story from the past, so that each event again becomes visually 'present.'"

The great abbey cupboard. That's a critical image to keep in mind when trying to make sense of the jumble of disparate ideas and images that one encounters in the works of Matthew and other medieval writers—especially in their maps.

* * *

MATTHEW HAD A passion for maps. He drew them throughout his adult life, following a number of traditional models, and those that survive provide a remarkably useful survey of the different ways in which educated medieval Europeans imagined and depicted the world.

One of the main sources from which Matthew received his geographical ideas was the hugely popular and influential *Etymologies*, by Saint Isidore of Seville: a vast compendium of ancient and medieval learning, written in the seventh century A.D. Throughout the Middle Ages and even into the Renaissance, Europeans considered Isidore one of their most trusted authorities. He began the geographical section of his *Etymologies* by situating his readers cosmically. "The earth," he wrote, "is placed in the central region of the cosmos, standing fast in the center, equidistant from all other parts of the sky." This age-old conception of the world—as a sphere that sat motionless at the center of the universe, with the moon, the sun, the planets, and the stars all revolving around it—was one that medieval authors often diagrammed in their works, and Matthew was no exception (*Figure 6*).

Medieval Europeans, even the most learned of geographers among them, are to this day often described as having believed that the world was flat.

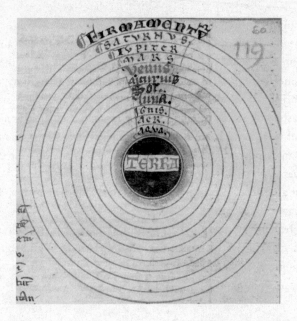

Figure 6. The medieval cosmos, by Matthew Paris (circa 1255).
The earth (*terra*) is fixed at its center, surrounded by the spheres of water,
air, fire, the moon, the sun, the planets, and the firmament.

But this simply isn't true. Thanks in large part to the labors of Arab astronomers and mathematicians, ancient Greek proofs of the earth as spherical had survived into the Middle Ages and were circulating in Europe—and at some point early in the thirteenth century an English scholar known as John of Holywood, or Sacrobosco, laid them out in an astronomical treatise appropriately titled *The Sphere*. For centuries afterward the work would be taught and studied in schools and universities around Europe. "If the earth were flat from east to west," Sacrobosco wrote, "the stars would rise as soon for Westerners as for Orientals, which is false. Also, if the earth were flat from north to south, and vice versa, the stars that were always visible to anyone would continue to be so wherever he went, which is false. But it seems flat to human sight because it is so extensive." Sailors certainly knew the world was round: a lookout at the top of a ship's mast, Sacrobosco pointed out, always catches sight of land before a lookout standing at the foot of the mast—"and there is no other explanation of this thing," Sacrobosco wrote, "than the bulge of the water." Copies of *The Sphere* almost invariably included a small drawing illustrating this concept (*Figure 7*).

Another source that would have helped determine Matthew's geographical outlook, and one that he had access to in the library at St. Albans, was

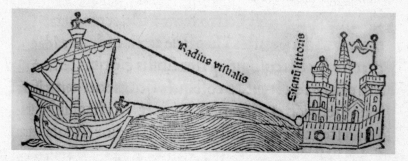

Figure 7. Proof that the earth is round: a sailor atop the mast of a ship catches sight of land before a sailor on deck. From a fifteenth-century edition of Sacrobosco's *The Sphere* (early 1200s).

the *Commentary on the Dream of Scipio*, by the fifth-century Roman writer Macrobius. The work—which for a full millennium after it was written would be used widely as a textbook in Europe—parsed *The Dream of Scipio*, a phantasmagoric musing on the world and its place in the cosmos, written some five centuries earlier by the Roman political philosopher Cicero. From a vantage point high up in the heavens, a character in Cicero's work had described the earth just as Isidore would later do, fixed at the center of the universe, but had also drifted in for a closer look. "You will observe," he declared, imagining the world as it would look from space,

> that the surface of the earth is girdled and encompassed by a number of different zones; and that the two which are most widely separated from each other and lie beneath opposite poles of the heavens are rigid with icy cold, while the central, broadest zone is burnt up with the heat of the sun. Two others, situated between the hot zones and the cold, are habitable. The zone which lies toward the south has no connection with yours at all. . . . As to its northern counterpart, where you yourselves live . . . the territory is nothing more than a small island, narrow from north

to south, somewhat less narrow from east to west, and surrounded by the sea that is known on earth as the Atlantic, or the Great Sea, or the Ocean.

Macrobius elaborated at considerable length on this division of the world in his *Commentary*, which contained a simple diagram. Today known as a zonal map, it showed the world as a circle, divided up into the five zones described by Cicero: two frigid zones, in the north and south; two temperate (or habitable) zones, closer to the center; and a torrid zone that wrapped around the earth's equatorial regions. Zonal maps were drawn and studied often during the Middle Ages, and Matthew, of course, produced his own version (*Figure 8*).

As Matthew's zonal map clearly shows, the northern temperate zone—which, following the practice of Arab geographers, Matthew placed at the

Figure 8. Zonal map, showing the earth's frigid, habitable, and torrid zones. South is at the top. Matthew Paris (circa 1255).

bottom of his map—contains the whole of the world as the ancients had described it and as medieval Europeans still knew it. Isidore of Seville succinctly described its makeup in his *Etymologies*. "It is divided into three parts," he wrote, "one of which is called Asia, the second Europe, the third Africa." To accompany this description, Isidore, or one of his early copyists, drew a rudimentary diagram, and for centuries after his death permutations of this diagram, known today as a T-O map, would adorn European encyclopedias, chronicles, religious texts, and travelogues (*Figure 9*). Matthew knew T-O maps well and drew a number of them in his work (*Figure 10*).

The standard T-O map places the world within a circular frame (hence the O). That frame represents the ocean as Cicero had described it: an all-encompassing body of water that washed every shore of the known world. Asia, the biggest continent, occupies the top half of the circle; Europe and Africa share the bottom. Dividing the three continents are two lines that

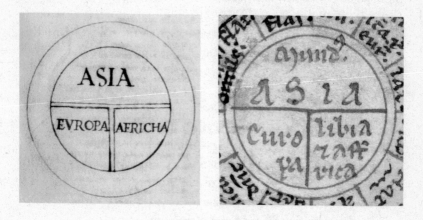

Figures 9 and 10. Left: A classic T-O map. *Right:* A T-O map drawn by Matthew Paris (circa 1255).

meet at a right angle in the middle of the map (hence the *T*). These represent three bodies of water: the Mediterranean, separating Europe from Africa; the River Nile, believed to separate Africa from Asia; and the River Don, in Russia, separating Europe from Asia.

As a whole, the T-O scheme concisely and effectively represents the world as medieval Europeans knew it. But at a certain level the scheme is disconcerting to the modern eye, because it puts east at the top of the map. Today north would be the natural choice, but that's an arbitrary convention, and in Matthew's time it had yet to come into being. East, in fact, had primacy of place in medieval Europe—which is why so many modern European languages still use a form of the word *orient* to describe getting one's bearings.

As the direction from which the sun rose, East represented the origin of things. (*Oriens*, the Latin root of *orient*, means "rising.") The Old Testament is built on this foundation. God planted the Earthly Paradise and its four great rivers "eastward in Eden," the Book of Genesis explains, and those waters nourished the world by flowing from East to West. "The glory of the God of Israel," reads the Book of Ezekiel, "came from the way of the east." The New Testament develops this theme: a star in the east announces the birth of Christ, who later dies on the Cross facing west. (*Occidens*, the Latin root of *occident*, means "falling" or "dying," and can refer to the setting sun.) The symbolic meanings of east and west in early Christian theology were set out clearly by the fourth-century church father Lactantius. God, he wrote,

> established two parts of the earth itself, opposite to one another and of a different character—namely, the east and the west; and of these the east is assigned to God, because He Himself is the fountain of light, and the enlightener of all things, and because He Himself makes us rise to eternal life. But the west is ascribed to that disturbed and depraved mind, because it conceals the light, because it always brings on darkness, and because it makes men die and perish in their sins.

* * *

THE EARLIEST SURVIVING T-O maps, which appear in eighth-century copies of Isidore's *Etymologies*, aren't invested with any such Christian symbolism. They're simple diagrams that seem based on a model that dates back at least to imperial Roman times (although no examples survive). They were a way of signifying the extent of not just the known world but also the world that the Romans aspired to rule: a world that the first Roman emperor, Augustus, described as "a global empire to which all peoples, monarchs, nations . . . consent."

After Christianity became the official religion of Rome, in the fourth century, the idea of Christendom came into being: a global Christian empire with Rome as its capital. In the centuries that followed, Christian mapmakers in Europe invested their simple T-O maps with increasingly complex layers of symbolism. Europe's Christian rulers, for example, began to be depicted sitting on thrones and holding T-O globes in their hands: an imperial pose that would fast become a Christian archetype, used for centuries to represent not only political rulers but also Christ himself. Matthew drew many different rulers, both ancient and modern, in this pose (*Figure 11*).

The very shape of the T-O map lent itself naturally to a specifically Christian kind of symbolism: the O called to mind a Biblical description of God sitting "enthroned above the circle of the earth," and the T called to

Figure 11. The Holy Roman Emperor Frederick II holding a T-O globe. Matthew Paris (circa 1255).

mind the *tau*, a mystical Greek letter that, according to Isidore, symbolized the cross. Inevitably, given these overtones, Christian mapmakers began to see the T in their T-O maps as representing not just the bodies of water that divided the three continents but also the cross that Christ had died on to bring them together (*Plate 2*). The symbolism was irresistible. What better way to convey the idea of the Trinity—a self-contained whole divided into three—than with a T-O map? "Most appropriate is this division of the earth into three parts," wrote the ninth-century archbishop of Mainz, whose work Matthew also had access to at St. Albans. "For it has been endowed with faith in the Holy Trinity and instructed by the Gospels."

The topography of the world also began to take on a distinctly Christian aspect on medieval world maps. "Paradise is a place lying in the eastern parts," Isidore told his readers. By Matthew's time most European geographical authorities gave the Earthly Paradise an actual location in the world; if it was earthly, after all, it had to be *somewhere*. Generally they placed it at the easternmost limits of the East, at the very edge of their maps, the part of the world where the sun rose and where time had begun. Jerusalem, for its part, gradually became not only the spiritual but geometrical center of the world, as dicated by the Bible. "This is Jerusalem," God declares in the Book of Ezekiel, "which I have set in the center of the nations, with countries all around her." The city even became identified with the Greek word *omphalos*, or "navel"—the geographical point on the earth where the gods made contact with the world to provide humankind with spiritual nourishment. This idea entered mainstream Christian thought in the fifth century, when an influential Latin translation of the Bible known as the Vulgate described the center of the world, in the Book of Ezekiel, as the *umbilicus terrae*: the navel of the earth.

Medieval geographers also begin to fill out their T-O maps with topographical and historical information, and by Matthew's time an elaborate and highly stylized kind of world map had come into being. These maps are often described collectively as *mappaemundi* (Latin for "maps of the world")—and, needless to say, one survives in Matthew's hand (*Figure 12*).

Following the T-O model, Matthew placed east at the top of his map. The west coast of continental Europe appears at the bottom left of the map, and above it, as one looks from west to east, are Germany ("Alemania"),

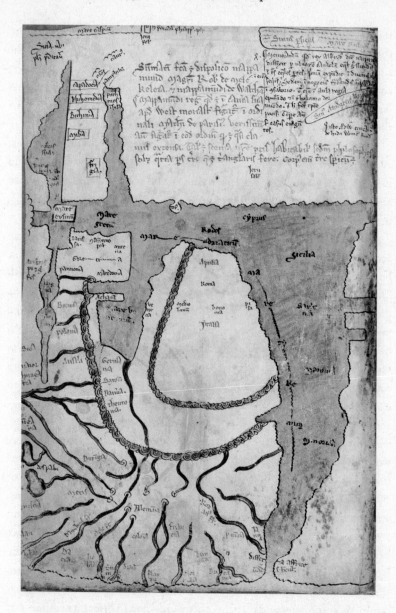

Figure 12. World map by Matthew Paris (circa 1255). East is at the top.
The darker portions of the map represent water. Asia is covered with text;
Europe is dominated by a stubby version of the Italian peninsula; and
a truncated North Africa appears at the right margin.

the Alps, and then a thick, stubby Italy. To the east and south (up and to the right) of Italy are Cyprus and Sicily, in the Mediterranean. Across the Mediterranean to the south is Africa, and to the east is Asia—on the west coast of which, occupying a lonely perch at the edge of the unknown East, is Jerusalem. Occupying the whole top portion of the map is Asia, which Matthew considered enough of a blank to fill in with notes to himself.

In terms of geographical accuracy, Matthew's map is a dud. It doesn't get even the basic contours of Europe right; it places legendary, biblical, and modern places side by side; and it doesn't try to come to terms with Africa and Asia. But geographical accuracy wasn't Matthew's goal in drawing this map, at least not as we now understand the term.

Remember that Matthew drew so that "what the ear hears the eyes may see." His maps were no exception. Like many religious writers of his time, he fully recognized the power of the image as a complement to the Word. "Know," John of Genoa wrote in a popular thirteenth-century religious dictionary known as the *Catholicon*, "that there were three reasons for the institution of images in churches. *First,* for the instruction of simple people, because they are instructed by them as if by books. *Second,* so that the mystery of the incarnation and the examples of the Saints may be the more active in our memory through being presented daily to our eyes. *Third,* to excite feelings of devotion, these being aroused more effectively by things seen than by things heard."

This description suits the function of the *mappaemundi* well. They, too, were designed for spiritual instruction and contemplation. They were devotional objects, guides to the divine cosmic order of things. Their picture of the world served as a backdrop onto which the various historical, religious, and symbolic coordinates of human history could be plotted: Adam and Eve in the Garden of Eden, Noah and the Ark on Mount Ararat, the Jews being dragged off to Babylon, Alexander the Great venturing into India, the Romans conquering Europe, Christ returning to Jerusalem, the Apostles spreading the Gospel to the most distant reaches of the earth, and much more. An "accurate" world map, in other words, had to orient its viewers not only in space but also in time.

Today the word *mappamundi* is used exclusively to refer to maps. But Matthew and his contemporaries used it more broadly. It could mean not only a visual depiction of the world but also a written description of it—a text very much like Matthew's *Chronicle*, that is. It traced the march of human history—scripted in advance by God at the beginning of time— from its origins in the East to its present state in the West. A map was a visual history; a history was a textual map; and each, like that great abbey cupboard, was stuffed full of disparate signs and symbols that, contemplated together, allowed believers to imagine seeing their places in history and the world as God could see them.

* * *

ONE OF THE notes that Matthew wrote on Asia on his world map records that the map was actually just "a reduced copy" of a *mappamundi* that he had seen elsewhere. He didn't say what map this was, but without a doubt it resembled some of the gorgeous thirteenth-century Christian *mappae-mundi* that do survive. Among them is the so-called Psalter Map, drawn in England in 1265 (*Plate 1*).

The Psalter Map has it all. It merges the image of an imperial Christ with many of the typical elements of medieval cartography: Christ hovers above the Earth holding a T-O globe in his hand, and below him is the three-part world in full, itself laid out according to the T-O scheme. At the top of the map, at the eastern limits of Asia, is the Earthly Paradise, just below the rising sun, with the faces of Adam and Eve clearly visible inside its walls. Europe appears at the bottom-left of the map, at the world's western edge—and at the center of everything is the holy city of Jerusalem, a bull's-eye target for Christians to set their sights on. Another *mappamundi* of the period, known as the Lambeth Palace map, depicts that very same vision of the world but takes its symbolism one step further (*Figure 13*). Here Christ not only surveys and embraces the world but also literally embodies it. His head appears at the top of the map, at the beginning of the East, alongside the Earthly Paradise. His hands appear at the north and south, representing an embrace of the whole world from north to south. Almost at the center of the map, where Christ's navel would be, is Jerusalem: the *omphalos*, the navel of the world. Finally, at the bottom of the map, in the west, are his feet, extending out into the ocean in the west and representing the end of both history and geography.

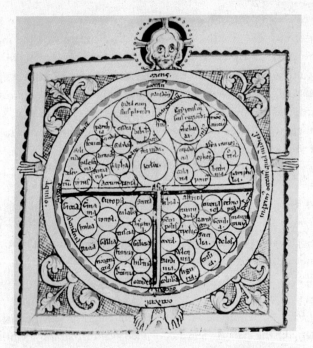

Figure 13. The Lambeth Palace Map (circa 1300). East is at the top.
The half circle directly under Christ is the Earthly Paradise; the large circle
in the middle is Jerusalem. Each other circle represents a region or a city.

The Psalter Map contains other features typical of the elaborate *map-paemundi* produced in Matthew's time. At its southern edge, for example, at the limits of Africa, are the monstrous races. Part human and part beast, the monstrous races had been famously cataloged in the first century by the Roman authority Pliny the Elder, in a vast and comprehensive work called the *Natural History*. This catalog had been embellished and popularized some two centuries later by another Roman writer, Julius Solinus, in his *Gallery of Wonderful Things*. Both works circulated widely in the Middle Ages and greatly influenced the ways in which Europeans imagined the distant parts of the world and their inhabitants.

Medieval artists often drew the monstrous races, and their fanciful depictions are enthralling to contemplate (*Figure 14*). As described by Pliny, Solinus, and others, the monstrous races included the Amazons (warrior women who cut off their right breast in order better to pull back the arrows in their bows); the Anthropophagi (cannibals); the Astomi (mouthless beings who derived most of their sustenance from sniffing apples); the Blemmyae (beings with no head or neck, whose faces were in their chests), the Cyclopes (the one-eyed race made famous by Homer); the Cynocephali (dog-headed beings who communicated by barking); the Ethiopians (whose faces had been burned black by the heat of the sun); the Panoti (whose ears were so large their owners had to hold them in their hands, and could use them to fly); and the Sciopods (a one-legged race with feet large enough to be used as umbrellas).

Reading about the monstrous races, and seeing them in illustrations and on maps, medieval readers were able to enjoy a form of lowbrow entertainment not unlike the one provided by modern tabloid accounts of vari-

Figure 14. Some of the monstrous races, believed by medieval Europeans to inhabit the world's distant places.

ous human monstrosities. But for Christian thinkers they posed a serious theological concern. What was their nature? Should they be considered human? This wasn't an idle question. If the monstrous were human, then they would have to be sought out and converted before the End of Days could come to pass. "This gospel of the kingdom," the Gospel of Matthew proclaimed, "shall be preached in all the world for a witness unto all nations; and then shall the end come."

The Psalter Map also contains another typical feature of late-medieval *mappaemundi.* At the northeast limits of Asia, behind a great wall, is the realm of Gog and Magog: a race of diabolical warriors, sometimes identified with one of the original tribes of the Jews. According to the Book of Revelation, the forces of Gog and Magog would at some point be united by the Antichrist and led on an apocalyptic rampage against the world's Christians. Popular medieval legend had it that during his great eastern campaign Alexander the Great had encountered Gog and Magog and trapped them behind a distant mountain range, to the north of the Caspian Sea—a story recounted, for example, in the widely read medieval *Romance of Alexander.* But everybody knew that Alexander's efforts had only forestalled the inevitable. Sooner or later, Gog and Magog, or the "enclosed Jews," as they were sometimes known, would burst out of captivity and begin to wreak havoc on the world.

The idea of the Apocalypse haunted Matthew, who believed, as many of his contemporaries did, that the arrival of the Antichrist and the end of the world were imminent. The *mappaemundi* told him this was so, in words and in pictures. From its starting point in the East—at the top of the map and at the beginning of time—human history had marched steadily westward: through the barbaric wilds of Asia, through the ever more civilized regions of Greece and Rome, through the Holy Land, and finally into Europe. All signs now suggested to Matthew that this long march was approaching its end—as the twelfth-century chronicler Bishop Otto of Freising had declared not long before Matthew was born.

> What great learning there was in Egypt and among the Chaldeans, from whom Abraham derived his knowledge! But what now is Babylon, once famous for its science and its power? . . . And Egypt is now in large part a trackless waste, whence science was transferred to the Greeks, then to the Romans, and finally to the Gauls and Spaniards. And let it be observed that because all human learning began in the Orient and will end in the Occident, the mutability and disappearance of all things is demonstrated.

At St. Albans, perched precariously at the very western edge of the world, in England, Matthew felt history and geography bearing down on him. The world would end in 1250, he decided, and he might well live to see the arrival of Gog and Magog.

* * *

As it was for most Latin Christians, Jerusalem was the focal point of Matthew's geographical imagination. But during the years that Matthew was writing his *Chronicle* and making his maps, Jerusalem and most of the Holy Land were in Muslim hands, despite more than a century's worth of Crusading efforts to win them back.

The First Crusade had been a success. Launched in 1095, it was the Latin West's response to a call for help sent by the Christians of the Byzantine Empire, who found themselves under threat from the Muslim armies of the Seljuk Turks. Earlier in the century the Turks had conquered Armenia and much of the Holy Land, including Jerusalem, and now seemed poised to move on Constantinople, Byzantium's capital. For a time the First Crusade—which culminated in the capture of Jerusalem, in 1099—helped stem the tide, and the mood among the Latins was triumphant. They established a series of outposts across the Holy Land and named this new Crusader state *Outremer* (a French name meaning "beyond the [Mediterranean] sea"). Christians had long been making pilgrimages to Jerusalem, but now soldiers and pilgrims and travelers began to flock there in unprecedented numbers. The twelfth-century British chronicler William of Malmesbury memorably captured the scene. "The Welshman left his hunting," he wrote, "the Scot his fellowship with lice, the Dane his drinking party, and Norwegian his warship. Lands were deserted of their husbandmen, houses of their inhabitants, even whole cities migrated."

These were heady times. But the Crusader presence in the Holy Land provoked Muslims throughout the Middle East to set aside many of their regional differences and fight back. Gradually, as the twelfth century progressed, Muslim armies won back many of the Holy Land's strategically important cities, and in 1187 the armies of Egypt, led by Saladin, retook Jerusalem.

For the Latins, it was a humiliating defeat. In the centuries that followed, dreaming of once again taking possession of Jerusalem, they would send Crusade after Crusade to the Holy Land. But they would never regain control of the city.

Recognizing that Christian pilgrims were Jerusalem's greatest source of income, Saladin and many of his successors allowed them to visit the city freely. As a result, by Matthew's time several standard pilgrimage routes from Europe to Jerusalem had come into being, and mapmakers had begun to include them on their maps, singling out important stopping points along the way. Matthew himself drew a set of such maps and included them in his *Chronicle (Figure 15)*. They're a valuable record of how medieval pilgrims made their way from Europe to the Holy Land, but they're also frustrating documents, because of how much they leave out. Matthew drew his maps as a series of destination cities for pilgrims, often separated by the French word *journée*—meaning "day," as in a day's travel, a usage that gave rise to the English word *journey*—but they reveal virtually nothing about what sort of places travelers would encounter in between those cities. They make no effort to depict Europe as a whole in a realistic way.

Why not? Because, once again, Matthew didn't have geographical accuracy in mind. He drew his itinerary maps primarily as contemplative guides for his fellow monks—ranks of the faithful often barred by their orders from making actual voyages to the Holy Land. "Going to Jerusalem," one monastic authority declared in the twelfth century, "is indicated for laymen but interdicted for monks."

This seems an odd proscription. Monks were devoted Christians, so why prevent them from making a pilgrimage to Jerusalem? There was no shortage of reasons. A monastery itself was considered a religious endpoint—a garden of spiritual delights, lovingly tended by those who had renounced the physical world. "This is an image of Paradise," one monk wrote in the twelfth century about the nature and function of a monastery: "it makes one think already of heaven." Setting off on a long and arduous journey in search of what one already had at home just didn't make sense.

Another problem was travel, which posed significant practical dangers. The world outside the cloister was a hazardous and ugly place. "Lord," one monk from Canterbury wrote in 1188 after making a passage across the Alps to Rome, "restore me to my brethren, that I may tell them that they come not to this place of torment." Worse yet, traveling the route to Jerusalem would force monks to run through an extended gauntlet of earthly vices—drink, desire, doubt—that could tempt them away from their chosen path. They'd already left all of that behind on their way to the monastery, so why suffer through it again? Even Jerusalem itself was best avoided. It was a city, Saint Anselm wrote, "not of peace but of tribulation."

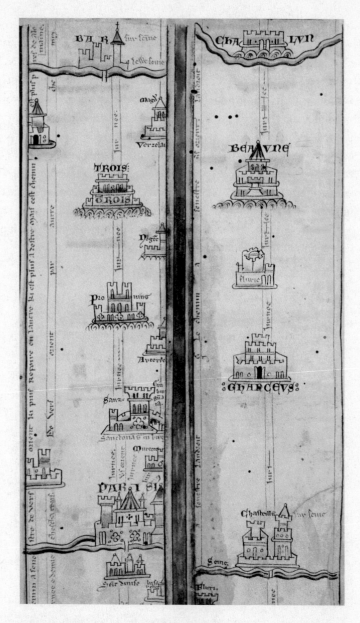

Figure 15. Pilgrimage itinerary through France. Paris is at the bottom left. Each castle represents a city, and each city is separated from the next by a day's travel—a *"journée,"* the root of *journey.* Matthew Paris (circa 1255).

Matthew drew his itinerary maps for mental travelers, not actual ones. They provided monks with a stop-by-stop guide for a meditative voyage— from an earthly reality in Europe to a spiritual summit in Jerusalem. "Who will give us wings like the dove," an anonymous Benedictine monk wrote in the twelfth century, "and we shall fly across all the kingdoms of the world, and we shall penetrate the depths of the eastern sky? Who will then conduct us to the city of the great King?"

Matthew's maps represented an effort to answer this call. They portrayed the world as God had created it, as God alone could see it when he surveyed it from above or held it in his hand. Human history had made the long march from East to West, and, God willing, a Crusade would once again win back Jerusalem. The course of events, as Matthew had laid it out on his maps and in his *Chronicle*, seemed clear.

Then came the Mongols.

◆ Central Asia ◆

CHAPTER TWO

SCOURGE OF GOD

In this year, that human joys might not long continue and that the delights of this world might not last long unmixed with lamentation, an immense horde of that detestable race of Satan, the Tartars, burst forth from their mountain-bound regions. . . . Overrunning the country, covering the face of the earth like locusts, they ravaged the eastern countries.

—Matthew Paris (1240)

*M*ATTHEW FIRST REPORTED that something was wrong in 1238.

Merchants from northern Europe customarily sailed each year to Yarmouth, a fishing port on England's east coast, to buy herring—but in 1238, Matthew noted, the merchants never came, and rotting fish piled up in the port's markets. That same year ambassadors representing the Assassins—a widely feared Muslim sect based in Iran and Syria—arrived in England and France, and appealed for help, Matthew wrote, "on behalf of the whole of the people of the East."

It was an extraordinary overture, given the relationship between Christians and Muslims at the time. But it was born of extraordinary circumstances. "A monstrous and inhuman race of men had burst forth from the

northern mountains," the ambassadors announced, "and had taken possession of the extensive, rich lands of the East." The invaders were called Tartars, and they came, Matthew reported, from "the Caspian mountains or the
adjacent places"—the northeastern limits of the world as it was portrayed
on the *mappaemundi*. Led by "a most ferocious man named Khan," the Tartars had overrun vast stretches of Muslim territory and had already ravaged
greater Hungary—lands on the southwestern side of the Ural Mountains,
that is, in present-day Russia. If Christians and Muslims didn't immediately put aside their differences and work together to repel the marauders in
Muslim territory, the ambassadors claimed, all of Europe itself would soon
be under attack.

Among those at the English court who heard the Assassins' appeal was
the bishop of Winchester. By Matthew's account, the bishop reacted with
contempt. "Let us leave these dogs to devour one another," he scoffed, "that
they may all be consumed and perish, and we, when we proceed against the
enemies of Christ who remain, will slay them and cleanse the face of the
earth, so that all the world will be subject to the one Catholic Church, and
there will be one shepherd and one fold." But Matthew wasn't so dismissive.
"Where have such people, who are so numerous, till now been concealed?"
he wrote. Who *were* these Tartars?

Grappling with such questions, Matthew and others looked to the Bible
and the *mappaemundi* for help. What they found there soon made apocalyptic alarm bells begin to sound in their heads. A race of diabolical warriors
from a mountainous region far off in the northeast, bent on the devastation
of the Christian West?

This sounded a lot like Gog and Magog.

* * *

MATTHEW DEVOTED LONG ENTRIES in his *Chronicle* to the Tartars.
Much of what he wrote derived from a report by a Hungarian friar named
Julian, one of the first Europeans to describe the invaders in any detail. During the mid-1230s Julian had left his homeland and traveled east in search
of converts to Christianity, and in 1237 he had located a community of
pagan ethnic Hungarians living near the Ural Mountains. To his delight he
had discovered that they spoke his language. But they had reported alarm-
ing news. An ominous force of Tartars—fearsome horsemen and archers,

Julian would later write, "with heads so great that they did not seem to fit their bodies"—had amassed some fifteen days to the east, and they intended to "devastate all the lands which they could subjugate."

Although Julian didn't know it, the invaders called themselves Mongols—and they were preparing to move into Western Europe. Julian's informants almost certainly described the Mongols to him not as Tartars but as Tatars, the name of one of the many tribes who formed part of the greater Mongol empire. But many educated Europeans quickly associated the name with *Tartarus*, a Latin name for the underworld, and soon began using the name Tartars widely: a fitting description for a race of warriors apparently unleashed from Hell.

The Mongols originally were a tribe from the region of Lake Baikal, to the north of Mongolia in modern-day Russia, but by the time Julian heard about them they had become a multiethnic federation of nomadic tribes ruled from the high Mongolian plateau: a region of bitterly cold winters, searingly hot summers, and vast open expanses of desolate terrain. The tribes had been united at the end of the twelfth century by a leader originally known as Temüjin, who in 1206 was declared their undisputed leader. Temüjin then proceeded to launch an astonishing series of military campaigns that, by the time of his death in 1227, put him in control of the largest contiguous land-based empire in history, one that extended from China in the east to the Caspian Sea in the west, and from Siberia in the north to northwest India in the south. Temüjin and the Mongols considered their campaigns of conquest to have been ordained by the supreme sky god, Tenggeri, and the whole world, they believed, was a Mongol empire-in-the-making. "Through the power of God," Great Khan Güyük would announce to the West in 1246, "all empires from sunrise to sunset have been given to us, and we own them." Those who refused to submit to Mongol rule were rebels against the divine plan, and punishment for this refusal was often the outright slaughter of whole cities and peoples.

The conquests led by Temüjin were legendary, and to celebrate them the Mongols posthumously bestowed on him the title Fierce Ruler, or Chingis Khan. Today, thanks to an imperfect Arabic transliteration of that name, he is widely known as Genghis Khan.

Chingis Khan recognized that his people, resilient horsemen accustomed to lives of hardship, deprivation, and perpetual motion, were natural warriors. They traveled with huge numbers of spare horses, and by using them in rotation managed to travel up to a hundred miles a day—a distance far greater than any other army of the time could travel. As nomads, they knew how to live off the land and the peoples they conquered, but during times of privation and hard travel they could sustain themselves by drinking the blood of their own horses—and, if necessary, by eating them. Such practices, coupled with the ferocity the Mongols displayed in battle, fed rumors in Europe of the Mongols as cannibals and savages. "The men are inhuman and of the nature of beasts," Matthew Paris reported, "rather to be called monsters than men, thirsting after and drinking blood, and tearing and devouring the flesh of dogs and human beings (*Figure 16*).

Figure 16. The Mongols through Western eyes: slaughtering, cooking, and feasting on their captives. Matthew Paris (circa 1255).

The Mongols' savagery was calculated. They wanted their reputation to precede them. Often, on the eve of an invasion, they would send advance word of their mission of conquest to their adversaries and would demand submission without a fight. Inevitably, many opponents would acquiesce, having already heard terrifying rumors about what would happen to them if they did not—and as a result, when the promised invasion actually did take place, the Mongols' ranks would already be swollen with captives. Some would be forced to fight as foot soldiers on the front ranks. Others would be enlisted as guides, interpreters, engineers, and spies.

* * *

MOST EUROPEANS HADN'T heard of the Mongols before Julian encountered them, but the Mongols had ventured into Eastern Europe before. In 1223, near the end of a raid that had swept around the southern end of the Caspian Sea and into the Caucasus, one of Chingis Khan's field armies had made a reconnaissance mission into what today is southern Russia. This was a standard practice: the Mongols regularly explored areas beyond their control, to gather information about areas of weakness they might exploit in later campaigns. Their incursion into Russia took the locals completely by surprise. "Unknown tribes came," one Russian chronicler wrote. "None knows who they are or whence they came, nor what their language is, nor of what race they are nor what their faith is." The Mongols demanded submission, the Russians refused, and on March 31, 1223, a major battle broke out near the Kalka River. Eighteen thousand Mongols and some five thousand allied troops took on a force of forty thousand Russians—and the Mongols won decisively. In celebration of their victory they placed a set of wooden planks on top of six captured Russian princes, created a makeshift table, and decked it out with a lavish celebratory feast. They then gorged themselves as their captives slowly suffocated under the weight of the feast.

As quickly as they had arrived, the Mongols disappeared. "We know not where they hid themselves again," the Russian chronicler wrote. In the uneasy calm after their departure, stories about what had happened in other parts of the Caucasus began to surface and reach points farther west. "The Tartars entered our land bearing the Cross in front of them," one Georgian writer reported to the pope, "and in this way, appearing to be Christians, they deceived us." Queen Rusudan of Georgia made a similar observation. "We took no precautions against them," she wrote, "because we believed them to be Christians." At the sight of the Cross-bearing Mongol armies, one Armenian chronicler reported, some locals had assumed them to be the armies of the Magi—the legendary Oriental kings of the Bible—and had greeted them with open arms. The locals were massacred on the spot.

The Mongols had learned well from their captive informants. For decades Christians in the area had been nurturing a religious fantasy, and the Mongols seem to have played it to their advantage. The fantasy was this: A wealthy and powerful Christian priest-king from the East had for some time been making his way toward the West. A direct descendant of the Magi, he was known as Prester John, and he was on a mission: to unite the Christians of East and West, to defeat Muslims armies everywhere he found them, and to take back Jerusalem for all time.

Although the armies who arrived in the Caucasus in 1223 turned out not to be the armies of Prester John, his legend lived on. In the centuries that followed, rumors about the great king would take on a life of their own, and the desire to locate him and his distant eastern kingdoms would play a critical role in motivating the European exploration of both Asia and Africa.

* * *

THE FIRST SURVIVING mention of Prester John dates from 1145. It derives from a Syrian prelate named Hugh, who had traveled to Europe that year to make an appeal for a new Crusade. Christian forces had successfully established themselves in the Holy Land at the turn of the century, but those forces were now quickly losing ground to advancing Muslim armies, and the Latins of Outremer desperately needed help. Hugh made his appeal personally to Pope Eugenius III on November 18, 1145, and during his visit also met with a bishop named Otto of Freising, who later recorded an account of his conversation with Hugh. It's in Otto's account of that conversation that Prester John makes his first appearance.

Hugh told Otto that the situation for Christians in the Holy Land was dire, but then went on to report news of a promising development in the East. "He related that not many years before," Otto recorded,

> a certain John, a king and priest who dwells beyond Persia and Armenia, in the uttermost East, and who, with all his people, is a Christian but a Nestorian, made war on the brother kings of Persians. . . . Putting the Persians to flight with dreadful carnage, Prester John—for so they are accustomed to call him—finally emerged victorious. [Hugh] said that after this victory the aforesaid John moved his army to the aid of the Church in Jerusalem, but that when he had reached the Tigris and was unable to transport his army across that river . . . he was forced to return home. It is said that he is a lineal descendant of the Magi, of whom mention is made in the Gospel, and that, ruling over the same peoples that they governed, he enjoys such great glory and wealth that he uses no

scepter save one of emerald. Inflamed by the example of his fathers who came to adore Christ in his manger, he had planned to go to Jerusalem but by reason aforesaid he was prevented—so men say.

If true, this was encouraging news indeed.

The story of Prester John would be told repeatedly in the decades that followed, but it might well have vanished from memory had not a curious letter—one of the great literary hoaxes of all time—begun to circulate in Europe in 1165, purportedly written by Prester John himself.

Who wrote the letter, and why, are unanswerable questions. All that can be said is that its author was probably a well-read western cleric who had spent time in the Holy Land. The letter may have been composed to generate support for a new Crusade or it may simply have been a joke. Whatever it was, it seems to have been accepted as genuine, and in the decades and centuries that followed, it would circulate widely, translated into many different languages—and would exert a lasting influence on the ways in which Europeans imagined the East.

Prester John introduced himself imperiously in the letter. "I, Prester John, who reign supreme," he announced, "exceed in riches, virtue, and power all creatures who dwell under heaven." He ruled over seventy-two vassal kings, he continued, and twelve archbishops; he fed thirty thousand soldiers each day at tables made of gold, amethyst, and emerald; he maintained a great army dedicated to protecting Christians everywhere; and he had dedicated himself to waging perpetual war against the enemies of Christ. He then proceeded to describe the manner in which he and his men rode into battle—and this description may well have been what the inhabitants of the Caucasus had in mind when, some sixty years later, they saw Cross-bearing Mongol armies advancing toward them. "When we ride forth to war," Prester John explained, "our troops are preceded by thirteen huge and lofty crosses made of gold and ornamented with precious stones, instead of banners, and each of these is followed by ten thousand mounted soldiers and one hundred thousand infantrymen. . . . When we go out on horseback on ordinary occasions, there is borne before us a wooden cross, without decoration or gold or jewels, so that we may be reminded of the passion of our Lord Jesus Christ."

Prester John was both specific and vague when it came to describing where he was from. "Our magnificence dominates the Three Indias," he wrote, "and extends to Farther India, where the body of St. Thomas the Apostle rests. It reaches through the desert toward the place of the rising of the sun." In the early Middle Ages, Europeans knew India as little more than a distant and mysterious part of the world somewhere to the east of the Holy Land, but by the twelfth century they had begun to divide it into three parts: Nearer India (the northern part of the Indian subcontinent, and points east); Farther India (the southern part of India, and the spice-producing regions of the Far East); and Middle India (Ethiopia and other African kingdoms, which in the T-O scheme were considered a part of Asia, since they were on the eastern side of the Nile). The division may seem odd to us today, but substitute "East" for "India" and you get something like our own peculiar way of dividing Asia: the Near East, the Far East, and the Middle East.

By the early 1200s the myth of Prester John had taken root in Europe and the Holy Land. Jerusalem by now was once again in Muslim hands, and the idea of yet another Crusade, the fifth, was taking shape. One of its proponents was the bishop of Acre, Jacques de Vitry—and he saw Prester John and the Christian armies of the East as its lynchpin. "I believe," he wrote in 1217, "that there are more Christians than Muslims living in Islamic countries. The Christians of the Orient, as far away as the land of Prester John, have many kings, who, when they hear that the Crusade has arrived, will come to its aid."

The Fifth Crusade was indeed soon launched, but it didn't go well. By 1221, its final year, it seemed doomed to fail, but Jacques remained optimistic, and that spring he broadcast good tidings. "A new and mighty protector of Christianity has arisen," he wrote. "He is King David of India, who has taken the field of battle against the unbelievers at the head of an army of unparalleled size." This King David, Jacques went on, was "commonly called Prester John." Some rumors suggested that this king must be the son or grandson of the original Prester John, but others told a different story: that Prester John had access to the Fountain of Youth and was immortal. "Know that we were born and blessed in the womb of our mother 562 years ago," Prester John claimed in one version of his letter, "and since then we have bathed in the fountain six times."

Jacques's pronouncements were Crusade propaganda, but they had roots in fact. In 1219, Chingis Khan had embarked on a campaign of conquest in Persia and had defeated powerful Muslim armies there. News of his victories had thrilled a sect of eastern Christians known as the Nestorians, who seem to have interpreted them as the fulfillment of a dream they had been clinging to for centuries: that an army of the Magi, led by a priest-king from India, would soon arrive and release them from Muslim domination.

Via the Nestorians garbled reports of Chingis Khan's victories reached the Holy Land and then Europe. Word had it that a great Christian army had arisen in the East, was decimating the Muslims in Persia, and would soon be marching on Jerusalem. Small wonder, then, that the Mongols were greeted as friends when they arrived in the Caucasus. Chingis Khan and Prester John appeared to be one and the same.

* * *

FRIAR JULIAN—THE HUNGARIAN missionary who stumbled across the Mongols in 1237, as they were preparing to enter Europe for a second time—harbored no such illusions. The Mongols, he decided, were not the armies of Prester John; they were nothing less than "the scourge of God." Abandoning his missionary efforts, he rushed home to warn his countrymen. On his way home he encountered two Mongol envoys who carried with them an order of submission intended for King Béla IV of Hungary translated, no doubt, by one of their captives, and it confirmed his worst fears. The order came from Ogodai, Chingis Khan's son and successor. "I am aware," Ogodai told Béla, "that you are a wealthy and powerful monarch, that you have under you many soldiers, and that you have the sole rule over a great kingdom. Hence it is difficult for you to submit to me of your own volition. And yet it would be better for you, and healthier, were you to submit willingly."

With the exception of King Béla himself, who may have tried to muster a defense of Hungary, nobody in Eastern Europe seems to have paid Julian's warning much mind. Not that it would have mattered much if they had. During the next four years, between 1237 and 1241, the Mongols swept through much of Russia, Poland, and Hungary, destroying entire cities, armies, and populations. After a rout in Silesia, one chronicler reported that Mongol soldiers had collected nine sacks full of their victims' ears and had sent them back to their capital, Karakorum, in the Mongolian steppe, as proof of their victory.

By 1241 the Mongols were only some forty miles south of Vienna and seemed poised to invade Western Europe. The region's ruling class, at last becoming aware of the threat, grew deeply worried. "If, which God forbid," Holy Roman Emperor Frederick II wrote to King Henry III of England, "they invade the German territory and meet with no opposition, the rest of the world will then feel the thunder of the suddenly coming tempest. . . . For they have left their own country, heedless of danger to their own lives, with the intention . . . of subduing the whole of the West, and of ruining and uprooting the faith and name of Christ." Henry Raspe, a German noble, interpreted the invasion in an even more apocalyptic light. "The dangers foretold in the Scriptures in the times of old," he wrote, "are now, owing to our sins, springing up and breaking out."

But then, once again, the Mongols disappeared. This time they made a violent retreat, burning and destroying everything in their path and leaving behind, as one chronicler memorably put it, "nobody to piss against a wall." Nobody knew why they had left or where they had gone, but this time everybody expected that they would soon be back.

Three years later, in 1245, Pope Innocent IV convened the first Council of Lyon—Christianity's thirteenth ecumenical council, a meeting in which bishops from all over Christendom met to settle matters of official doctrine and practice. Hundreds of high-ranking members of the Latin clergy and European nobility all converged on Lyon for the event. High on the agenda, everybody agreed, was the need to come up with what the pope called "a remedy against the Tartars."

* * *

ONE POSSIBLE OPTION soon emerged. A Russian bishop named Peter reported that at their encampments in Russia the Mongols were encouraging visits from European ambassadors. Heartened by this news, Pope Innocent IV appointed three small groups of envoys, two led by Dominicans and one by a Franciscan, and dispatched them on missions to find the Mongols. The Dominicans were to travel through the Middle East, in search of army outposts there, the Franciscans were to make their way through

Eastern Europe and Russia and into the eastern steppe beyond, in an effort to find the Great Khan.

The medieval European exploration of Asia was about to begin.

What did these early envoys know about where they were going? Almost nothing. But they did set out with a number of geographical preconceptions. As educated thirteenth-century Christians, they would have been familiar with biblical descriptions of Asia, including those that mentioned the Earthly Paradise and Gog and Magog. They would have understood Asia as the largest and most symbolically important part of the known world and would have considered Jerusalem to be its spiritual center. The monstrous races at the margins of the earth, Alexander the Great in India, Prester John in the Three Indias—they would have heard stories of them all. Their conception of the world, in other words, was the one laid out on the *mappaemundi.*

The leader of the mission sent to find the Great Khan was John of Plano Carpini, an overweight monk well into his sixties who had been one of the original companions of Saint Francis of Assisi, the founder of the Franciscan Order. John was appropriately daunted upon setting out. "We feared," he wrote to the pope upon his return, in a report that has come to be known as *History of the Mongols*, "that we might be killed by the Tartars or other people, or imprisoned for life, or afflicted with hunger, thirst, cold, heat, injuries, and exceeding great trial almost beyond our powers of endurance—all of which, with the exception of death and imprisonment for life, fell to our lot in various ways, and in a much greater degree, than we had conceived beforehand."

John and his companions departed Lyon on Easter Sunday, April 16, 1245. With them they carried two letters from the pope, addressed to "the King of the Tartars." The first letter, dated March 5, introduced John and his party as friends, laid out the principles of Christianity, and expressed the naïve hope that the Tartars, upon hearing these principles, would cease their hostilities and convert. The second letter, dated March 13, was more of a diplomatic overture.

> It is not without cause that we are driven to express in strong terms our amazement that you, as we have heard, have invaded many countries belonging to Christians and to others and are laying them waste in a horrible desolation. . . . Following the example of the King of Peace, and

desiring that all men should live united in concord in the fear of God, we do admonish, beg, and earnestly beseech all of you that for the future you desist entirely from assaults of this kind and especially from the persecution of Christians, and that after so many and such grievous offences you conciliate by a fitting penance the wrath of Divine Majesty. . . . On this account we have thought fit to send to you our beloved son [Friar John] and his companions, the bearers of this letter. . . . When you have had profitable discussions with them concerning the aforesaid affairs, especially those pertaining to peace, make fully known to us through these same friars what moved you to destroy other nations and what your intentions are for the future.

What the pope and Friar John didn't know was that the Mongols expected only one thing from foreign envoys: offers of absolute submission.

For ten months, John and his companions made their way north and east through Europe, and then early in the winter of 1246 they began slogging through western Russia to Kiev. John was now "desperately ill," as he would later report, and had to be "carried in a vehicle through the snow in the bitter cold." What he saw along the route to Kiev didn't make him feel any better. "When we were journeying through that land," he wrote, "we came across countless skulls and bones of dead men lying about on the ground. Kiev had been a very large and thickly populated town, but now it is reduced almost to nothing."

The envoys made it to Kiev but only stayed a few days. On February 3, 1246, they set out for "the barbarian nations" of the East.

* * *

THREE WEEKS LATER they found the Mongols—or, rather, the Mongols found them. "On the first Friday after Ash Wednesday," John wrote, "we were putting up for the night as the sun was setting, when some armed Tartars came rushing upon us in a horrible manner, wanting to know what kind of men we were." John explained that he and his companions had been sent by the pope with the message that "all Christians should be friends of the Tartars and be at peace with them." Upon hearing this, the Mongols changed their attitude toward the friars, evidently for a simple reason: they used the same word for *submission* as they did for *peace*, and therefore understood the friars to be carrying a message of surrender. They provided the friars with fresh horses and rode east with them for the next month, until they reached a great tent city on the Volga River: the imperial encampment of Batu, a grandson of Chingis Khan and the ruler of Mongol-occupied Russia and Eastern Europe.

At Batu's camp the friars finally learned why the invasion of Western Europe had never happened. Great Khan Ogodai had died in 1241, and Mongol armies on campaigns of conquest around the world had been summoned back to Karakorum to elect a successor. Güyük, another of Chingis Khan's grandsons, had been chosen, and his enthronement was due to take place in Karakorum in the summer of 1246.

Batu ordered the friars to attend. And so John and his companions once again found themselves heading east. "Together with the two Tartars who had been assigned to us," he wrote, "we left with many tears, not knowing whether we were going to death or to life."

The missionary envoys now had to ride as hard and fast as the Mongols themselves. Urged on by their escorts, and relying on the Mongols' postal relay stations for fresh horses and supplies, they rose early each morning, changed mounts several times a day, and stopped to eat and rest only at night. During the daylight hours they rode through more scenes of death and desolation. "We came across many skulls and bones of dead men lying on the ground like dung," John wrote, adding that he saw "innumerable ruined cities and demolished forts and many deserted towns." Day after day, week after week, month after month, the friars made their way through the desert steppes of Central Asia—across the entire length of present-day Kazakhstan, over the Altai Mountains, and deep into Mongolia. It was a journey of almost three thousand miles, through terrain that John could only describe as "more wretched than I can possibly say," and the traveling took its toll. "We were so weak we could hardly ride," John wrote. "During the whole of that Lent our food had been nothing but millet with water and salt, and it was the same on other fast days, and we had nothing to drink except snow melted in a kettle." Such travel was routine for the Mongols, but it was an ordeal that would have tested the hardiest of Europeans. How the elderly, fat, and sick Friar John survived it boggles the mind.

On July 22, the envoys finally arrived at their destination: a huge tent city set up a half day's ride south of Karakorum. The city had been set up to receive the many dignitaries flooding in to submit and pay tribute to the new Great Khan—more than four thousand of them by Friar John's count, including noblemen from Russia and Georgia, ten Muslim sultans, an ambassador of the caliph of Baghdad, and the leaders of many Asian tribes. "So many gifts were bestowed by the envoys," John recalled, "that it was marvelous to behold—gifts of silk, samite, velvet, brocade, girdles of silk threaded with gold, choice furs, and . . . more than five hundred carts, which were all filled with gold and silver and silken garments." The friars found many Russians, Hungarians, and others in the camp—Latin and French speakers, he explained, who "had been among the Tartars, some for thirty years, through wars and other happenings, and who knew all about them, for they knew the language and had lived with them continually." For several weeks, as the friars waited for the enthronement ceremony to take place, they gathered a wealth of information from these men about the customs, beliefs, military tactics, and intentions of the Mongols. But they also picked up disheartening news. "Men who knew" told them that this new Great Khan intended "to raise his banner against the whole of the Western world."

After the enthronement John at last was able to present the Great Khan with his letters of introduction. The response he received came some days later, in the form of a menacing order of submission that he was ordered to carry back to the pope.

Having taken counsel for making peace with us, You Pope, and all Christians have sent an envoy to us, as we have heard from him and as your letters declare. Wherefore, if you wish to have peace with us, You Pope, and all kings and potentates, in no way delay to come to me to make terms of peace, and then you shall hear alike our answer and our will. The contents of your letters stated that we ought to be baptized and become Christians. To this we answer briefly that we do not understand in what way we ought to do this. To the rest of the contents of your letters—that you wonder at so great a slaughter of men, especially of Christians and in particular Poles, Moravians, and Hungarians—we reply likewise that this also we do not understand. . . . Because they did not obey the word of God and the command of Chingis Khan and the Khan . . . God ordered us to destroy them and gave them up into our hands. . . . If you accept peace and are willing to surrender your fortresses to us, You Pope, and Christian princes, in no way delay coming to me to conclude peace. . . . But if you should not believe our letters and the command of God nor hearken to our counsel, then we shall know for certain that you wish to have war.

John had no choice but to agree to deliver the khan's letter to the pope, and so he and his companions headed back across the steppe to the west, once again enduring great hardships along the way. "We traveled throughout the winter," John wrote, "often sleeping in the desert on the snow, except when we were able to clear a place with our feet. When there were not trees but only open country we found ourselves many a time completely covered with snow." When at last they arrived back in Christian territory, they were celebrated as heroes, and as they made their way back through Russia to France they enthralled local audiences with stories about where they had been and what they had seen. John's official report to the pope, composed in the latter half of 1247, after he had arrived back in Lyon, became avidly sought after in Europe—as did the good friar himself, who seems to have embarked on something of an author tour. "He wrote a great book about the deeds of the Tartars," a Franciscan chronicler named Salimbene recorded in 1247, "and caused this book to be read aloud whenever he found it too laborious to tell the story, as I myself saw and heard; and when the readers were astonished or did not understand, he himself expounded or discussed matters of detail."

In the coming centuries Friar John's report would become the most widely read of all the early accounts of the Mongols, thanks in large part to a Dominican friar named Vincent of Beauvais, who included excerpts from the report in his mid-thirteenth-century *Speculum historiale*, or *Historical Mirror*, one of the most popular and influential encyclopedias of the late Middle Ages. John made passing references to Prester John and monstrous races in his report, but this was secondhand information, picked up from others, and he gave it only a cursory treatment.

*   *   *

WHAT JOHN HAD to say about the Mongols and their intentions did not bode well for Western Europe. But he did offer up one hopeful bit of news. Nestorians in the service of the Great Khan, he reported, "told us that they firmly believed he [the khan] was about to become a Christian, and they have clear evidence of this."

Similar messages from other quarters were also making it to the Church. One of the pope's Dominican missionaries returned from his mission with a report that the Great Khan's mother was a Christian—a daughter of Prester John, no less. The pope and King Louis IX of France were also both shown an encouraging letter, from an Armenian who had spent time in Samarkand in 1248, that reported surprising news: great numbers of Christians lived throughout the Mongol empire; the Mongols had recently given support to a Christian ruler in India in his war against Muslims there; and the Great Khan and all of his people had converted to Christianity. There was talk, too, that Sartaq—the son of Batu, who ruled the Mongols' European territories—was also a convert.

The most heartening development of all came at the end of 1248. Two Nestorian envoys, a certain David and Mark, arrived in Cyprus with a proposal for King Louis, who was making final preparations on the island for the launch of the Seventh Crusade. David and Mark brought with them a letter from a Mongol prince named Eljigidei. Uncharacteristically, the prince made no demands for submission and instead wished Louis well. More important than the letter itself was what David and Mark told King Louis in person. Eljigidei and many other Mongol princes had converted to Christianity, they announced, and now Eljigidei, outraged by recent indignities inflicted by Muslims upon Christians in the Holy Land, had decided to invade Baghdad the following summer. In fact, he was interested in forming an alliance of sorts. Would the Crusaders be willing to attack Egypt that summer, to prevent the Muslim armies there from marching to defend Baghdad?

Once again, it seems, the Mongols were shrewdly manipulating Christian religious susceptibilities—and once again it worked. Louis greeted David and Mark's proposal with enthusiasm. He forwarded the letter to his mother in France, who passed it along to King Henry III in England,

and soon enough rumors began to circulate that the Mongols had not only converted but also agreed to join the upcoming Crusade.

Such rumors motivated the first expressly evangelical mission to the Mongols, launched in 1253 by the Franciscan friar William of Rubruck. William had spent time on Cyprus and in Egypt in the company of King Louis during the ill-fated Seventh Crusade, and after the venture had ended in failure he had turned his attention to the Mongols. Like others, he had heard rumors about the conversion of the Mongol prince Sartaq—and he decided he would try to find him.

*　　*　　*

WILLIAM BEGAN HIS journey by traveling first from the port city of Acre, in the Holy Land, to Constantinople, in early 1253, and then on to Soldaia, a merchant city on the Black Sea coast of the Crimean Peninsula (today known as Sudak) that had become an important port of exchange between Asia and Europe. In Soldaia he made preparations for his upcoming journey, gathering supplies and whatever information he could about the Mongols from local officials and merchants, and while there also delivered at least one public sermon, at a church called Saint Sophia. Among those who may have heard him preach that day was a Venetian merchant, who owned a house in the town: Marco Polo, the uncle of the Marco Polo who later in the century would himself set out in search of the Mongols and beccome the most celebrated of all European travelers to Asia.

William left Soldaia on June 1. For the next two months he and his companions traveled toward Sartaq's camp on the Volga. They fell in with the Mongols not long after leaving, and were immediately disappointed. The Mongols evinced little interest in Christianity and Christians. Instead, they seemed only to care about gifts, which William, as a mendicant friar who had taken a lifetime vow of poverty, was in no position to offer.

After two months on the road William reached Sartaq's camp. Sartaq turned out not to be much of a convert at all. "Whether Sartaq believes in Christ or not I do not know," William would write in a report he submitted to King Louis at the end of his journey. "What I do know is that he does not wish to be called a Christian: in fact, my impression is rather that he makes sport of Christians."

Assuming that William carried letters of submission, Sartaq ordered him to travel farther along the Volga to meet his father, Batu. But William fared no better with Batu than he had with Sartaq. Upon meeting Batu, William began to lecture him on the eternal damnation that awaited him if he did not convert to Christianity. The sermon didn't go over well. Batu gave "a slight smile," William remembered, "and the other Mo'als [Mongols] began to clap at us in derision." According to another account of the meeting, Batu also addressed William directly. "The nurse begins first to let drops of milk fall into the child's mouth," he told him, "so that the sweet taste may encourage the child to suck; only next does she offer him the nipple. Thus you should have persuaded us in simple and reasonable fashion, as this teaching seems to us to be altogether foreign. Yet you threatened [us] at once with everlasting punishment."

Batu then ordered William and his companions to travel to Karakorum for an audience with the Great Khan, now a recently elected grandson of Chingis Khan's named Mangku—and so it was that William found himself making the same journey across Central Asia that Friar John had made before him. The Mongol escort assigned to accompany him offered little comfort at the outset of the journey. "I am to take you to Mangku Chan," he told William. "It is a four-month journey, and the cold there is so intense that rocks and trees split apart with the frost: see whether you can bear it. . . . If you prove unable to bear it, I shall abandon you on the way."

William survived the journey. He and his companions ended up spending more than six months in Mangku's camp and in nearby Karakorum, during which time they mingled with an international cast of characters that included Persians, Turks, Chinese, Armenians, Georgians, Russians, Hungarians, Germans, Frenchmen, and even an Englishman. Most were probably captives from various Mongol campaigns of conquest; all worked for the Mongols in some capacity. The court also swarmed with priests and elders of different religions, among them great numbers of Nestorians—some of whom told William that Mangku was indeed about to convert. But when William at last had his audience with the Great Khan, any hopes he was harboring about the Mongols' joining a new Crusade were dashed. In an order of submission that he ordered William to carry back to King Louis, he alluded to the visit that Louis had received on Cyprus from David and Mark. "A man named David came to you, as though an envoy of the Mo'als," Mangku told Louis. "But he was a liar."

William arrived back in the Holy Land in 1255, after a punishing return journey across the Central Asian steppe, and delivered his report to King Louis soon after. He described what he had seen during his travels in great detail, bringing the Mongols and their customs to life far more vividly than had Friar John. When it came to some of the Europeans' traditional beliefs about the East, he injected a new note of skepticism, based on empirical investigation. He gently took issue with Isidore of Seville, for example, concerning the Caspian Sea. On his journey, William wrote, he had rounded the Caspian to the north and had discovered it to be a landlocked body of water. This meant that "what Isidore says, to the effect that it is a gulf extending inland from the ocean, is incorrect." He suggested, too, that the monstrous races might not actually exist. "I enquired about the monsters or human freaks who are described by Isidore and Solinus," he told King Louis, "but was told that such things had never been sighted, which makes us very much doubt whether [the story] is true." Nowhere was he more skeptical than when it came to Prester John. "The Nestorians called him King John," he explained, "[but] only a tenth of what they said about him was true. For this is the way with the Nestorians who come from these parts: they create big rumors out of nothing."

William had to admit, however, that as an evangelical enterprise his mission had been a failure. He had spent six months among the Mongols but had managed almost no conversions. "We baptized there," he wrote, "a total of six souls."

* * *

FRIAR JOHN AND Friar William had both returned to Europe with deeply discouraging news. The Mongols had not converted to Christianity; they were bent on a program of world conquest; and they intended to wipe out those who would not submit to their rule. But there was a silver lining. Those who had suffered the most at the hands of the Mongols, at least so far, were Muslims and Byzantines, Latin Christendom's great foes and rivals. Not only that, it was too early to declare defeat in the attempt to convert the Mongols; as a monotheistic people with no organized priesthood or national faith, the Mongols and their leaders were surely susceptible to persuasion. And even if the Mongols themselves didn't convert, the large numbers of schismatic Christians that Friar John and Friar William had discovered living in Asia might themselves be convinced to join forces with the Latins to make a grand pincer move on the Holy Land from the east and west.

The Mongols had done the Europeans another favor. By seizing control of and governing a vast expanse of territory that stretched all the way from China to the edges of Europe, they had opened up a new highway to the East. This highway represented an unprecedented opportunity for Europe's missionaries and diplomats—and its merchants. In the fifty years since Chingis Khan had united the nomadic tribes of the Mongolian steppe, he and his successors had plundered and exacted tribute from some of the richest and most powerful nations in the world. By the middle of the thirteenth century they had amassed fabulous stores of wealth and were quickly developing an appetite for luxury goods—which is why, in 1260, "in the hope of a profitable venture," two Venetian merchants named Niccolò and Maffeo Polo loaded a ship with expensive jewels and set sail from Constantinople to Soldaia.

This itself was not an unusual move. Venetian merchants had established a trading colony in Soldaia by the time the Polo brothers set out. It was the port where Friar William had stopped and preached in 1253, and where Niccolò and Maffeo's brother Marco owned a house. But after reaching Soldaia, Niccolò and Maffeo did decide to do something new and different: they decided to try their luck farther east, among the Mongols. And so, in 1260, just as Friar William had done seven years earlier, they left Soldaia and headed north toward the Mongol encampments on the Volga.

They would be gone for nine years.

+ China +

CHAPTER THREE

THE DESCRIPTION OF THE WORLD

*From the time when with His hands our Lord God formed Adam,
our first parent, down to this day, there has been no man, Christian
or pagan, Tartar or Indian, or of any race whatsoever, who has known
or explored so many of the various parts of the world and of its great
wonders as . . . Messer Marco Polo.*

Prologue, Book of Marco Polo (circa 1300)

WHEN NICCOLÒ AND Maffeo Polo finally reappeared, in Acre in 1269, they had with them a magnificent gold tablet and a letter. The tablet, which granted the brothers provisions, lodging, and safe passage anywhere they traveled in Mongol territory, came from "the Great Khan of all the Tartars, whose name was Kubilai." Kubilai presided over a vast and powerful kingdom known as Cathay (China), located at the very "ends of the earth, in an east-north-easterly direction," and the brothers claimed to have visited him there. The letter they carried also came from the Kubilai. In it he addressed the pope directly but made none of the customary Mongol demands for submission. Instead he sent greetings and requested a hundred learned Christians be dispatched to his court, where he hoped to see them "argue and demonstrate plainly to idolaters and those of other persuasions that their religion is utterly mistaken." He also made a memorable promise.

On the day when we see this, I too will condemn [the idolaters] and their religion. Then I will be baptized, and all my barons and magnates will do likewise, and their subjects in turn will undergo baptism. So there will be more Christians here than there are in your part of the world.

It wasn't until 1271 that the Polos managed to get an audience with the pope, the newly elected Gregory X, but when they did, he received the letter with enthusiasm. He couldn't spare a hundred men, but did what he could: he summoned two of his most learned Dominican friars, bestowed on them "plenary authority to ordain priests and bishops and grant absolution as fully as he could himself," and dispatched them, along with the Polo brothers, back to the Great Khan, bearing gifts and salutations. This time Niccolò decided to bring along his seventeen-year-old son, Marco.

The Dominicans, evidently made of less sturdy stuff than Friar John and Friar William, soon got cold feet, conferred on the Polos "all of the privileges and letters they had received," and promptly turned back. The Polos made their way east on their own, hampered by "snow and rain and flooded rivers," and eventually, after three and a half years of traveling, rejoined the Great Khan in Cathay—where they would remain for the next seventeen years.

The young Marco Polo blossomed in the Mongol court. He mastered four local languages, proved himself wise and trustworthy well beyond his years, and gradually became a special envoy of the Great Khan, who "showed him such favor and kept him so near his own person that the other lords were moved to envy." Having noticed that Marco had a special gift for observation, the Great Khan sent him on important missions to many far-flung realms of the Far East, and Marco not only conducted his official business with great skill but also always returned to delight the Great Khan with colorful and detailed descriptions of the places and peoples he had seen.

Eventually the Polos decided to return home, and reluctantly the Great Khan allowed them to go. He provided them with two more tablets, to ensure safe passage, and then, because of the dangers of traveling overland through warring Mongol territories to the west, placed them on a seagoing mission he was sending to Persia. He had selfish reasons for doing this too: the Polos had already spent time sailing in his service between Persia and

Cathay, he knew, and could help serve as guides in the "strange seas" around India.

For the better part of the next two years the Polos sailed through the 12,700 islands of the Indian Ocean, the extent and marvels of which defied full description. In Persia they disembarked among the Mongols, who once again gave them gold tablets guaranteeing safe passage, and from there they returned first to the Black Sea and then to Constantinople. Finally, in 1295, twenty-four years after having set out, they arrived back in Venice—the first medieval Europeans to have lived among the Mongols in Cathay and India.

Or so goes the story that Marco related upon his return.

* * *

THE MARCO POLO story survives in a bewildering variety of versions and languages. For convenience's sake, modern Polo scholars often refer to any edition of it simply as the Book. Some 150 manuscripts of the Book survive from before the advent of printing, in the mid-1400s, and are written not only in Latin, the language of educated European readers, but also in a number of local languages and dialects, among them Franco-Italian, Northern French, Tuscan, and Venetian, German, Czech, Irish, and Italian. No two versions of the text are exactly the same, some are dramatically different, and, to complicate matters, they all seem to derive from a highly problematic source: a lost original purportedly composed by a certain Rusticello of Pisa, an Italian writer of lowbrow chivalric romances. Rusticello claimed to have met Marco Polo in a Genoese prison, in 1298, and to have collaborated with him there on the writing of the work.

Questions about Marco Polo and his Book abound. Did he and Rusticello really meet? Did Rusticello faithfully record his story? Did later copyists preserve the substance of the original account? Did Marco actually make it to China? Did he even exist? These and other questions have provoked animated speculation and debate, and the result, when it comes to the study of the Book itself, is what one exasperated Polo expert has described as an "undecipherable enigma." Most scholars agree, however, that much of the Polo story is real, and that Marco did reach China.

The Polos may well have been the first Latin Christians to visit Kubilai at his court in Cathay, but they weren't the first Westerners to make it there.

Records from the court survive, and in a chronicle of the year 1261 an entry describes the arrival, on June 6, of an embassy representing a nation called the Fa-lang. The word is a transliteration of *Farang*, itself a rendering of *Frank*, and was often used at the time, in the Near and Far East, to refer to any European people living to the west and north of Muslim territory. "These people," the entry reads,

came and presented garments made from vegetable fabrics and other presents. These envoys had traveled three years from their country to Shang tu. They reported that their country is in the Far West beyond the Uighurs. In their country there is constant daylight and no night. It is evening there when the field mice come out of their holes. If somebody dies there, then Heaven is invoked, and it might even happen that the person is restored to life. Flies and mosquitoes are born from wood. The women are very beautiful, and the men usually have blue eyes and blond hair. There are two oceans on the route from there, one which takes a month to cross, the other a whole year. Their ships are so big that they can hold between 50 and 100 men. These people presented a wine beaker made from the eggshell of a sea bird. If one poured wine into it, the wine became warm immediately. . . . The emperor was very pleased that these people had come from so far and gave them liberal gifts of gold and textiles.

This passage is captivating on several different levels. No mention of any earlier visit by Europeans to China appears in any medieval source, Eastern or Western. These visitors therefore may indeed have been some of the very first—although other Western traders surely reached the Far East at about the same time, but either failed to get an audience with the Great Khan or chose to keep their new sources of goods secret. The passage does corroborate the idea that European merchants began to venture east in the second half of the thirteenth century, after the Mongol invasions of Europe and the Near East, and it's also a useful reminder that the West appeared every bit as distant and strange and marvelous to those in the East as did the East to those in the West. Centuries before medieval Europeans arrived, in fact, the Chinese had been telling themselves stories about the monstrous beings that lived at the world's margins, including those in the distant, mysterious West. One ancient and much-loved work known as *The Classic of Mountains and Seas*, referred to such races as the Loppy Ears (who "have such big ears they flop down onto their shoulders"), the Feathered Folk (who "can fly, but not very far"), the Hairy Folk (covered in hair "like a pig"), the Mushroom People (whose aspect was like "a meat fungus"), and the Progenyless Folk (a boneless race who, because they eat only air, "are clear-headed and live a long time") (*Figure 17*).

Figure 17. Monstrous beings believed by the ancient Chinese to inhabit the world's distant places. Note the similarities to Europe's monstrous races (*Figure 14*).

The Fa-lang account is also a reminder that for every great surviving story of travel and discovery, like that of the Polos, countless others are lost. How did the Fa-lang travelers get to China? What made them go? What of the earliest Mongols to arrive in Europe, for that matter? What stories did they bring home? We'll never know.

Which brings us back to Marco Polo. What matters most about him is not just that he was one of the first Europeans to travel beyond Central Asia to the Far East. It's that he—or somebody, at least—put into writing what he found there.

*   *   *

MARCO IS NEITHER the hero nor the subject of his own story. Unlike Friar John and Friar William before him, he provides almost no information about what he himself underwent, thought, or felt on his travels. What little narrative that does appear in the Book—namely, the version of events set out at the beginning of this chapter—is limited almost wholly to its prologue, which in one modern English translation takes up only the first thirteen of more than three hundred pages. The rest of the Book is not a travelogue or an adventure story but a strikingly impersonal survey of "all the great wonders and curiosities" of the East, laid out, as the prologue announces, "in due order." The Book aimed to do much more than just recount the details of Marco's travels, in other words. It aimed to provide a newly full picture of the world.

Despite such a grand aim, the Book is written in an almost comically long-winded and disorganized style. "Now let us leave these regions and turn to the Black Sea," the author announces at one point.

It is true that there are many merchants and others who know it well; but still there are a great many more who do not know it. So for their sake it is worthwhile to set down the facts in writing. Let us begin, then, with the mouth of the Black Sea, the strait of Constantinople. At the entrance to the sea, on the western side, stands a mountain called the Faro. But, now that we have embarked on this topic, we have had second thoughts about setting it down in writing; for, after all, it is very well known to many people. So let us drop the subject and start on another one.

The Book declares right from the outset that its function is to instruct and entertain. Marco will record "all the things he had seen and had heard by true report, so that others . . . may learn them," a job he will undertake "to afford entertainment for readers." But one of the many curious things about the Book is that it lacks just the sort of instruction and entertainment that Europeans had come to expect when it came to the wonders of the East: descriptions of diabolical warrior tribes, the monstrous races, mythical beasts, famous sites from antiquity, the legendary topography of the Bible, and so on. What made Marco's account so different—so very hard to make sense of—was that the Far East it describes did *not* contain such things. Instead, Marco matter-of-factly describes it as wealthier, more powerful, more populous, more extensive, more technologically advanced, and more civilized than anything Europeans had ever imagined.

* * *

"LET ME BEGIN with Armenia." With that brief sentence Marco begins the main body of his Book—and he then proceeds methodically to introduce Europeans to the wonders of the Far East in a sweeping account that would captivate readers for centuries to come. He describes the route overland from Europe to Cathay; he opens up a new window on the astonishing wealth and civilization of the peoples under the dominion of the Great Khan; he fills in blank spaces throughout the Three Indias; and he paints a beguiling picture of the Indian Ocean, describing it as a vast maze of islands awash in gold, gems, and exotic spices. Today the title of Marco's Book is often given as *The Travels of Marco Polo*, creating expectations that the work is a kind of a personal travelogue. But it's something else entirely. During its early history the Book often bore the title *The Description of the World*, which in many ways is a much more fitting title. It suggests that the Book was read and understood initially not as a blow-by-blow account of one man's travels but as a kind of verbal *mappamundi*, composed to flesh out Europe's skeletal understanding of the East.

It's clear that Marco did compose his Book with maps in mind. In the case of the islands and seas of the Indies, he specifically acknowledges his debt, noting that he has relied not only on his own observations but also on "the maps and writings of the practiced seamen who ply in these waters." Many of his descriptions of the route across Central Asia also seem drawn from a kind of map—specifically, itinerary-like descriptions known to have been made of the stages of the vast Mongol postal relay system. Internal evidence in the Book, too, suggests that Marco relied on new exploratory surveys that Kubilai Khan ordered made of his Chinese territories. For his broad conception of the East, which he laid out far more expansively than any Christian encyclopedia or *mappamundi* of his time, Marco appears to have relied on now-lost maps prepared by Chinese or Persian cosmographers working at the Mongol court.

Marco's description of the world doesn't include anything about Europe, but the omission isn't not surprising; he didn't feel it necessary to describe places that were well-known to his readers. The East is where he focuses his attention—and he delivers his descriptions in a terse style that seems almost designed to provide European mapmakers with material they could use in legends on their maps. Describing some of the provinces of Manzi (southern China), for example, he writes,

The first of these, a very splendid and wealthy one, is called Ngan-king. The people are idolaters [Buddhists], using paper money and subject to the Great Khan. They live by trade and industry. They have silk in abundance and manufacture cloths of gold and silk fabrics of all sorts. They also have game in plenty. They burn their dead. There are many lions here. And there are many rich merchants, so that the province yields a heavy tribute and a big revenue to the Great Khan. . . . As there is nothing else worthy of note, let us turn to the very splendid city of Sian-yang-fu. . . . It is an important center of trade and industry. The people are idolaters, use paper money and burn their dead. They are subject to the Great Khan. They have silk in plenty and weave cloths of gold and silk of all sorts. Game too is plentiful.

Embedded in the short passage above are three elements of Marco's Book that greatly impressed European readers: the power and wealth of the Great Khan, the material resources and sophisticated industries under his control, and the previously unsuspected masses of people still living outside of the reach of Christianity and Islam. Fittingly, for a man who lived much of his life in the Mongol court, Marco's depiction of the Great Khan is reverential in the extreme, but it also plays on the European stereotype, evident in the legend of Prester John as well, of what an Oriental king should be. Kubilai is the richest and most powerful ruler the world has ever known. He presides benevolently over the largest empire ever to have existed. He lives in Khanbalik (Beijing), the capital city of Cathay, in the largest palace ever seen, the interior of which is adorned with gold and silver, and the main hall of which can accommodate more than six thousand diners. In the now-ruined town of Shang-tu, or Xanadu, he maintains a huge summer palace of marble and gold, set in the midst of carefully tended parks and springs and game (which, thanks to Marco's description, gave rise to Coleridge's poem "Kubla Khan" and its famous opening lines: "In Xanadu did Kubla Khan / A stately pleasure-dome decree.") The Great Khan rules over entire kingdoms of Buddhists, Christians, Muslims, and Jews. He affords all of these religions equal treatment and celebrates all of their festivals, but he regards only one as "truest and best": Christianity. This prompts Marco to make a rare and pointed editorial aside about Kubilai's famous request for a hundred learned Christians. "If men had really been sent by the Pope with the ability to preach our faith to the Great Khan," he writes, "then assuredly he would have become a Christian. For it is known for a fact that he was most desirous to be converted."

No city in the Great Khan's vast dominion was greater than Kinsai (today Hangzhou), a Chinese city under Mongol control in the province of Manzi. The city, Marco writes, is "the finest and most splendid city in the world" and as such well deserves its name, which, he claims, means "City of Heaven." Crisscrossed by a vast network of streets and canals and bridges ("like Venice," one Venetian Polo manuscript notes), the city has ten principal marketplaces. Three times a week, forty to fifty thousand merchants arrive at each one, bringing with them an impossible abundance of game, fruits and vegetables of all kinds, great quantities of fish, and much more. Surrounding the markets are "shops in which every craft imaginable is practiced and every sort of luxury is on sale, including spices, gems, and pearls."

Marco finds this same sort of abundance in the great Chinese port city of Zaiton (today Quanzhou), the capital of another of Manzi's provinces, all of which are subject and pay tribute to the Great Khan. Like Kinsai, Zaiton overwhelms Marco with its riches, and his description of them would exert a profound pull on subsequent generations of Western merchants and explorers as they began to ponder sailing from Europe to the East. The city, he writes, is "the port for all the ships that arrive from India laden with costly wares and precious stones of great price, and pearls of fine quality. . . . And I assure you that for one spice ship that goes to Alexandria or elsewhere to pick up pepper for export to Christendom, Zaiton is visited by a hundred. . . . I can tell you further that the revenue accruing to the Great Khan from this city and port is something colossal."

Marco goes out of his way, too, to note that the wealth of Manzi is not only commercial but also cultural. The province itself, like much of the rest of China, is home to great numbers of craftsmen, doctors, astrologers, and magistrates—the foot soldiers of civilization, in European terms, and the kinds of professionals absent from previous accounts of what was thought to be the primitive East. Marco describes most of the inhabitants of Manzi as idolaters but digresses at one point to recount how, while traveling in the region, he and Maffeo discovered the presence of a strange eastern sect of

Christians—probably Nestorians. The Polos convinced the Great Khan to recognize them and to offer them the privileges he offered to other Christians, Marco writes, and later learned that in all some seven hundred thousand Christians lived "here and there" in Manzi. The supposed presence of so many Christians in the Far East, along with the abundance of riches to be found there, would make the region an object of particular fascination for later generations of Europeans.

* * *

THE FIRST ISLAND Marco describes after leaving Cathay and Manzi is one that he calls Cipangu—that is, Japan. (The name derives from the Chinese *Jin-pön-kuo*, or "Land of the Rising Sun," and would have pronounced in Italian as "Chipangu.") Marco's description of the island, which he doesn't claim to have visited, is the first on record to have reached Europe.

Cipangu, Marco writes, lies some 1,500 miles east of the Chinese mainland in the Sea of Cathay—a body of water that he claims contains 7,448 islands, most of them rich in gold, gems, spices, and exotic trees and plants. Cipangu is very large, he reports. Because of its distance from the mainland, and the difficulties involved in reaching it by sea, its people are wholly independent; they neither rule nor are ruled by any other nations. They are idolaters, just like the inhabitants of Cathay and Manzi, and they exhibit a particularly barbaric streak: they consider cooked human flesh "the choicest of all foods."

What's most notable about Cipangu is its wealth. Awash in precious stones and gold, the island is a veritable Promised Land for European merchants.

They have gold in great abundance, because it is found there in measureless quantities. And I assure you that no one exports it from the island, because no trader, nor indeed anyone else, goes there from the mainland. That is how they come to possess so much of it—so much indeed that I can report to you in sober truth a veritable marvel concerning a certain palace of the ruler of the island. You may take it for a fact that he has a very large palace entirely roofed with fine gold. . . . All in all I can tell you that the palace is of such incalculable richness that any attempt to estimate its value would pass the bounds of the marvelous.

After describing Cipangu, Marco takes his readers on a phantasmagorical journey south and then west through what he calls "islands of the Indies." These include not only the 7,448 islands of the Sea of Cathay but also the 12,700 islands that can be found in the body of water to the southwest: the Indian Ocean. On this journey he alights on island after island, finding a profusion of native peoples, pearls, gems, and precious metals, along with forests rich in spices, rare trees, monkeys, screeching parrots, and other exotic animals. The natives he meets often wear almost no clothes, because of the extreme heat; some are docile and friendly, others are cannibals. He provides no recognizable itinerary in his narrative, but he does progress gradually south and then west through the Indies, very much as though he is tracing a finger around the ocean rim of a circular *mappamundi*. Loosely speaking, that progression represents a voyage through parts of the Philippines, Malaysia, and Indonesia, followed by a passage through Indonesia's Strait of Malacca, well known today as the quickest sea route from China to India.

Much of Marco's description of his wanderings in this ocean wilderness has a formulaic quality to it: here you find gold and spices, there you find idolaters and cannibals. But something very unusual happens when Marco arrives at what he calls Lesser Java (probably Sumatra): a giant inhabited island that he claims has a circumference of more than two thousand miles. When introducing the island, Marco pauses to make an observation that he predicts will "surprise everyone." It concerns not the island's size or its riches or its cannibals but rather its location.

"The truth is," he writes, "that it lies so far to the south that the Pole Star is not visible there."

* * *

Only a single map appears in all of the surviving manuscripts of Marco's Book—a rudimentary world map dating from sometime in the fourteenth century (*Figure 18*). The map almost certainly had nothing to do with Marco himself and is probably the work of the scribe who copied the manuscript. It's a strange thing to behold—it has a formless, amoebic aspect—but in the fourteenth century it would have been instantly recognizable to anybody who had studied geography. It was a zonal diagram, and it illustrates one of the most important initial ways in which Marco's Book was received:

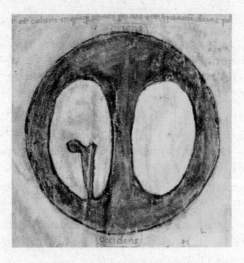

Figure 18. The only world map known to have been included in
a manuscript of Marco Polo's Book (fourteenth century).

as evidence that it actually was possible to enter the torrid zone, cross the
equator, and find habitable land in the southern hemisphere.

Whether such feats were possible had long been the subject of theological and scholarly debate. The authorities of ancient Greece and Rome had
generally agreed that land probably did exist somewhere on the south side
of the torrid zone, perhaps placed there as a counterbalance to the lands in
the north. These were often called the Antipodes, named after their presumed inhabitants, the Antipodeans—literally, those with their feet on the
opposite side of the earth. Some ancient theories even imagined the globe
as divided into four equal quarters, each surrounded by ocean and each containing a continental landmass, just one of which was the known world. In
this school of thought the Antipodes existed not just to the south but also
to the west. It was the place on the far side of the globe where you'd arrive
if you could travel from the known world straight through the center of the
earth. Because they lay on the south side of the torrid zone, the Antipodes
were considered unreachable, and as such they became a metaphor for inaccessibility and extreme distance.

The name Antipodes was also used to refer to lands anywhere in the
southern hemisphere, and the question of their existence vexed early Chris-

tians. If, as the Bible suggested, God had situated Adam and Eve and the Earthly Paradise in Asia; if, as the Bible also suggested, "the whole earth" was peopled" by Noah's three sons, Shem (Asia), Ham (Africa), and Japeth (in Europe); and if crossing the torrid zone from the north to the south was impossible—if all of these things were true, then the extent of human habitation *had* to be limited to the northern temperate zone.

It all made sense in theory, but in practice nobody had been able to confirm or deny the existence of the Antipodes. What if they did exist? If so, what if they were inhabited? How was a good evangelist, enjoined by the Bible to carry the Word across "all the earth," supposed to reach the Antipodeans and bring them into the Christian fold?

One of the earliest Christian authorities to express an opinion on the subject was Saint Augustine. Writing early in the fifth century, in his *City of God*, Augustine was outright dismissive. "But in regard to the story of the Antipodes," he declared,

> that is, that there are men on the other side of the earth, where the sun rises when it sets for us, who plant their footprints opposite ours, there is no logical ground for believing this. . . . Even if the world is held to be global or rounded in shape, or if some process of reasoning should prove this to be the case, it would still not necessarily follow that the land on the opposite side is not covered by masses of water. Furthermore, even if the land there be exposed, we must not jump to the conclusion that it has human inhabitants. For there is absolutely no falsehood in the Scripture, which gains credence for its account of past events by the fact that its prophecies are fulfilled. And the idea is too absurd to mention that some men might have sailed from our part of the earth to the other and have arrived there by crossing the boundless tracts of ocean, so that the human race might be established there also by descent from the one first man.

For Augustine, the Antipodes' existence was doubtful. But Isidore of Seville, writing a century after Augustine, disagreed. "Apart from these three parts of the world," he wrote, referring to Asia, Africa, and Europe, "there exists a fourth part, beyond the ocean, which is unknown to us."

Figure 19. T-O map showing a fourth part of the world lying west of Europe and Africa across the ocean (tenth-century). The continent is identified as India and as the site of the Earthly Paradise.

Isidore, of course, would become one of the most trusted geographical authorities in the Middle Ages, and some medieval cartographers therefore felt compelled to append this fourth part of the world to the standard T-O scheme (*Figure 19*).

What the zonal Marco Polo map (*Figure 18*) depicts should now be apparent. East is at the top. On the left is the known world, occupying much of the northern temperate zone, and on the right, occupying much of the southern temperate zone, is a giant amorphous continent: the Antipodes. Why would an anonymous fourteenth-century scribe have inserted this particular map into Marco's Book? Because of Lesser Java, the great island that Marco had described as lying "so far to the south that the Pole Star is not visible there." Those versed in geography and astronomy would have understood what the absence of the Pole Star meant: Marco had sailed into the torrid zone, crossed the equator, and arrived at a giant inhabited land-mass in the southern hemisphere. Perhaps the Antipodes, and the Antipo-deans, really did exist after all.

When Marco finally returned home, in 1295, his stories got a mixed reception. According to one contemporary source, which dates from not long after Marco's death in 1324, plenty of people considered much of what he had to say as "past all credence." But a Paduan professor named Pietro d'Abano felt differently. In 1310, having heard of Marco's travels, D'Abano sought Marco out, discussed matters of geography with him, and declared himself impressed. Marco, he wrote, was "the most extensive traveler and diligent explorer that has ever been known."

*   *   *

MANY PEOPLE ALSO doubted Marco's stories about Kubilai, the Great Khan for whom Marco claimed to have worked. But Kubilai was indeed a historical figure. For an extraordinary thirty-four years, from 1260 to 1294, he presided over much of China and other Mongol-occupied territories. In theory and by title he ruled over the entire Mongol empire, but that empire had in fact begun to split apart by the time he took power, gnawed at from within by factional rivalries. In 1258 the Mongols had captured Baghdad and seemed on the verge of taking over the entire Islamic world, with West-ern Europe apparently next on the list, but just two years later the tide began to turn. The Mamelukes of Egypt defeated the Mongols in the Holy Land that year and drove them out of the Middle East, and the Mongols would never return.

With their power waning, and facing increasingly successful Muslim resistance to their presence in the Near East, the Mongols began to make increasingly friendly overtures to the Christian West. In 1262, Hülegü, the ruler of the Mongols in Persia, touted himself as a "kindly exalter of the Christian faith." In 1274 several Mongol envoys were baptized in Lyon, an event much discussed and celebrated in the West in the decades and cen-turies to come. And in 1287 Arghun, a Mongol ruler in Persia, dispatched a Nestorian envoy named Rabban Sauma to Rome with greetings for the newly elected pope, Nicholas IV. The arrival of an envoy from "the King of the Tartars" surprised some cardinals in the pope's inner circle, but Rabban

Sauma assured them that the conversion of the Mongols to Christianity was well under way. "Know ye," he told the cardinals,

> that many of our fathers in times past entered the lands of the Turks, the Mongols, and the Chinese and have instructed them in the faith. Today many Mongols are Christian. There are queens and children of kings who have been baptized and confess Christ. The Khans have churches in their camps. And as the King is united in friendship with the Catholics and proposes to take possession of Syria and Palestine, he asks your aid for the conquest of Jerusalem.

Not surprisingly, the pope greeted Rabban Sauma warmly and sent him home with a letter "confirming his patriarchal authority over all the Orientals." Whether Arghun had any real interest in submitting to Nicholas's authority isn't clear, but he did at least make noises that must have pleased the pope. In 1288 he promised to be baptized in Jerusalem if a Mongol-European alliance ever managed to take back the city; in 1291 he had his son baptized with the Christian name Nicholas; and in that same year—alluding to the mission of Rabban Sauma, and citing the authority of Kubilai himself—he proposed a genuine joint military action against the Muslims occupying the Holy Land. "When we have taken Jerusalem from this people," he wrote to the King Philip IV of France, "we will give it to you."

Such is the context in which Kubilai purportedly made his famous request, via the elder Polos, that the pope dispatch a hundred learned Christians to China. Maybe Kubilai made the request or maybe he didn't, but one way or another, Christian missionaries did reach the Far East before the end of the century. In 1294, even before Marco had returned to Europe, a Franciscan known as John of Monte Corvino arrived in Khanbalik, the capital of Mongol-occupied Cathay, where he lived out the rest of his life building churches and baptizing Christians. The Franciscans made a special effort to establish a presence in the Far East, and by 1318 had set up at least thirty-four monasteries in areas controlled by the Mongols, with at least four in Cathay itself. Many of their order not only traveled extensively in Asia but also wrote memorable personal accounts of their voyages, and for generations afterward these accounts, like Marco's Book, would exert a profound influence on the ways in which Europeans imagined the East.

It was missionaries who initially seem to have taken Marco's Book the most seriously. In 1314 Francesco Pipino, a Dominican friar from Bologna, produced an important Latin translation, and in his introduction he laid out the reason he had done so.

I am of the opinion that the perusal of the Book by the faithful may merit an abounding grace from the Lord; in contemplating the variety, beauty, and vastness of God's Creation, as herein displayed . . . the hearts of some members of the religious orders may be moved to strive for the diffusion of the Christian faith, and by divine aid to carry the name of our Lord Jesus Christ, forgotten among such vast multitudes, to those blinded nations among whom the harvest is indeed so great, and the laborers so few.

Missionaries and merchants alike shared this optimism about the harvest open to them in the Far East, and during the first few decades of the fourteenth century they traveled back and forth between Europe and Asia with some regularity—enough so that by 1340 a merchant's handbook could report that "according to what the merchants say who have used it," the route from the shores of the Black Sea all the way to Cathay was "perfectly safe, whether by day or night." Even among those who didn't travel, a kind of Mongol chic developed, born of a growing interest in the Far East. Nobles commissioned lavish editions of Marco's Book for their libraries; Italian parents began to name their children after Mongol Khans; Mongol captives become the most popular class of slave. On one memorable occasion, in 1344, King Edward III of England held a tournament in which all of the participants gaily bedecked themselves in Mongol garb.

Then, in the middle of the century, apparently in the space of just a few years, contacts between Europe and Asia dried up. There were several precipitating factors. One—momentous but not often discussed—was the mingling of Asian and European rats. Thanks to expanding trade and travel between East and West under Mongol rule, plague-infected populations of Asian rats managed to reach past the previously isolating expanses of the

Central Asian steppe, and the result in Europe was the Black Death—an epidemic that ravaged the continent, reducing its interest in travel and its appetite for trade. The resurgence of Muslim power in the Holy Land and the Near East was another factor. By 1291 the Muslims had expelled not only the Mongols but also the Latins from the Middle East. The Mongol rulers who remained in Persia began converting to Islam, and by the middle of the century they were actively discouraging all travel between Europe and Asia. A third development, in 1368, was perhaps the most significant of all: after decades of humiliating occupation, the Chinese finally overthrew the Mongols and pushed them back out into the steppes. Expelled along with the Mongols were the Western missionaries and merchants who had been living in their midst.

Europeans wouldn't return to the Far East until the 1500s. In the interim, long after the Mongols had been pushed out of China and their empire had dissolved, European Christians held on to the vision of the Far East that Marco Polo and the early missionaries had set out in their writings. It was a domain of unimaginable wealth and civilization, rich in natural resources and human industry, inhabited by not only an almost limitless number of potential converts to Christianity but also large communities of Eastern Christians who might easily be brought into the Latin fold. And presiding over it all was an immensely powerful Great Khan who—if only he could be found—would welcome their return.

• Eastern China and Japan •

CHAPTER FOUR

THROUGH THE OCEAN SEA

*Let us sail westward to the island that is called the Promised Land of the
Saints, which God will give to those who come after us, at the end of time.*

—Anonymous, *The Voyage of Saint Brendan the Abbot* (circa A.D. 750)

*M*ARCO POLO's *Description of the World* provided Europeans with an
alluring new vision of the East, one that would fix itself in the minds of
merchants, missionaries, and monarchs for centuries to come. It also suggested the astonishing idea, at least to those well versed in geographical theory, that the southern hemisphere might be both reachable and habitable.
But it brought about another profound change in the way many Europeans
looked at the world. By taking them out into the island archipelagoes of the
Sea of Cathay and the Indian Ocean—at the very margins of the earth as
it appeared on the *mappaemundi*—the *Description* drew their minds to the
uncharted oceanic spaces that lay beyond them to the east and to the south.
In short, it helped lure them out to sea.

When Marco first took his readers out into the Sea of Cathay, he made
a point of explaining to them that the waters he was entering were not as
unfamiliar as they might sound. "When I say that this sea is called the Sea

of Cathay," he wrote, "I should explain that it is really the ocean. But as we say 'the sea of England' or 'the sea of Rochelle' [the Aegean], so in these parts they speak of 'the Sea of Cathay' or 'the Indian Sea,' and so forth. But all these names really apply to the ocean."

As asides go, this is something of a bombshell. Until this point Marco has taken his readers on a relatively straight path overland to the East, revealing to them an Asia far greater in extent than anything they had ever imagined. But now, as he leaves the known world and heads out to sea, he warps that line. The world is round, the Sea of Cathay and the Sea of England are one—and looking east out over the horizon from the Indies, one can almost see the West.

Marco stopped short of claiming it was possible to sail from Asia to Europe, but by so directly juxtaposing the Sea of Cathay and the Sea of England he forced his readers to contemplate the idea. As a theoretical possibility, this was nothing new to thirteenth-century Europeans. "A man could go around the world," one French geographical authority wrote in 1246, "as a fly makes the tour of an apple." At the time it was written, this remark was nothing but an abstract proposition, but only half a century later, thanks to Marco and the early missionaries, it began to seem more and more like a statement of actual possibility. If Asia indeed spread out so much farther to the east than had previously been imagined, then by definition the ocean that separated Asia from Europe had to be that much smaller.

* * *

READERS DIDN'T HAVE to wait long for a travel writer to expand Marco Polo's vision of the East. In a fourteenth-century book known today as *Mandeville's Travels*, one of the most popular and influential of all medieval geographical works, an author calling himself Sir John Mandeville took his readers on a full circuit of the globe.

Copies of *Mandeville's Travels* began appearing in Europe sometime between 1356 and 1366. Describing himself as a knight from England, Sir John claimed to have spent thirty-four years traveling throughout the known world, during which time he had witnessed many of the wonders of the East firsthand. When he finally returned to England, Sir John found his compatriots so hungry for information about where he had been that he decided to write his book to satisfy their appetites.

Sir John's identity is a mystery. No record of anyone by that name who could have possibly written his book has ever been found. Several real-life figures have been suggested as possibilities, but making a definitive identification has proved impossible. Whoever he was—and for simplicity's sake in this account he will simply be called Sir John—at least one thing can be said with certainty about him: he didn't make the journey that he said he did. Instead, as scholars have documented amply, he cobbled his account together by borrowing freely from a variety of ancient and contemporary geographical sources.

The result was an exercise in armchair travel and geography that, although often fanciful and misguided, provided as full a picture of the world as was possible in the middle of the fourteenth century. An abundance of accurate geographical information was available in the works that literate Europeans could be expected to know—the very works that Sir John relied upon in composing his book—but an abundance of inaccurate information was also available. The challenge of distinguishing truth from falsehood—especially improbable truth from improbable falsehood—was an impossible task.

More accessibly and concisely than any other writer of his time, Sir John gathered together the travel lore of his day and presented his readers with a vision of the world in which the traditional elements of medieval geography (the contents of that great abbey cupboard) rested side by side with the newly revealed wonders of Asia. As it happened, his book appeared just as the Mongols were about to be expelled from China, and just as the wall was coming back up between East and West. As a result, for more than a century and a half *Mandeville's Travels* would stand as one of the most up-to-date summaries of the East available in the West—a guide to what European travelers might expect to encounter there if they ever managed to find their way back.

Sir John constructed his itinerary artfully, moving from west to east. He begins at the very western edge of the known world—at St. Albans, where Matthew Paris had lived and worked a century before. The book starts out as a kind of itinerary guide, taking readers through the regions they might pass through on a pilgrimage to Jerusalem. Sir John spends considerable time in the Holy Land but then moves into Asia, where he encounters just what any fourteenth-century reader in Europe would expect him to:

fabulous beasts, strange peoples, monstrous races of various kinds, miraculous natural phenomena, Gog and Magog, Prester John, and the Great Khan. But after reaching Cathay and the Indies, Sir John doesn't turn back, as Marco had done. Instead he makes an audacious move. Having traveled to the very eastern limits of Asia, he sets his sights on the place where "the earth begins": the Earthly Paradise itself. Disarmingly, he admits he hasn't been there himself; such a feat, he explains, is impossible.

> You should realize that no living man can go to Paradise. By land no man can go thither because of the wild beasts in the wilderness, and because of the hills and rocks, which no one can cross; and also because of the many dark places that are there. No one can go there by water either, for those rivers flow with so strong a current, with such a rush and such waves, that no boat can sail against them. There is also such a great noise of waters that one man cannot hear another, shout he never so loudly. Many great lords have tried at different times to travel by those rivers to Paradise, but they could not prosper in their journeys; some of them died through exhaustion from rowing and excessive labor, some went blind and deaf through the noise of the waters, and some were drowned through the violence of the waves. And so no man, as I said, can get there except through the special grace of God.

Sir John does recount what the inhabitants of regions near the Earthly Paradise have told him about the place. It sits atop land so high, he says, that it escaped the ravages of Noah's flood. A great wall surrounds it, thickly overgrown with moss and brush. Deep inside the wall—behind a region of perpetually raging fire—lies a giant well, the source of the four biblical rivers of Paradise: the Phison (Ganges), the Gyon (Nile), the Tigris, and the Euphrates. Rushing down from the heights, the rivers flood the adjacent Indian lands—creating the thousands of islands found there—and then sink below the surface of the earth, where they flow thousands of miles underground before rising to the surface again in their respective portions of the known world.

Soon after this description of the Earthly Paradise, Sir John brings his book to a close. But elsewhere in his account he has already taken readers even farther east, on a voyage through what he describes as "those countries

girdling the roundness of the earth and sea." It's here, in a memorable anecdote designed to illustrate the roundness of the earth, that he brings his
readers full circle.

> I have often thought of a story I heard when I was young, of a worthy
> man of our country who went once upon a time to see the world. He
> passed India and many isles beyond India, where there are more than
> 5,000 isles, and traveled so far by land and sea, girdling the globe, that
> he found an isle where he heard his own language being spoken. For he
> heard one who was driving a plough team say such words to them as he
> had heard men say to oxen in his own land when they were working at the
> plough. He marveled greatly, for he did not understand how this could
> be. But I conjecture that he had traveled so far over land and sea that,
> circumnavigating the earth, he had come to his own borders.

Those isles environing all the roundness of the earth and of the sea. What Sir
John is describing is the East as it appears on the elaborate Christian *mappaemundi:* a three-part known world, surrounded by an island-studded ring
of ocean, with the Earthly Paradise lying somewhere at the very beginning
of the East. Although Sir John had made his journey to the outskirts of the
Earthly Paradise by traveling overland to the east, the implication of his
worthy-man anecdote was obvious. Sail west from Europe far enough out
into the ocean, and sooner or later you'll reach Paradise—which is exactly
what had happened, centuries earlier, to an Irish monk named Brendan.

*   *   *

ANYBODY WHO HAS spent time on the west coast of Ireland knows what
to expect. Rain and fog can move in for days at a time, and with them comes
an atmosphere of bleakness and obscurity so pervasive that the island really
does seem to occupy a place at the ends of the earth. But without warning
the skies can clear, and when they do they unveil a breathtaking sight: a vast
soul-replenishing expanse of lapis-colored ocean, out of which rise up small
islands glowing a rich emerald green.

Such moments are what drew the early monks of Ireland out into the
ocean. The island was converted to Christianity in the fourth and fifth
centuries, and by the sixth century it was home to scores of monastic

communities. These were busy places—too busy for the liking of some monks, who sought out a more solitary way of life. Elsewhere this might have meant heading for the mountains or the desert. In Ireland it meant heading out to sea.

So began centuries in which the Christian monks of Ireland improbably bobbed and floated and rowed and sailed their way out into the northern Atlantic—first setting themselves up on the scores of small islands just off Ireland's west coast, then moving up in the Faroes, to the north of Scotland, and finally even making their way north and west to Iceland. The monks, it seems, believed that the waters that they were exploring could ultimately take them all the way to the East. One account of the life of the earliest and most famous of them, Saint Columba, records that seven "children of the king of India" sailed across the sea to Ireland to receive baptism from him.

Sometime in the eighth century, an anonymous Irish monk composed a work known today as *The Voyage of Saint Brendan the Abbot*. This Brendan was a real person: early Irish annals show him to have been a sixth-century monk who built churches, established a monastery, and traveled widely.

The story of Brendan's famous journey, as recounted in the *Voyage*, goes something like this. One evening, as Brendan is tending to the business of his monastery, a monk by the name of Barrind, who has just returned from a long sea voyage, arrives before him and collapses in great spasms of tears and prayer. When pressed by Brendan to explain himself, Barrind describes having reached an island that he calls the Promised Land of the Saints. "A fog so thick covered us that we could scarcely see the poop or prow of the boat," he tells Brendan. "But when we had spent about an hour like this, a great light shone all around us, and there appeared to us a land wide, and full of grass and fruit. . . . We could not find the end of it."

Inspired by Barrind's story, Brendan gathers together a crew of fellow monks, oversees the construction of small sailboat, and launches an expedition in search of Barrind's island (*Plate 3*). So begins a circular sequence of mystical island hopping that Brendan and his companions are fated to repeat annually for seven years. Among other adventures, they find island paradises full of talking birds; they celebrate Easter on a rocky outcropping that turns out to be a whale; they sail for four days around a mysterious "pillar in the sea" made of "bright crystal" (a giant iceberg); they venture so far north they arrive at the point where the sea coagulates—and then they do it all over again. And again. When their seven years are up, Brendan and his companions sail for forty days and finally reach the Promised Land of the Saints, which appears exactly as Barrind described it: hidden behind a great fog, eternally awash in the light of Christ, apparently limitless in extent. The monks explore the island for another forty days, and then make their way home, where Brendan soon expires.

The Voyage of Saint Brendan the Abbot would enjoy an unusually wide readership in Europe during the Middle Ages. Like the works of Marco Polo and Sir John Mandeville, it was translated into a great number of languages, and more than a hundred manuscript copies still survive today. The work exerted a pull on medieval geographers, who by as early as the twelfth century had begun including the islands of Saint Brendan—or just Saint Brendan's Isle—in their descriptions of the world. "There is in the ocean," one authority recorded in 1130,

> a certain isle agreeable and fertile above all others, unknown to men but discovered by chance and then sought for without anyone being able to find it again, and so called the "Lost Isle." It was, so they say, the island whither once upon a time St. Brendan came.

By the next century Saint Brendan was making routine appearances on *mappaemundi*, among them the huge late-thirteenth-century Ebstorf Map, where it appears to the west of Africa, accompanied by the following legend: "The Lost Island. St. Brendan found this. It has been found by no man since he sailed from it."

* * *

AT JUST ABOUT the time the Ebstorf Map was being made, a new and radically different style of map was beginning to circulate in Europe: the marine chart. It too would change the way Europeans looked at the world, and help draw them out to sea.

The marine chart, also known as the portolan, makes its first definitive appearance in the historical record on July 2, 1270. That month King Louis IX of France set sail for the Holy Land, at the head of the Eighth Crusade. The expedition got off to an inauspicious start. A storm struck almost immediately, blowing King Louis's ship off course and dispersing his fleet. When the heavy weather finally passed, the king anxiously demanded that he be shown the location of his ship—and his crew produced a map on which they were able to point out that they were not far from Cagliari, a town on the coast of the island of Sardinia.

This account appears in a life of King Louis written by a Benedictine chronicler. The author refers to the map as a *mappamundi*, but no map known today by that name could have helped sailors explain their whereabouts to the king. What they must have shown him, and what his landlubber of a biographer probably didn't have the vocabulary to describe, can only have been a marine chart. As it happens, the oldest surviving marine chart dates from almost exactly the same time, approximately 1275, which makes it possible to imagine with some specificity what Louis's chart looked like. Known as the Carte Pisane (*Figure 20*), the chart was made in Italy, as were all of the earliest surviving marine charts (*Figure 21*).

Since antiquity, European sailors in the Mediterranean and the Indian Ocean had taken written sets of instructions with them on their voyages. These instructions, some of which still survive, allowed sailors to recognize coastal landmarks and navigate safely from port to port. The best-preserved of these guides is the first-century *Periplus of the Erythraean Sea.* "Before the harbor," reads a typical entry, describing Egypt's Red Sea coast, "lies the so-called Mountain Island, about two hundred stadia seaward from the very head of the bay, with the shores of the mainland close to it on both sides." It's possible that sailors in antiquity and the early Middle Ages also relied on some form of illustrated chart to accompany such descriptions, but if they did none has survived, and no unambiguous reference to an actual chart exists from before the thirteenth century. The Carte Pisane therefore arrives like a bolt out of the blue, revealing a new kind of world: fully formed and entirely without precedent.

The world of the early marine charts is a revelation. Looking at it after puzzling over the fuzzy, figurative world of the *mappaemundi* is like putting on a pair of glasses and having everything snap into focus. The effect is most dramatic when you look at Italy. Declared by Pliny the Elder to have the shape of an oak leaf, and routinely depicted by Matthew Paris and other medieval mapmakers as simply a stubby rectangular peninsula (*Plate 4*),

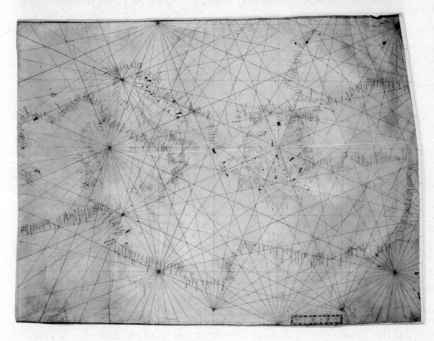

Figures 20 and 21. Top: the Carte Pisane, the oldest surviving European marine chart (circa 1275). Drawn on a single piece of vellum, the chart shows North Africa, Europe, and the Mediterranean. Italy is at the center. *Bottom:* An anonymous marine chart (fourteenth century) displaying the characteristic spare design of the Italian school: rhumb lines, coastlines crowded with names, and empty continental interiors. Europe is at the top left; North Africa is at the bottom; the Holy Land is on the right.

Italy on the marine charts is transformed into what we know it to be today: a boot (*Plate 5*).

Almost two hundred marine charts survive from before 1500—a number that represents only a tiny fraction of the charts that must have actually circulated. Most charts were used under punishing conditions at sea, not contemplated dreamily on land, and they eventually wore out, were lost, or were discarded when newer versions became available. The charts that do still survive are primarily decorative copies, made for the personal libraries of wealthy shipowners and merchants.

By the early fourteenth century most marine charts had taken on a standard look. Usually drawn on a single rectangular piece of parchment, they showed the Mediterranean Basin, the Black Sea, and, in some cases, a limited portion of the Atlantic coasts of Europe and North Africa. They provide details only along the coasts; what lay inland naturally mattered little to sailors who consulted the charts only to orient themselves at sea and to locate ports and natural coastal features. Today the charts are easiest to make sense of by looking at them with north at the top, but in fact they were designed to be turned in all directions, and looked at from all angles, depending on a ship's heading. This explains why place-names on the charts always follow the contours of the coasts.

The most instantly recognizable feature of all the marine charts is the elaborate web of lines that covers them. These are rhumb lines, or loxodromes: fixed lines of direction that radiate out from a wind rose at their center (*Figure 22*). A small diagram designed to allow sailors to determine course headings, the wind rose takes its name both from its floral shape and from the fact that around its petals it showed the various winds that early European sailors used to describe directional headings. To sail north, for example, might be to sail with Africus (the south wind), and to sail south might be to sail with Boreas (the north wind). But actually to make use of the rhumb lines and wind roses that appeared on their charts, sailors needed something else: a mariner's compass.

Where and when the mariner's compass was first invented is unclear. Both China and Europe claim the invention as their own; each may in fact have come up with it independently. References to the device begin to appear in both regions at about the same time. In twelfth-century China,

Figure 22. Wind rose from the marine chart of Jose Aguiar (1492).

writing about commercial voyages between Canton and Sumatra (just the sort of voyages Marco Polo would make a century later), a certain Chu Yü noted that "in dark weather," when the stars are no longer visible, sailors navigate by looking at "the south-pointing needle." In Europe, in 1187, Alexander Neckam of St. Albans recorded a similar practice. "Mariners at sea," he wrote, "when, through cloudy weather in the day which hides the sun, or through the darkness of the night, they lose the knowledge of the quarter of the world to which they are sailing, touch a needle with a magnet, which will turn round till, on its motion ceasing, its point will be directed towards the north."

Such early compasses were rudimentary instruments, often involving nothing more than a magnetized needle that floated in a bowl of water or balanced on a pin and pointed north—a good backup when all else failed, but certainly not a practical tool for use at sea. By 1269, however, just before King Louis set sail on the Eighth Crusade, an experimental scientist named Petrus Peregrinus had devised a more sophisticated version of the compass, one much better suited to the challenges of life at sea, by inserting a magnetized needle into a solid box and covering it with glass or crystal. "Divide the cover first into four parts and subdivide these into 90 parts," he proposed. "Mark the parts north, south, east and west. . . . Then turn the vessel until the needle stands in the north and south line already marked on the instrument. . . . By means of this instrument you can direct your course towards cities and islands and any other place wherever you wish to go by land or sea." Peregrinus, however, was a scholar, not a sailor—and sailors are a notoriously superstitious bunch. Many thirteenth-century sailors, one contemporary source records, refused to have anything to do with the new invention, considering it "an infernal spirit."

Wherever the compass came from, and however sailors began to adopt it, its practical uses were immediately obvious. Consider the situation that confronted King Louis and his crew after the passing of the storm that had blown them off course. Their vessel sat at the center of a featureless circle of ocean, the circumference of which was the horizon. With clouds still overhead, no sun or stars would have been visible. But even the most rudimentary of magnetic needles would have allowed his ship's navigator to determine the direction of the winds that had blown the ship off course, and thus to turn it back where he knew land would be. And once Sardinia came into view they would have been able to compare its coastal features with what appeared on their chart—and then, using their compass and their wind rose, would have been able to chart a course for the port of Cagliari.

By allowing sailors to fix the points on their horizon, the mariner's compass allowed them to take precise bearings and plot sight lines from one point of land to another. This, in turn, allowed them to draw their remarkable charts. The invention—or at least the arrival—of the mariner's compass in Europe in the late twelfth century therefore may explain the sudden appearance of the marine chart in the following century.

One thing that marine charts *didn't* do was take account of the curvature of the earth. This meant that the bigger the area that appeared on a chart, the more distorted it would be. But because the region that appears on the early marine charts amounted to only a small portion of the globe, this distortion was inconsequential—especially when compared with the many other advantages that marine charts offered sailors. By far the greatest was this: using their charts and their compasses, and without worrying about the visibility of the stars, sailors could set sail with increasing confidence across long stretches of open water.

It was a development that would have momentous consequences. By the 1270s European merchants had begun slipping through Muslim shipping lanes in the Straits of Gibraltar and out into the Atlantic, on trading voyages to Portugal, Spain, England, and the Netherlands. These voyages, which skirted the western edge of the known world, opened up new oceanic horizons for European sailors, to the south and the west—and, in 1291, prompted two brothers from Genoa, Ugolino and Vadino Vivaldi, to come up with an ambitious plan. The brothers fitted out two galleys with food, water, and supplies; gathered up a crew (including two Franciscan monks); and, in May of 1291, not long before Marco Polo returned from his epic journey, set sail for the open ocean. An entry describing their intended destination survives in the Genoese annals for the year 1291 and records that the brothers planned to sail "through the Ocean Sea to parts of India and to bring back useful merchandise from there."

What's not recorded is what the Vivaldi brothers intended to do after passing through the Straits of Gibraltar. Did they intend to turn south and hug the west coast of Africa, nosing their way into the torrid zone in search of an eastward passage around Africa to India—a passage clearly visible on zonal maps of the world? Or, emboldened by their compasses and by stories of ocean wanderers like Saint Brendan, whose island makes an appearance on some early marine charts, did they intend to reach India by sailing across the ocean to the west?

Nobody knows, because the brothers never came home.

• The North Wind •

CHAPTER FIVE

SEEING IS BELIEVING

Every point of the earth is the center of its own horizon.

—Roger Bacon (1267)

By the early fourteenth century, marine charts were no longer just for sailors. The Genoese chart maker Petrus Vesconte, for example, saw no reason why the techniques he used to make his charts shouldn't also be applied to the making of a new kind of world map. So he gave it a try.

The result is a fascinating hybrid—and a cartographical first (*Figure* 23). The map's overall scheme is that of a traditional *mappamundi*: it shows a round world divided into three parts, centered on Jerusalem, and oriented with east at the top. Africa and Asia are as vaguely depicted as ever, but the contours of Europe, the Mediterranean basin, and the Black Sea are those that appear on the marine charts. Vesconte also covered his map with a network of rhumb lines. This was nothing more than a symbolic gesture, given that he had very little new information about Asia or Africa to present, but it sent an unambiguous message: with the help of the compass and the marine chart, a better picture of the world is possible.

That message soon made its way beyond the mapmaking world. In 1321 the Christian propagandist Marino Sanudo sent Vesconte's new hybrid world map, along with several of his regional marine charts, to both Pope John XXII and King Philip of France. Sanudo included them all in

Figure 23. The hybrid world of Petrus Vesconte (1321), bringing together features of the *mappaemundi* and the marine charts. Asia is at the top; Europe is at the bottom left; Africa is on the right.

his *Liber secretorum*, or *Book of Secrets*, a work in which he enthusiastically and elaborately laid out the case for a new Crusade. "Whosoever exercises the leadership of the Crusade must wholeheartedly follow the directions as proposed in the *Book of Secrets*," Vesconte would later write to King Philip. "The Crusade leader should study and pay close attention to the map of the world, and pay very careful attention to the maps showing Egypt, the Mediterranean, and the Holy Land. If these precautions are followed, with the help of God, this venture will come to a victorious conclusion."

The worlds of the marine chart and the *mappamundi* also come together in one of the most lavish and famous of all surviving medieval world maps: the Catalan Atlas (*Plate 6*). Drawn in 1375 by Cresques Abraham, a Jewish mapmaker who lived and worked on Majorca, the map is another fascinating hybrid. In the west it shows Europe, North Africa, the Mediterranean, and the Black Sea in the precise style of the marine charts; in the south it presents a newly full picture of northern Africa, incorporating the geographical knowledge of the Majorcan Jews, who were able to travel between Christian Europe and Muslim North Africa; and in the east it resembles a fancifully drawn *mappamundi*, rich with illustrations and descriptions of such stock medieval characters as Prester John, the Three Magi, and Gog and Magog. But the Catalan Atlas is important for another reason. It's one of the first world maps to make a serious effort to expand the traditional vision of the East to include the places described by Marco Polo, Sir John Mandeville, and the early missionaries: Cathay and Manzi, and the dazzling new world of islands that lay beyond them to the east and the south.

In grafting this vision of the Far East onto a map that also brought together the established worlds of the marine chart and the *mappamundi*, the Catalan Atlas provides a rich illustrated survey of the state of European geographical learning and ideas at the end of the fourteenth century. But other ways of depicting the world were beginning to emerge—including one that had occurred during the previous century to an English Franciscan named Roger Bacon.

* * *

BACON WAS A polymath: a theologian, a philosopher, a scientist, and a geographer. During the mid-1200s he spent time teaching at the University of Paris, one of the great centers of medieval learning. During his time in the city, sometime in the late 1250s or early 1260s, Bacon encountered Friar William of Rubruck, who had recently arrived there from the Holy Land after delivering his report on the Mongols to King Louis. Bacon pored over the report with great interest, and the two men evidently spent much time discussing what William had seen and heard on his travels. Bacon was especially concerned about the imminent advent of the Antichrist. Like Matthew Paris and others, he wondered about the relationship of the Mongols to Gog and Magog, and therefore paid very careful attention to what William had to say about the Mongols: where they based themselves, how fast they could travel, what their strengths and weaknesses were, what religions they seemed to favor, what their imperial ambitions and political intentions seemed to be. He listened with a mixture of fascination and dejection, too, as William told him about the unexpectedly broad geographical extent of the East, and the vast numbers of non-Christians living there. He asked lots of questions. What were the true contours of the Caspian? How far off were the Gates of Alexander? Where *exactly* were Karakorum and Tibet and Cathay? Where could communities of Christians be found?

Neither William nor anybody else could answer these questions with any precision. In Bacon's view this was a serious problem. How could Christians—Latin Christians—win the struggle against the world's infidels and idolaters, and bring the Word to the ends of the earth, if they didn't have a good idea of what the earth looked like? It was time for a new kind of map—and Bacon had some ideas about how to make one.

In 1266, he got a chance to put those ideas into practice. Pope Clement IV had ascended to the papacy just the year before, and upon taking the job had found himself overwhelmed by the many challenges of his new office, among them the question of how to deal with the Mongols in the east and the Muslims in the south. So Clement wrote a letter to Bacon, whose learning he admired, and made an urgent request. Could Bacon in all haste please send "writings and remedies for current conditions"?

Perhaps Clement expected something along the lines of a policy paper: a small brief containing advice that would be of practical use to the world's busiest statesman. What Bacon sent Clement a year later was something altogether different: an astonishing half million words or so, covering . . . just about everything.

The work, known as the *Opus majus*, or *Great Work*, was an encyclopedic argument for a new, experience-based approach to learning, designed to provide Christians with the tools required to understand—and, implicitly, control—the world. Drawing not only on the ancient authorities of Greece and Rome but also on his own experiments and research, as well as the more sophisticated contemporary writings of Muslim and Jewish thinkers, Bacon tackled subjects as varied as alchemy, astronomy, biology, calendar reform, cosmology, geography, grammar, linguistics, mathematics, and moral philosophy. Guiding him throughout was a single principle. "All the sciences are connected," he would later write; "they lend each other material aid as parts of one great whole, each doing its own work, not for itself alone, but for the other parts."

Knowledge was a relational web of information for Bacon. And so was geography itself. Understanding the places of the world, he told Clement, required analyzing and collating the firsthand accounts of "those who have in great measure traveled over the places of the world." Such observations, properly synthesized into a new picture of the world, would be invaluable on many levels. Detailed geographical knowledge was critical for missionary efforts. "Men without number," he wrote, "have failed to succeed in the most important business of Christendom, simply because they did not understand the immense differences between the regions of the world." Knowledge of the world was also critical for a proper understanding of the scriptures, which abounded with geographical references.

> He who has gained a good idea of places, and has learned their location, distance, height, length, width, and depth, and has tested their diversity in heat and dryness, cold and humidity, color, savor, odor and beauty, ugliness, pleasantness, fertility, sterility, and other conditions, will be pleased very greatly by the literal history [as laid out in the Bible], and will be able easily and admirably to gain an understanding of the spiritual meanings. For there is no doubt that corporeal roads signify spiritual roads.

Travelers like Friar William and Friar John could provide much of this sort of information. But how could anybody pin down exactly where they had been? Answering this question, Bacon felt, required looking at the world from a new perspective. The earth was a sphere at the center of the cosmos, so why not start there? Why not consider it a mathematically definable object that could be mapped according to the principles of geometry? The thought sent Bacon, who had a special expertise in the field, into a rapture. "The whole truth of things in the world," he told the pope, "lies in the literal sense, as has been said, and especially of things relating to geometry, because we can understand nothing fully unless its form is presented before our eyes."

Bacon was laying down a challenge to the pope. Forget what you think you know about the world. What does it *really* look like?

* * *

THIS WAS NOT a question that could be asked in isolation. Understanding the world meant understanding the cosmos—and that meant turning to Aristotle.

Much of the literature of ancient Greece had been lost or forgotten in Europe during the early Middle Ages, but Muslim and Jewish scholars had preserved and studied it carefully for centuries, some of them in the great centers of learning that had been established in Spain. By the end of the twelfth century, through contacts with the Muslims and the Jews, Christian scholars in Europe began to translate the work of various important Greek authors, one of the first and most significant of whom was Aristotle.

Aristotle was an empiricist. He looked at the world around him and tried, in a systematic and orderly fashion, to describe it. When it came to describing the cosmos, he observed how the land, the sea, and the skies seemed to fit together, and then came up with a system that explained everything he saw. The visible cosmos, he felt, could be divided into two regions, the material and the celestial. The material region consisted of four elements: earth, water, air, and fire, each of which occupied its own sphere. But these spheres were not perfectly self-contained: land is found above water, after all, water is found in the air, air is found trapped in land and water, and fire can burn on land and in the air.

The heaviest of the material elements was earth. This was obvious: if you dropped a stone it would travel in a straight line down through air and water until it hit land. Water came next, pooling as it did above land but below the air. Fire was the lightest element: flames rose straight upward through the air, became invisible, and gathered in a sphere at the outer limits of the material region. Earth, water, air, and fire: they all wrapped around one another in concentric, if imperfect, spheres, drawn in toward a common center by a sort of universal gravitational field. As the heaviest element, earth—and thus the earth—by definition came to rest at that center.

Above the material region was the celestial region: another set of concentric spheres, each fitted smoothly within the next, and all pressed in tightly around the material spheres. The moon, the sun, and the planets each had their own sphere, which revolved at its own pace around the material region, and above them all was the firmament: a single sphere in which all of the stars were fixed. At the outer circumference of all the cosmic spheres, enveloping everything else, was an invisible and rapidly spinning sphere that Aristotle called the *primum mobile*, or "first moved"—so named because its rotation set all of the other spheres in motion. And beyond that?

Things got a little fuzzy. All Aristotle could say was that it was a region of "neither place nor void nor time."

Aristotle laid out the architecture of his whole system with grace and concision. But he provided few mechanical details, and his successors in antiquity, notably the second-century Greek astronomer and mathematician Claudius Ptolemy, had to exert themselves manfully to explain, say, why the planets behaved erratically, apparently jumping forward and backward as they progressed around the world. To explain such inconsistencies, Ptolemy proposed an ingeniously convoluted system of planetary motion; the logic was somewhat tortuous and the math complex, but the model predicted the movements of the planets better than any other.

Such was the picture of the cosmos that arrived fully formed in Europe in the late twelfth century. It's this very picture that Matthew Paris would diagram just a few decades later (*Figure 6*). The model appealed greatly to an emerging school of Christian thinkers and logicians known as the scholastics, who, since the beginning of the century, had been trying to reconcile medieval theological ideas with the secular teachings of the ancients. By the thirteenth century the scholastics had managed to transform Aristotle's physical model of the cosmos into a specifically Christian one: the *machina mundi*, or "world machine." Like so many other elements of medieval thought, traces of this model of the cosmos still survive today; we allude to its workings whenever we talk about finding "our own element," being in "seventh heaven," or having a "sphere of influence."

The fit between Aristotle's cosmology and Christian theology seemed so very apt. Relying only on observable phenomena and commonsense deduction, Aristotle showed the world to be at the center of the cosmos—which, of course, was exactly where the scriptures had also revealed it to be. The settling of the various material elements into their separate spheres corresponded nicely with the biblical account of the creation, too ("And God said, Let the waters under the heaven be gathered together unto one place, and let the dry land appear: and it was so.") As for that most distant region of "neither place nor void nor time" that Aristotle had placed beyond the *primum mobile*, what better way to characterize the ineffable, all-encompassing mind of God? "This quiet and peaceful heaven," wrote Dante, who had been schooled in the scholastic tradition, "is the abode of that Supreme Deity who alone doth perfectly behold Himself."

The fit was good, and easy to diagram (*Figure 24*)—but it wasn't perfect. As the thirteenth century wore on and the scholastics tried to explain the idiosyncratic workings of the world machine, they found themselves confronting a vexing incongruity. If the earth was supposed to be surrounded by water (like a yolk surrounded by the white of an egg, as some of the scholastics described it), and if the whole system involved a series of concentric spheres with the earth at its center, then reason dictated the whole earth should be submerged. So how could it be that the known world was exposed to the air?

One way of answering the question was to take the Bible at face value: the land was kept dry by a divine miracle, and so no explanation was necessary. Sacrobosco adopted this approach in his *Sphere*. Water, air, and fire,

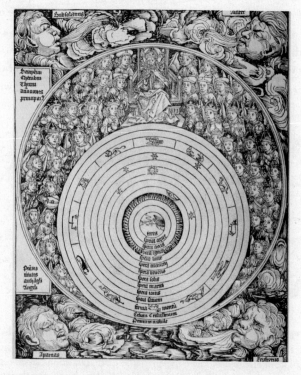

Figure 24. The Christian cosmos (1492). Earth is at the center, the *primum mobile* is at the circumference, and presiding over everything, in a region beyond space and time, are God and the heavenly host.

he wrote, "surround the earth on all sides spherically, except in so far as the dry land stays the sea's tide." How exactly the dry land did this, however, he didn't say.

Some thinkers demanded an explanation. Even if land had initially been exposed to the air by a miracle, there had to be a physical reason that it now stayed that way. The scholastics grappled with this question during the thirteenth century and eventually came up with a theory to explain it—one that would remain popular until well into the sixteenth century, when Nicholas Copernicus would finally break with tradition and propose that the earth revolved around the sun.

The theory had a fetchingly circular logic. The land remained free of water because it became lighter where God had caused it to be mixed with the air. As a result the earth's center of gravity had shifted from the exact center of the cosmos; in effect, it had bobbed to the surface of the ocean on one side of the watery sphere, just as an apple put underwater in a bathtub bobs to the surface (*Figures 25 and 26*).

It was a neat solution. Like that apple in the bathtub, most of the earth was covered with water, but a portion of it—the known world, which one

Figures 25 and 26. Two different views of the center of the cosmos.
Left: the earth at the exact center, surrounded by the sphere of water, then air, then fire (detail from Figure 24). *Right*: the off-center earth. Drawn in classic zonal form, the earth here is the smallest circle. It has bobbed to the surface of the water (shaded with wavy horizontal lines), exposing part of itself to the air and making life on land possible. From a 1499 edition of Sacrobosco's *The Sphere* (early 1200s).

scholastic writer called the "front face of the earth"—floated above the surface. The theory seemed to resolve another important debate. If the earth had bobbed to the surface of one side of the watery sphere, then, by definition, the farther that one sailed away from the shores of the known world, the deeper the ocean had to get, which meant that on the opposite side of the globe from the known world no land could possibly be exposed. The Antipodes, therefore, must not exist.

* * *

BACON HIMSELF didn't tackle the question of the Antipodes, but he did try to figure out the ratio of exposed land to water on the earth. Only once he was able to determine the actual size of the known world, he recognized, could he begin to render it geometrically as a part of the whole globe. But the ancient authorities were divided on the subject, he told the pope. Ptolemy had proposed a ratio of one to six, whereas Aristotle had proposed more than one to four.

Bacon sided with Aristotle. Citing ancient accounts of far-flung voyages, he argued that the known world from west to east had to stretch around at least half the circumference of the globe, and possibly more. "Aristotle," he wrote, "suggests that the sea between the west of Spain and the eastern edge of India is of no great extent . . . [and] Seneca informs us that this sea can be crossed in a few days if the wind is favorable." Bacon also turned to an unusual religious source to back his views up: the Bible's Book of Esdras (also known as Ezra), a controversial work generally rejected by Roman Catholics but accepted by Orthodox Christians. "Esdras tells us in his fourth book," Bacon reported, "that six parts of the earth are inhabited, while the seventh is covered with water. Nobody should question the authority of this passage by claiming that this book is apocryphal and of dubious authority; everyone knows that the saints of old used this book constantly, to confirm the sacred truths, and even used this book's pronouncements in the divine office."

There was more. Not only did land cover much of the world's northern habitable zone, it seemed to do the same in the southern hemisphere. Bacon found evidence for this in Pliny the Elder's *Natural History*, which records an anecdote about Taprobane—a giant island "southeast of India," where it extended 1,250 miles from north to south. "Some men from this place came to Rome in the reign of Claudius," Bacon wrote. "They were astonished to find that their shadows fell to the north, and that the sun traveled to the south." The cause of the men's astonishment was obvious to Bacon: Taprobane, or at least part of it, lay below the equator.

Bacon, then, imagined that the known world extended around much of the globe from east to west, and beyond the equator to the south. In this view a relatively narrow channel of ocean divided Spain from India. Bacon drew a curious little picture to illustrate this channel in his *Opus majus*: a dumbbell-shaped diagram, oriented with west at the top, delineating the global extent of the ocean and visually emphasizing the proximity of ends of the West to the beginnings of the East (*Figure 27*).

Bacon then presented Clement with his most original geographical proposal: a technique for mapping the world using mathematical coordinates.

Since these zones and their famous cities cannot well be described by words alone, a map must be used to make them clear to our senses. I shall, therefore, first present a map of our quadrant [quarter of the globe], and on it I shall label the important cities, each in its own place, with the distance from the equator—what we call the latitude—of the city or region.

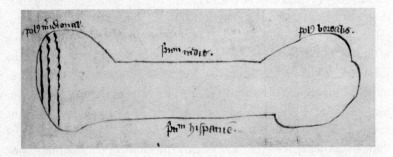

Figure 27. Diagram by Roger Bacon showing the small size of the sea separating East from West (circa 1267). West is at the top, and the north pole at the right; everything inside the diagram is water. The circles at the ends of the diagram are the two polar regions, where, according to Bacon, cold causes a "natural pile-up of water" that exposes land elsewhere on the globe. Between the poles is the narrow sea channel (the Atlantic) that Bacon believed lay between the beginning of India ("*principium Indie*") and the beginning of Spain ("*principium Hispanie*").

I shall also label them according to their distance from the east or west, what we call the place's longitude. In my assigning of zones, and likewise of latitude and longitude, I shall make use of the prestige and experience of the wisest scholars. To locate each city in its proper place [on this map] by its longitude and latitude, which have already been discovered by my authorities, I shall use a method by which their positions may be shown by their distances north and south, east and west. The device is this: parallel to the equator . . . a straight line is drawn. This intersects another straight line [running north to south], from the point corresponding to the number of degrees of latitude of the place. . . . This procedure is both easier and better [than anything now in use], and a map drawn in this way is quite capable of representing to the senses the location of any point in the world.

The "authorities" from whom Bacon drew his coordinate data were Arabic astronomers. But they hadn't compiled the latitude and longitude of various cities because they wanted to determine their location on the globe. At the time Bacon was writing, latitude and longitude were astronomical tools, not geographical ones, and were used to help predict how, at any given time and place, the planets and the stars would align themselves with a particular city—information that was necessary for, among other things, the casting of horoscopes. Bacon, however, realized that these geographical coordinates, scattered in tables that pertained to the movements of the heavens above different earthly regions, could be combined in such a way as to begin to develop a picture of the earth. But he, too, connected the places of the earth with what bore down on them from above. "This is the first axiom of our study," he told the pope; "every point on the earth is the apex of a pyramid that transmits the power of the heavens."

On Bacon's map—which, regrettably, is lost—the midpoint between north and south was easy to determine. It was the equator: the zero-degree point, located anywhere around the circumference of the globe, the place where the Pole Star disappears below the horizon and where day and night always last twelve hours each. But where did east or west begin? Here there was no natural zero-degree point, but choosing one was essential for deriving a uniform table of longitudes—that is, relative east-west distances from the same zero-degree point. Because Bacon relied largely on Islamic authorities for his data, he chose the midpoint along the equator that they had used in their tables: a mythical city called Aryne. The choice was an obvious and logical one, based on nothing other than mathematical necessity and already existing data. But its meaning would not have been lost on anybody familiar with the *mappaemundi* in vogue at the time: Jerusalem was no longer at the center of the world.

Bacon brought his proposal to a close with a lament. In drawing his map, he said, he had relied on the coordinate tables compiled primarily by his Arabic sources. These were the best available, he continued, but they didn't go far enough. "We sorely need more accurate ones," he wrote, "since the latitudes and longitudes of the Latin-speaking world and its cities have not yet been established. Indeed, they never will be, except under an apostolic or imperial decree, or the support of some great ruler willing to offer his backing to philosophers."

If Bacon had read his sources a little more closely, he might not have made that claim. Instead he might have noticed occasional references to an ancient geographical text containing just the kind of information he was looking for. In *On the Science of the Stars*, for example, translated into Latin in the twelfth century, the astronomer al-Battani spoke of "the system proposed by Bartholomeus and confirmed by the ancients, in which the places and regions of the world are noted according to latitude and longitude." This Bartholomeus, al-Battani went on to note, had laid out this system in "his book about the form of the world, called *Ieraphie*."

Who was this Bartholomeus? And what was his book? Bacon could have found the answer in another treatise translated into Latin at about the same time: *On the Operation of the Astrolabe*, by Ibn al-Saffar. Bartholomeus, it turned out, was none other than Claudius Ptolemy (whose name had been corrupted by its passage into Arabic and then Latin), and his book seemed actually to have been called the *Geography*. Ptolemy, al-Saffar explained, "collected the latitudes and longitudes of all known regions in a book called *Gerapphie*."

By the middle of the thirteenth century more such references to Ptolemy in the works of the Arabs were circulating in Latin translation. In discussing Ptolemy and his astronomical works one writer made passing mention of the "book of his that constructs the form of the habitable earth." Another spoke of the "book in which he named cities, islands, and seas," and yet another mentioned "his book titled *Mappamundi*." Ptolemy himself made a passing remark in his *Almagest*—a much-studied and translated astronomical treatise—that seemed to confirm that he had indeed written a stand-alone study of world geography. "What is still missing in the preliminaries," he wrote, after laying out tables of astronomical phenomena, "is to determine the positions of the noteworthy cities in each province in longitude and latitude. . . . But since the setting out of this information is pertinent to a separate, cartographical project, we will present it by itself."

But whatever this geographical work had been called, and whatever it had contained, everybody who mentioned it seemed to agree on one point: it was lost.

PART TWO

NEW WORLD

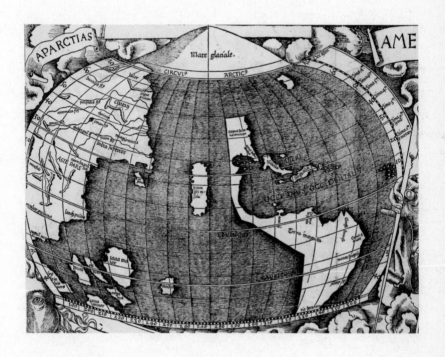

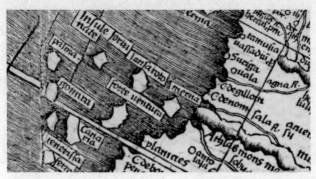

• The Canary Islands and northwest Africa •

CHAPTER SIX

REDISCOVERY

*Who can doubt that Rome would not rise up again if she but began
to know herself?*

—Francesco Petrarch (1341)

ON JULY 1, 1341, two sailing vessels and a small warship picked up the
ebb tide out of Lisbon harbor and headed southwest into the Atlantic, carrying with them "horses, arms, and various machines of war, built for taking
cities and castles." Chartered and outfitted by the king of Portugal, Alfonso
IV, the vessels sailed under the joint direction of Genoese and Florentine
mariners, and their mission, according to the sole surviving account of the
voyage, was to find "these islands we generally call 'rediscovered'"—a reference to two islands visited sometime earlier in the century by a Genoese
sailor named Lanzarotto Malocello, who seems to have been searching for
traces of the Vivaldi brothers' expedition.

The expedition didn't sail off into a void. In all likelihood, the ships'
navigators brought charts along that were much like the one drawn just two
years earlier, in 1339, by the Majorcan chart maker Angelino Dulcert, the
earliest surviving map to show Malocello's two islands. On Dulcert's chart
the islands appear not far off the southern coast of present-day Morocco,
at the time the farthest point south along the African coast to be charted by

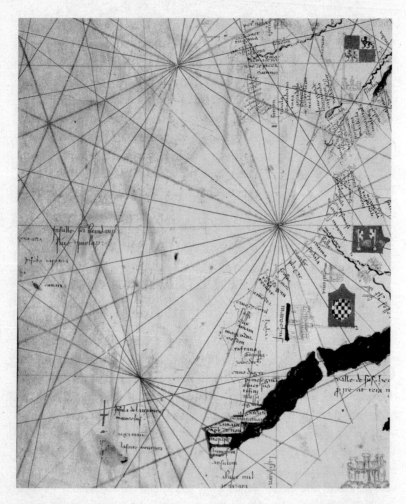

Figure 28. The islands of the Atlantic, from the marine chart of Angelino Dulcert (1339). Portugal and Spain are at the top right. Below them is the northwest corner of Africa. Offshore at the bottom, marked by a cross, are the two islands discovered by Lanzarotto Malocello. To their northwest are some of the Atlantic islands of legend: Saint Brendan's Isle and the Fortunate Isles, not yet associated with Malocello's islands.

European sailors (*Figure 28*). Somewhat to the northwest of them the chart also depicts Saint Brendan's Isle—the migration of which from the world of the *mappaemundi* to that of the marine chart demonstrates the broad imaginative pull of the Brendan legend at the time. Next to Malocello's two islands Dulcert drew the red cross of Genoa, as a tribute to their discoverer, and gave one of them the name Insula de Lanzarotus Marocelus. The island is known today as Lanzarote, one of the Canary Islands, but at the time of its discovery some Europeans assumed it to be one of the legendary islands known to the ancients as the Fortunate Isles.

Medieval Europeans knew the legend well. It dated back at least to the Greeks, who believed the gods sent many mortal heroes to the Fortunate Islands for an afterlife of gentle weather, agricultural abundance, and eternal bliss. The Greeks described the islands hazily as somewhere off in the ocean to the west of the Straits of Gibraltar, but in Roman times various authorities began to locate them with more precision. Writing in the first century, Plutarch reported that sailors had recently discovered two Atlantic islands some distance off the coast of Africa and noted that it was generally believed that these islands—blessed with gentle winds, moderate rains, and fertile soil—were the Fortunate Isles. Pliny the Elder got more specific in his *Natural History*, placing the islands a few hundred miles off the coast of North Africa and giving them individual names. "About the Fortunate Isles," he recorded, "Juba has ascertained that . . . the first island reached is called Ombrios . . . that the second island is called Junonia, and that there is a small temple on it built of only a single stone; and that in its neighborhood there is a smaller island of the same name, and then Capraria . . . and that in view from these islands is Ninguaria, so named from its perpetual snow, and wrapped in cloud; and next to it one named Canaria, from its multitude of dogs of a huge size." By the middle of the fourteenth century the Fortunate Isles were making regular appearances on the *mappaemundi*, roughly where Plutarch and Pliny had located them—and where Lanzarotto Malocello had rediscovered them.

After leaving Lisbon, King Alfonso's sailors picked up a favorable breeze. For five days they let it carry them south, toward the region where Dulcert's chart had put Malocello's islands, and finally they arrived at a chain of volcanic islands: rocky, heavily forested, dauntingly mountainous places, rich in supplies of fresh water and set amid surprisingly calm seas. They visited thirteen of them, and on six encountered a collection of primitive peoples who appeared never to have had contact with European civilization. These peoples, who would be wiped out entirely by European slave raids and colonial ventures, today are known as the Guanches.

Early in their visit Alfonso's sailors made for one of the biggest of the islands, and as they approached, a large throng of native men and women came to the water's edge to meet them. The islanders seemed interested in trade, but the sailors, made apprehensive by the language barrier, anchored a short ways offshore and held their distance. This prompted several young men to swim out to greet them—a gesture the Europeans rewarded by taking the men prisoner. Later the sailors grew bolder and raided a different portion of the island, where they discovered evidence of lives led in tranquility and abundance: vegetable gardens, fig trees, and palms; stockpiles of dried figs, wheat, and barley; houses artfully constructed of stone and wood (which the sailors broke into); and a small temple (which they looted).

The sailors explored some of the other islands, too. One seemed to be covered with nothing but "numerous trees, large and rising straight up to the sky." On another they found great numbers of pigeons and birds of prey. Another consisted of "extremely high rocky mountains, mostly covered with clouds, where the rains are frequent." From a distance they saw yet another island that appeared to rise some seventy miles in the air and turned white at its summit. After exploring the islands for some time, the sailors gathered up a sampling of what meager products they had to offer—a few animals, some goat and seal skins, natural wood dyes, fish oil, and tallow—and set out for home, bringing with them four of their prisoners. The account of the voyage that survives describes them in some detail.

The four men they took with them were young, beardless and handsome; they wore a loin cloth and had a belt of cord around the hips from which hung thick, long strips of palm leaves . . . with which to cover their shame in front and behind, unless the wind or something else lifted it. They are not circumcised and have long blond hair down to their waists, and they cover themselves with their hair, and walk with bare feet. The island from which they were taken was called Canaria, the most inhabited of them. They cannot understand anyone else's language, as they were spoken to in various languages. In height they do not pass ours. They are strong-limbed, lively, and very robust and of great intelligence, in as much as one can judge. We speak to them through gestures, and with gestures they respond, in the manner of the dumb. . . . They were shown gold and silver coins but did not recognize them. Nor with perfumes of any kind, nor necklaces of gold, etched vases, sabers and swords of any sort, the like of which it seems that they have never seen or heard.

No description exists of the sailors' return journey to Portugal, or of how they were received upon their return. But word of what had been discovered traveled fast. A second voyage to the islands was soon launched, this time by the Catalans, and by 1344 the news of the islands' discovery had reached Pope Clement VI, who decided he had the authority to grant the islands to a Franco-Spanish prince named Louis de La Cerda. After La Cerda died, in 1348, the rights to the islands once again became an open question, and during the remainder of the century both the Catalans and the Portuguese would repeatedly send missions to the islands to stake their claims.

It's worth pausing for a moment to consider the many dynamics at play in this little-known episode. With the financial backing of a king from the Iberian Peninsula, a maritime expedition led by Italian sailors sets off in search of islands rumored to exist somewhere out in the Atlantic. Guided by maps that contain a blend of the real and the imaginary, they find a series of islands, on which they encounter a pagan society previously unknown to Europe. They take prisoners, search for riches, loot and plunder, and finally return home, bearing samples of commercial goods, including human captives, that they believe might be of interest to merchants in Europe. Word of the islands spreads fast and reaches the pope, who considers himself empowered to grant ownership of them to whomever he sees fit—and who, when he does so, sets in motion a race among different Iberian monarchs to explore, exploit, and Christianize the region. These are the very dynamics, of course, that would assert themselves time and again in the Atlantic as the European Age of Discovery got under way.

The reason that the details of Alfonso's expedition have survived at all is that Florentine merchants in Seville interviewed at least one of its members after his return, and in November 1341 sent back to their colleagues in Florence a description of what they had learned. Although the original correspondence is lost, it seems to have been little more than a routine accounting of the expedition's commercial discoveries. But the correspondence arrived in Florence at a pivotal historical moment, just as the city was becoming an important center of geographical learning. The age of the humanists had just begun.

* * *

HUMANISM IS A maddeningly slippery term, but in the context of four-teenth-century Italy, where the movement was born, it has a very specific meaning: it describes a surge of interest in the works of the ancient Greeks and Romans, the development of a new kind of critical method for study-ing those works, and the gradual emergence of a program of education and cultural renewal based on classical thought.

As the early Italian humanists saw it, the fall of Rome in the fifth cen-tury A.D. had pushed Europeans into a centuries-long period of intellectual decline and social decay, and the only way to reverse the trend was to revive the wisdom and learning of antiquity. The humanists felt that Italians, liv-ing as they did in the former heartland of Roman civilization, were the natural heirs to this heritage—if only they would devote themselves to the difficult job of bringing it back.

The prime mover of early Italian humanism was Francesco Petrarch. Born in 1304 as the son of a Florentine law clerk, Petrarch traveled far and wide for much of his life in search of classical texts, many of which had lain ignored or little-known for centuries in the monastery libraries and private collections of Europe. He met with great success, bringing to light a remarkable number of forgotten texts, and as he realized the vast extent of what had been lost he grew hot with anger. "Each famous author of antiq-uity whom I recover," he fulminated,

> places a new offence and another cause of dishonor to the charge of ear-lier generations who, not satisfied with their own disgraceful barrenness, permitted the fruit of other minds, and the writings that their ancestors had produced by toil and application, to perish through insufferable neglect. Although they had nothing of their own to hand down to those who were to come after, they robbed posterity of its ancestral heritage.

This was an idea that Petrarch would return to repeatedly in his writ-ing over the years. His own era, he argued, was one of intellectual "darkness and dense gloom," whereas the ancients had lived in a golden age of learning. But he held out hope for the future. "There will follow a better age," he wrote at the end of his Latin epic poem *Africa*. "This sleep of forgetfulness will not last for ever. When the darkness has been dispersed, our descendants can come again into the former pure radiance." During the course of his career Petrarch defined the three-part view of European cultural evolution that dominates historiography today: first comes Antiquity, a glorious era of classical learning; then come the Middle Ages, long, dark centuries when the learning of the ancients is forgotten or lost; and finally comes the Renaissance, when that learning is revived.

Immense practical challenges confronted anybody in the fourteenth century trying to study the literature and history of antiquity. Students of the classics today can consult critical editions of the texts prepared by specialized scholars—a collective task of constant revision, emendation, and reconstruction that has been ongoing for centuries. These scholars do their work in collaboration with careful editors and reputable publishers, and the works themselves come helpfully supplemented with user-friendly overviews that discuss the sources of texts, the lives of authors, and the historical and literary contexts in which those texts appeared. For ease of reading and reference, they contain title pages, tables of contents, geographical glossaries, maps, diagrams, pictures, explanatory glosses, footnotes, bibliographies, indexes, and recommendations for further study. They use modern punctuation and are divided into chapters, sections, and paragraphs. Petrarch had none of this critical apparatus to help him. Instead he often had to grapple with untitled, undifferentiated, unreliably copied, and incomplete masses of text. Sometimes no author would be mentioned at all in the manuscripts he worked with; at other times the attribution might be wrong, improperly copied or mistakenly added by a scribe. Pliny or Seneca might be mentioned as the author of a text without any hint of what we now know: that there were two of each, Pliny the Elder and Pliny the Younger, and Seneca the Elder and Seneca the Younger. Trying to determine the actual age of a text was no easy task, and even distinguishing between medieval and ancient Latin authors was often problematical.

Petrarch embraced the challenge. What was necessary, he decided, was a sort of literary archaeology—an unprecedented and painstaking effort to locate, dig up, sift through, sort out, clean up, save, compare, contrast, correct, and study the scattered textual remains of antiquity.

It was an inspiring idea. But early on in his studies Petrarch identified a very basic problem. To appreciate classical literature and history, he realized, he needed to understand the geographical context in which they had been written and had taken place. This presented a host of difficulties. How could one study a history of the Roman campaigns in Gaul, for example, without knowing what the ancients had considered Gaul to be? Without a firm understanding of the geography of the ancient world, how could one fully appreciate and draw inspiration from epic poems such as Virgil's *Aeneid* and Homer's *Odyssey* and *Iliad*, which involved allegorical stories of travel in specific geographical settings?

The shards of ancient geography that Petrarch began to gather and piece together created a picture that was confusing at best, alien at worst, and in many respects almost impossible to square with contemporary reality. The names of ancient towns, rivers, mountains, lakes, seas, countries, and regions had all changed. Old cities had been destroyed, and new ones had arisen. Whole countries, peoples, and empires had come and gone. Parts of the world known to the ancients had dropped off the medieval map, and regions explored since the fall of Rome had been added. Even when continuity actually did exist, it was often hard to discern: scribal errors, compounded over generations by copyists who knew nothing of geography or antiquity, had spawned mutant spellings that varied greatly from manuscript to manuscript, and some had become entirely indecipherable. It was a mess—and Petrarch wanted to clean it up. He described his motivations at the outset of a biographical sketch of Julius Caesar. "About to write of the exploits of Julius Caesar in Gaul," he wrote, "I believed I should start by describing the [geographical] site of Gaul, where the events had taken place; its description itself has been made in such a confusing manner by some that it prevents an understanding of the facts."

But Petrarch didn't limit his focus to Gaul. He had something much more ambitious in mind: a full picture of the world as the ancients had known it.

* * *

PETRARCH WORKED TOWARD this goal in several ways. One of his first steps was to seek out whatever geographical information he could find in the encyclopedic and historical writings of the Romans. The most obvious source to turn to was Pliny the Elder's *Natural History*, a work not well-known in Italy at the time but commonly consulted in other parts of Europe. Reading the *Natural History*, Petrarch noticed that Pliny made several mentions of the work of a first-century geographer almost entirely forgotten in the fourteenth century: Pomponius Mela, a Roman citizen who had lived and worked in Spain. Petrarch made the search for Mela's work a part of his book-hunting expeditions, and sometime in the 1330s, possibly in papal circles in Avignon, he found what he was looking for: a copy of Mela's slender *De chorographia*, or *On Chorography*. Today Mela's little book is considered the oldest extant Latin treatise on geography and, like Marco Polo's Book, is often referred to as *The Description of the World*.

Written far more concisely and elegantly than Pliny's rambling *Natural History*, Mela's *Description* immediately appealed to Petrarch, who made a copy and soon began circulating it among his friends. Along with Pliny's *Natural History* and Solinus's *Gallery of Wonderful Things*, Mela's *Description* would become one of the most popular and widely consulted works of ancient geography in the coming two centuries—and almost every copy made during that period can be traced back to this copy that Petrarch made for himself.

In the decades that followed Petrarch relied on his copies of Pliny and Mela for help in deciphering the places and regions mentioned in classical literature and history. Although his original copy of the *Description* is lost, his copy of the *Natural History* survives, and its geographical sections demonstrate just how methodically Petrarch pursued his studies. He annotated the manuscript heavily, in effect creating a kind of rudimentary geographical glossary for himself that ran alongside Pliny's rambling narrative. When Petrarch came across the names of places and geographical features he would write them in margins of the manuscript, creating an easy-to-scan set of names that he could consult as he encountered unclear geographical references in other works. He also devised a simple system for signaling to himself the kind of geographical object that his marginal annotations referred to. Around the names of mountains he drew a little box that rose to a kind of peak above the word; under the names of rivers he drew a line that rose up vertically to the left of the word; to the left of the names of whole regions he drew a little vertical line; and over the names of cities he drew a line that descended vertically on the left.

Petrarch also marked up many of his manuscripts with notes about geography, one of the most lengthy and revealing of which appears in his copy of Virgil's *Aeneid*. The poem, an epic conceived of as a Roman analogue to the Greek epics of Homer, describes the ocean wanderings of the Trojan citizen Aeneas, who flees Troy after its fall to the Greeks. Aeneas sails to Italy, where he and his companions settle and become the ancestors of the Romans. Virgil intended the *Aeneid* to be read as the national poem of the Roman people, and Petrarch and the early humanists understood it as such, giving it a central place in the classical revival they were trying to engineer. Unpacking its geography therefore was a job of critical importance.

The long, revealing note that Petrarch made in his copy of the *Aeneid* appears at the point in the narrative when Aeneas and his sailing companions first catch sight of Italy. A harbor reveals itself to them, above which lies something that Virgil identifies as the temple of Minerva. "The harbor had been formed into the shape of a bent bow by waves blown from the east," the text reads. "It was hidden by projecting rocks which foamed with salt spray, and from the towered crags its two walls, like drooping arms, ran steeply down. The temple lay back from the shore."

What was this place? Petrarch threw himself into researching the question and eventually decided that Virgil had to be describing the Italian coast near Otranto, a port town at the southeastern tip of Italy, situated at the heel of the boot. He explained his reasoning at considerable length.

Many things cause errors concerning the knowledge of places, among them: the inaccessibility of regions to men of our age; the change of names, the rarity and lack of clarity of authors, and sometimes the dissent among them; but above all the lack of intellectual curiosity and the laziness of those who care for nothing that isn't right before their eyes. Not only the general reader but also scholarly commentators neglect to pause over these things. As for us, as much as we have been able, through a quite scrupulous survey of not only the works of authors, especially cosmographers, but also descriptions of the world and certain very ancient maps that have come into our hands, we have discovered that this place is located at the very corner of Italy, above or beyond Otranto, and is called the castle or camp of Minerva. When one crosses from [the east side of the Adriatic Sea] to the Italian coast, this is the first place that one comes to. . . . . As concerns this place, either we simply accept the term Temple of Minerva such as it is, because it was the first place sighted [by Aeneas], or, wanting to give the place a name, he [Virgil] used the word *temple* for *castle*, transposing one place for another, as was his custom. And indeed there is another place known by this name the Temple of Minerva, although Pomponius calls it a promontory on a different shore.

These remarks are illuminating on many levels. They lay bare how deeply Petrarch immersed himself in the study of ancient geography—and just how confusing and daunting the whole enterprise must have been. They show Petrarch deploying geography in the service of literary criticism, a defining element of the early humanist movement, and they show him playing different ancient texts and sources against one another, deliberately trying to expose and then resolve their contradictions—another humanist hallmark. By referring explicitly to "cosmographers," too, Petrarch departs from medieval tradition and grants to geography a newly independent status, making it no longer simply a subordinate element of other intellectual disciplines. And he breaks with tradition in another way as well: his remarks show that he was trying not only to revive ancient geography on its own terms but also to understand its correspondence to present-day reality.

Petrarch's reference to "very ancient maps" is particularly telling. Medieval *mappaemundi* couldn't possibly have supplied Petrarch with the kind of detail he needed to resolve the geographical questions he was trying to answer. The far more likely alternative is that he was referring to marine charts, which he may have considered to be a part of a tradition that extended right back to antiquity. In other contexts Petrarch is known to have relied heavily on marine charts to help him work out problems of ancient geography and to devise a geographical framework for his own literary creations. He is the earliest known writer, in fact, to have described Italy as having the shape of a boot—a reference he simply could not have made without having seen a marine chart. "Powerful on the sea and on land," he wrote of Italy in one of his letters, "by your very aspect you appear destined for empire, as if you would like to strike the world with your heel. Like a spur you stretch out your Otranto and oppose to the Nordic waves two-headed Brindisi."

Petrarch's reliance on marine charts comes across most clearly in his one explicitly geographical work, a pilgrimage guide titled *Itinerary to the Sepulcher of Our Lord Jesus Christ*. The great majority of the work describes the route south through Italy on the way to the Holy Land, and in effect it amounts to the textual equivalent of Matthew Paris's itinerary maps: a geographical description of a spiritual journey. This idea—that maps made armchair travel possible—was one that he clearly cherished. "I decided," he wrote in one of his letters, "not to travel just once on a very long journey by ship or on horse or on foot to those lands, but many times on a tiny map, with books and the imagination."

* * *

WITH BOOKS AND THE IMAGINATION: the phrase sums up one of the main ways in which Petrarch approached the study of geography. But it wasn't the only one. A widely traveled man, Petrarch knew his way around Europe better than almost all of his contemporaries, and he prided himself on his expertise. "I have been around almost all of the most distant borders of those regions," he wrote at the beginning of his life of Caesar, "either for leisure and the sole aim of seeing and learning, or on business." Petrarch felt that his travels qualified him uniquely for the job of deciphering the puzzles of ancient geography, and he felt that his knowledge of ancient geography could help him expand his understanding of the modern world. Remote places exerted a special pull on him. The search for knowledge about them became a powerful metaphor for the humanist project: the vast, collaborative, and never-ending attempt to see and understand the world as a whole. When he learned that his friend Philippe de Vitry, after an extended stay in Italy, had referred to his time there as an "exile" from France, he wrote to chastise him for his geographical small-mindedness. "Once India used to appear not too distant to you," he wrote. "At one time with eager mind you used to take measure of Thoprobanes [Taprobane] and whatever unknown places exist in the Eastern Ocean. At other times, you used to sigh for

Ultima Thule [probably Iceland, and signifying the northwestern limits of the world]. . . . In my opinion, you have forgotten that man, who, when asked where he was from, answered that he was a citizen of the world."

Medieval writers and geographers were generally content to place such quasi-mythical places as Taprobane and Thule wherever there was room for them at the margins of their maps. But Petrarch sought a new kind of precision. If such places really did exist, where *exactly* were they? Could they be rediscovered? Might finding them again, like rediscovering the lost texts of antiquity, help usher in that much-hoped-for new golden age? Petrarch never was able to pin down the location of Taprobane and Thule to his satisfaction, but he had better luck when it came to the Fortunate Isles. "Within the memory of our fathers," he wrote in 1346, just five years after King Alfonso's expedition to the islands discovered by Lanzarotto Malocello, "the warships of the Genoese penetrated to them."

How he had learned about the expedition, Petrarch didn't say, but chances are that he got word of it from somebody close to Pope Clement VI, whom he knew personally and had spent time with in Avignon. Word about the expedition reached the pope in the early 1340s, and scholars in his entourage had quickly associated the newly discovered islands with the Fortunate Isles, evidently after consulting Pliny's *Natural History*. Louis de La Cerda, the nobleman to whom the pope granted authority over the islands, certainly had Pliny in mind when he wrote to the pope in 1344. De La Cerda made a point of identifying himself as the "prince of Fortune" and then listed the islands under his control as "Canaria, Ningaria, Pluviaria, Capraria, Iunonia, Embreonea, Atlantia, Hesperide, Cernent, Gorgonide, and Galeta." The first six of these names correspond directly to Pliny's names for the Fortunate Isles, and the rest correspond to islands that Pliny located in nearby waters—islands, presumably, that La Cerda intended to locate and bring under his control.

La Cerda's letter reveals that an important new trend was under way in Europe: the Latin church and the European nobility were beginning to rely on ancient geographical ideas to help them develop a modern program of Atlantic exploration and imperial expansion. Conveniently for Petrarch and his friends, the trend dovetailed with the emerging humanist effort to revive the learning, power, and geographical reach of Rome.

* * *

ONE OF PETRARCH'S most important disciples was the Florentine author and scholar Giovanni Boccaccio. Like Petrarch, Boccaccio is today celebrated as one of the founding fathers of Italian literature and is best known for his poetry in the Italian vernacular. But Boccaccio also made a name for himself as an authority on ancient geography—and it was he, in fact, who recorded the sole surviving account of King Alfonso's expedition to the Canary Islands.

Boccaccio's most important contribution to the humanist study of geography was a work almost entirely forgotten today: a reference dictionary, widely circulated and relied upon in his day, titled *On Mountains, Forests, Springs, Lakes, Rivers, Swamps or Marshes, on the Names of the Sea*. In preparing his dictionary Boccaccio pored over ancient geographical texts, medieval *mappaemundi*, marine charts, and the accounts of contemporary travelers—and, not unlike Petrarch, he staggered away from them all reeling at the immensity of the mess. The ancient authors contradicted one another in a dizzying number of specifics; scribal errors had muddled everything; and the regions of the world known to the ancients and the moderns didn't fully overlap. These weren't new problems, but for the most part ancient and medieval writers hadn't grappled directly with them. Instead, they had simply recorded everything that they had heard reported or rumored about the world. This wasn't good enough for the humanists.

By marking up his copy of the *Natural History*, Petrarch had created a geographical glossary of sorts, but it was unwieldy and intended only for his personal use. Boccaccio's *On Mountains* represented the next logical step: a single easy-to-use volume devoted to the geography of antiquity that could be copied and consulted by anybody interested in classical history and literature. Boccaccio combed through ancient texts with great care as he prepared his dictionary, but he also examined modern sources. He looked to marine charts as a way of resolving inconsistencies he found in classical texts, and he spent considerable time comparing the geographical accounts of recent travelers with those of ancient authorities. In the work of either Marco Polo or William of Rubruck, for example, he read that the Caspian Sea was landlocked—a claim contradicted by such authorities as Pliny, Pomponius Mela, and Isidore of Seville, all of whom wrote that it emptied out into the ocean in the north. The bodies of water that the modern and ancient writers described just didn't seem to be the same. "I discovered that there are two Caspian Seas," Boccaccio wrote. "One is located in the middle of the land and has no connection with the sea, while the other flows from the ocean." Rather than take sides on the issue, he decided to include two separate entries on the Caspian in his dictionary. "Which one of these opinions is true?" he asked. "I leave it to be examined by those more diligent than I, since I dare not remove my confidence in the ancients, nor can I deny the eyewitness testimony of the moderns." This problem—how to square ancient and modern descriptions of the world—would consume humanist geographers for centuries to come.

During the 1350s, Boccaccio maintained a journal in which he jotted down notes to himself and copied out classical quotations and excerpts that he thought might be useful to him in his studies and his writing. In most respects the journal represents an idiosyncratic collection of the work of classical writers, but at one important point in the middle of the journal, in an entry that dates to the 1350s, Boccaccio recorded a contemporary development: King Alfonso's expedition to the Canaries. Summarizing in Latin the account of the expedition sent back to Florence from Seville, Boccaccio produced a short description of the episode that in subsequent years would take on a life of its own among Florence's humanist scholars, who knew it under the title *On Canaria and Other Islands Recently Discovered in the Ocean Beyond Spain.*

Boccaccio didn't just boil down and translate the merchants' account. Instead he recorded those parts of the account that to him seemed to make it clear that the islands being described were the same as Pliny's Fortunate Isles. Both Boccaccio and Pliny, for instance, write of a chain of mountainous islands, some inhabited and some uninhabited, located off the coast of northwest Africa. Both mention the presence of simple human dwellings and a small temple. Both describe the islands as being rich in figs, palms, and birds. Both single out three islands for special attention: a perpetually rainy island, an island entirely covered with strikingly tall trees, and a giant mountainous island apparently capped with snow: the island today known as Tenerife, which rises to more than twelve thousand feet above sea level. Boccaccio's use of the name Canaria, too, is an obvious interpolation based on Pliny's use of the name; the island's inhabitants certainly wouldn't have used it. Boccaccio and Pliny both describe the native inhabitants of their islands in the same way: as a happy, intelligent, communally minded people who live in a setting of natural abundance, far from the cares of civilization.

Boccaccio wasn't the first to make such connections. His reference at the outset of *On Canaria* to "these islands we generally call 'rediscovered'" suggests that plenty of others were thinking along similar lines. Merchants and scholars alike had rediscovered a part of the world not visited since antiquity, and no doubt others would soon be found. Boccaccio summed up the mood when he described the progress made by Alfonso's sailors as they explored the Canary Islands. "The more they advanced," he wrote, "the more of them they discovered."

• Claudius Ptolemy •

CHAPTER SEVEN

PTOLEMY THE WISE

World cartography . . . enables one to show the positions and general configurations [of places] purely by means of lines and labels.

—Claudius Ptolemy (circa A.D. 130–150)

$\mathcal{P}$ETRARCH WANTED TO do more than revive the culture and learning of Rome. He wanted to return the city to its place as Christendom's world capital. Pope Clement V and his entourage had decamped to Avignon in 1309—and ever since, in Petrarch's view, the papacy had been held "captive," and Rome had been left "lamenting for her bridegroom." For years Petrarch hectored the Avignon-based popes about returning, and in his *Life of Solitude*, in a chapter titled "How the Catholic Faith was of old diffused through well-nigh the entire world, but is now reduced through the negligence of the great," he summed up his feelings. Christendom had become only "the image and shadow" of what it had once been—the "real empire" created when Constantine had embraced Christianity in the fourth century. "The ancient Roman Empire," Petrarch wrote, "lacked only a small portion of the East, while we, alas, lack nearly everything except a small portion of the West." On the authority of Saint Jerome, who had lived in the fourth and fifth centuries, Petrarch also reported that Christendom's geographical reach had even once exceeded Rome's. "Jerome affirms," he noted, "that besides France and Britain, countries of our region, Africa and Persia and the East and India and all the barbaric lands worshipped Christ."

But that was ancient history. Now the Mongols and the Muslims and the Byzantines had all encroached on Rome's eastern territories, which had to be won back. And why limit the Crusades to the liberation of the Holy Land? Why not make their object the liberation and Christianization of all of Rome's former possessions? Here was a powerful way in which classical studies could serve the interests of the Church. The study of ancient geographical writings could reveal the extent of Rome's former reach, and could show missionaries and Crusaders where to go in order to win those territories back.

Yoking classical studies to the Christian mission presented an ethical dilemma, however. The ancients had been pagans, and much of what they had to say—about religion, philosophy, ethics, law, and other subjects—simply didn't accord with Christian doctrine. Fortunately, geography presented few such challenges. Ancient descriptions of the world generally contained no troublesome theological or moral messages. They could be appropriated with ease into the Christian body of learning, as had been done with Aristotle's vision of the cosmos.

For guidance in their attempt to reconcile classical writings with Christian doctrine, many early humanists and Church scholars turned to the work of Cassiodorus, a sixth-century Roman statesman who in midcareer had retired from public office to become a Christian monk. In his retirement Cassiodorus had amassed a great library of important Greek and Roman texts—works that, he contended, could greatly enrich Christian learning if studied properly. Cassiodorus listed many of these texts in an encyclopedic work titled *Institutions of Divine and Secular Learning*, which in the fourteenth century would become invaluable to the humanists in their hunt for lost classical texts.

One passage in the *Institutions* would have immediately caught the attention of any humanist with an interest in ancient geography. "Read through geographical writings," Cassiodorus instructed his audience, "so that you know the location of each place you read of in holy books. . . . If you are fired with interest for this noble subject, you have the book of Ptolemy, who described every place so clearly that you might almost think that he was an inhabitant of all regions. Thus, although you are in one place (as monks ought to be) you may traverse mentally what others in their travels have collected with a great deal of effort."

It was only the briefest of descriptions, but it spoke volumes. Here was firsthand testimony from a venerable Latin authority that confirmed what the Arab commentators had suggested. Ptolemy, it seemed, really *had* written a treatise on geography.

*   *   *

ALMOST NOTHING IS known about Claudius Ptolemy, the man, except for what his works reveal about him. As best can be deduced, he was born in about A.D. 100, pursued his professional life in or around the northern Egyptian city of Alexandria, and died at about the age of seventy, having written important treatises on not only astronomy and geography but also astrology, music, and optics. His name provides a few additional hints about his identity: Claudius is Roman, and Ptolemy is Greek, which suggests, based on the naming conventions of the time, that he was a Greek-speaking citizen of the Roman Empire, born into the Hellenic community that thrived in Alexandria during the second century.

These were Rome's glory days. The Roman Republic had swollen into the Roman Empire—which now encompassed much of Europe, North Africa, and the Middle East, and would reach its greatest territorial extent during Ptolemy's lifetime. Beyond the limits of the empire, Roman traders and soldiers and surveyors had fanned out in all directions, heading south into Africa, north into the inhospitable lands of the Germanic tribes, and west to the British Isles. They were moving east as well. Taking advantage of seasonal monsoon winds, Roman ships plied the waters of the Indian Ocean, carrying silver and gold to India and beyond, and returning home with cargoes of gems, spices, hardwoods, ivory, and silk. (The silk trade, Pliny the Elder grumbled in the first century, involved vast amounts of labor and expense and travel for a singularly frivolous purpose—"to enable the Roman maiden to flaunt transparent clothing in public.")

By Ptolemy's time the Roman Empire seemed on the verge of extending right to the fabled island limits of the globe. Plutarch and Pliny both reported that sailors in the first century had reached the Fortunate Isles; Pliny told of a Roman mariner who at about the same time had been blown off course on his way to Arabia and reached the legendary island of Taprobane; and the Roman general Gnaeus Julius Agricola, also writing in the first century, claimed that while sailing north of Great Britain he had caught sight of Ultima Thule. To many patriotic Romans, the empire seemed destined to encompass not just the three parts of the known world but also the many islands in the ocean beyond it—including the Antipodes themselves. Writing in the first century B.C., in the *Aeneid*, Virgil had famously prophesied this course of events. The reach of the empire, he wrote, would some day extend south and east beyond the limits of the known world to a distant land somewhere in the southern hemisphere—a land, he wrote, that "lies beyond the stars, beyond the paths of the year and the sun, where Atlas the Heaven Bearer turns on his shoulder the firmament, studded with blazing stars."

Such was the context—imperial reach, geographical ambition—in which Ptolemy wrote his *Geography*. He presented second-century Romans with a picture of the world they knew, and that they intended to rule.

* * *

IN WRITING THE *Geography*, Ptolemy intended to do something dramatically different from what writers like Pliny and Pomponius Mela had done in their works, which described the world in narrative, anecdotal form. Such narratives had their uses, but Ptolemy considered them little more than "chit chat about places." He set out to write something more practical and scientific. Building on the traditions of Greek mathematical geography, a centuries-old discipline not widely studied or understood in the Latin-speaking world of his day, Ptolemy wrote the *Geography* to teach his readers how to map the world. Hence the work's rarely cited Greek title: *Geographike hyphegesis*, or *Guide to Drawing a World Map*. No other treatise devoted to cartography has survived from antiquity.

Ptolemy begins the *Geography* by announcing that he intends to discuss only world cartography, not regional cartography. "The goal of regional cartography," he explains, "is an impression of a part, as when one makes an image of just an ear or an eye; but [the goal] of world cartography is a general view, analogous to making a portrait of the whole head." Regional cartography, he explains, focuses on small-scale geographical features that are easily observed by a single person ("harbors, towns, districts, branches of rivers, and so on"), and as a result it is the province of artists. World cartography, on the other hand, depicts an object that no one person can possibly inspect as a whole: the earth. Mapping the world therefore becomes an abstract geometrical problem, and the only way to solve it, Ptolemy writes, is to employ "the mathematical method" and to give special consideration— as Roger Bacon would try to do more than a thousand years later—to "the proportionality of distances for all things."

The first challenge for the world cartographer was to determine the size of the earth; only when that was established could the known world be assigned its proper proportion of the globe. Here Ptolemy turned to the calculations of his Greek predecessors, many of whom had produced reasonably accurate estimates of the earth's circumference. They did so by marshalling the power of a simple geometrical principle—one apparently first applied to the measurement of the earth in the third century B.C. by the scientist and philosopher Eratosthenes.

Eratosthenes used the process of deduction to determine the earth's size. First he chose a city in Egypt where at the solstice the sun cast no shadow at noon: Syene (Aswan). Then he chose another city that he presumed lay five hundred miles due north of Syene: Alexandria. If the sun was directly overhead in Syene, he knew, it could not also be overhead in Alexandria; it had to strike the city and cast a shadow at an angle. By determining what that angle was and then multiplying it by the distance between the two cities, Eratosthenes realized, he could come up with an estimate of the earth's circumference. The angle turned out to be a little more than 7 degrees, or about one-fiftieth of a full circle, which meant that the circumference of the earth had to be fifty times the distance between Syene and Alexandria: twenty-five thousand miles, which turns out to be almost exactly the true circumference of the earth.

For reasons he didn't explain, Ptolemy rejected Eratosthenes' estimate. Other Greek mathematicians had made different estimates, based on different calculations, and Ptolemy adopted one of them: an estimate (we now know) that made the globe about 18 percent smaller than it actually is.

With the size of the earth established to his satisfaction, Ptolemy moved on to determining the size of the known world, or, as the Greeks called it, the *oikoumene*. This time he looked to the stars instead of the sun. Like most Greek astronomers, he understood the stars all to be embedded in the firmament: the single fixed sphere on the outer edge of the spherical cosmos that rotated daily around the earth. As anybody looking up at the night sky could see, this sphere appeared to revolve around a single invisible axis, or pole. The stars near the pole moved in tiny circles, whereas those far from it traced wide arcs across the sky.

This axis became the starting point of the Greeks' system for mapping the firmament. They called its top the *north celestial pole* and its bottom the *south celestial pole*—the latter of which, because they lived in the earth's northern hemisphere, they couldn't see. With the north and south celes-

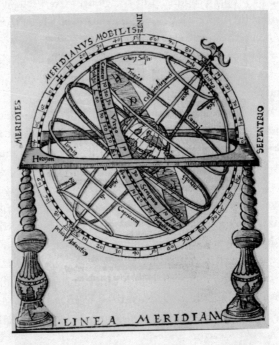

Figure 29. An armillary sphere (1524), showing the correspondence of the celestial and the terrestrial zones. The north celestial pole is at top right; the south celestial pole is at bottom left. The five parallel circles that surround the pole are the Arctic Circle, the Tropic of Cancer, the equator, the Tropic of Capricorn, and the Antarctic Circle.

tial poles as fixed reference points, the Greeks then laid grid lines over the firmament that allowed them to plot the locations and movements of the stars. At various north-south points along the pole they sliced horizontally through the sphere, creating a series of small circles near the poles, and, at the middle, a single *great circle*—that is, a circle that girdled the full circumference of the sphere and passed through its center. For obvious reasons, all of these circles were called *parallels*, and the one that passed through the middle was called the *equator*. The Greeks also made repeated vertical slices through the sphere, always starting at one pole, cutting through the center of the sphere, and ending at the other pole. This created a series of great circles, equal in circumference, that were not parallel to one another but rather divided the sphere into vertical sections, much like the wedges of an orange. These circles they called *meridians*—a name, like *equator*, that alluded to the circle's passing through the center of the earth. To teach these concepts, and to predict the motions of the heavenly spheres, ancient and medieval astronomers often designed elaborate models known today as armillary spheres (*Figure 29*).

Because the Greeks considered the firmament and the earth to be spheres that shared a center, and because they considered the earth to occupy the exact center of the cosmos, they realized that the celestial axis must also pass right through the middle of the earth. This meant that the earth, too, had a north and a south pole and an equator—and that the net of celestial parallels and meridians could be cinched in tight around the earth itself. It was this grid of parallels and meridians, called lines of latitude and longitude when applied to the earth, on which Ptolemy based the mapping system he laid out in the *Geography*. Every celestial and terrestrial location, he recognized, could be described as a unique point of intersection between a line of latitude and longitude (*Figure 30*).

Determining terrestrial latitude was relatively easy. If you looked at the night sky from a point along the equator, the celestial poles aligned themselves with the horizon, which is why the Pole Star dipped below the horizon when Marco Polo crossed over the equator near Lesser Java. The equator thus represented the zero-degree point of latitude. Everything else fell into place after that. By using a quadrant or an astrolabe—instruments that allowed observers to measure the height of celestial objects above the

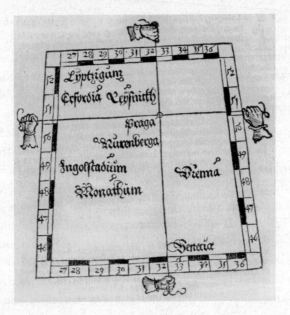

Figure 30. Latitude and longitude, explained in easy-to-understand graphic form (1524).

horizon—you could record how high the north celestial pole sat in the sky above the horizon (*Figure 31*). Then, as Sacrobosco would later illustrate in his *Sphere*, all you had to do to get your latitude was to measure the angular distance between those two points (*Figure 32*).

Certain latitudes had a special astronomical significance, and the Greeks gave them names. These included the Arctic Circle, in the northern hemisphere (approximately 66 degrees north), above which there was at least one day during the year when the sun never set and another when it never rose; and its counterpart in the southern hemisphere, the Antarctic Circle (approximately 67 degrees south). They also included the Tropic of Cancer (approximately 23 degrees north) and the Tropic of Capricorn (approximately 23 degrees south)—the northernmost and southernmost latitudes at which the sun could appear directly overhead at the summer and winter solstices. Wrapped around the globe together, these four lines of latitude created the climate belts that would appear on so many medieval maps of the world: the two frigid zones lay above the Arctic Circle and below the

Figure 31. Instruments used to observe the heavens and measure distances on land (1533). The quadrant is second from the left.

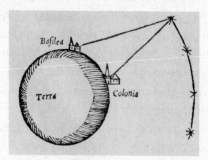

Figure 32. Diagram showing the perceived height of the Pole Star at two different latitudes (1490). The angle created where the two sight lines meet equals the degree difference in latitude between the two observation points. Anybody with an estimate of the earth's circumference can easily convert that degree difference into an actual distance.

Antarctic Circle; the two temperate zones lay between the Arctic Circle and the Tropic of Cancer, and between the Tropic of Capricorn and the Antarctic Circle; and the equatorial torrid zone lay between the Tropic of Cancer and the Tropic of Capricorn.

As Roger Bacon would later discover, determining longitude was far more difficult. First, you had to arbitrarily designate a place where east and west began—the zero-degree line of longitude, often called the prime meridian. The prime meridian effectively sliced the globe into eastern and western halves, each of which could be divided into 180 degrees of longitude. But trying to determine exactly how far a given place on the earth lay from the prime meridian posed great challenges. The world wasn't wrapped in a perfectly flat network of roads that would allow for the easy measurement of east-west distances. The roads that did exist meandered in all directions through terrain that might include mountains, valleys, and large bodies of water. Units of measurement varied dramatically from place to place, too, which meant that a cartographer trying to estimate distances routinely had to execute a series of error-inducing conversions, using figures that were already unreliable to begin with. Compounding the difficulty was the fact that accounts of voyages to far-off lands tended to refer to distance only in terms of days traveled and approximate direction. "Concerning the land journey from Garame to the Aithiopians," Ptolemy wrote in one typical passage, describing a distance calculation in Africa made by the cartographer Marinos of Tyre, "he says that Septimius Flaccus, who made a campaign out of Libye, reached the Aithiopians after leaving the people of Garame after marching south for three months."

The difficulty of determining distances traveled over land paled in comparison to determining distances traveled at sea. Here landmarks didn't exist, and winds and currents made sailing in a straight line impossible. "Concerning the sail between Aromata and Rhapta," Ptolemy wrote, still discussing Marinos's description of Africa, "he says that a certain Diogenes, who was one of those who sailed to India, returning the second time, was driven back when he got to Aromata by the *Aparktias* [north] wind and had Troglodytike on his right for twenty-five days, and [he then] reached the lakes from which the Nile flows."

When it came to longitude, the stars didn't seem to offer much help at sea *or* on land; the north celestial pole occupied exactly the same place in the night sky when viewed from the same latitude anywhere in the world. The most obvious difference in the sky when viewed from two places at once was the location of the sun. It took the sun twenty-four hours to make the circuit of the earth, so when it was noon at the prime meridian it was six o'clock in the evening at 90 degrees east, midnight exactly halfway around the world, and six in the morning at 90 degrees west. Longitude, in other words, could be considered a measurement of time. Ptolemy thought of it that way; he drew lines of longitude on his world map at 15-degree intervals, the space in between them representing the distance traveled by the sun in an hour.

Given the difficulty of accurately determining longitude, it's not surprising that when Ptolemy tried to map the world he did a better job of estimating latitude. He can't be blamed for this: he worked with bad data and had no way of accurately recording the time in two different places at once. The result was that he overestimated the east-west extent of the known world by some 40 degrees, and thus shrank the size of the ocean between India and Spain.

*   *   *

AFTER ESTABLISHING THE size of the globe and casting a net of latitude and longitude over it, Ptolemy's next challenge was to define the boundaries of the known world: Europe, Asia, and part of Africa. "It is agreed by absolutely everybody," he wrote, "that the dimension from east to west of the *oikoumene* . . . is much greater than that from north to south." He ran his prime meridian through the Fortunate Isles, a reasonable enough decision given that the islands were considered to represent the western limit of the known world. To define its limit in the east he studied the work of cartographers and astronomers who had preceded him, read the accounts of ancient and contemporary travelers, and eventually decided to extend his Asia exactly halfway around the globe, to 180 degrees of longitude. (This didn't imply that the continent ended there, just that its known part did.) Because Ptolemy considered Thule to be the northern limit of the known world, he placed it just below the Arctic Circle, at 63 degrees north; and because he believed that none of the ancient Greeks or Romans had ever reached the equator, he made Tapbrobane the southernmost place with a name on his map, situating it at approximately 4 degrees north of the equator. He knew that Africa descended into southern latitudes, however, and even admitted that habitable regions might exist there, so he showed the continent extending 16 degrees south of the equator. He didn't try to map what lay below that point, but he did claim that Africa at that latitude connected up with a giant unknown land that then extended far to the east and also joined Asia. This had an important consequence: on Ptolemy's map of the world the Indian Ocean appears surrounded by land.

Ptolemy didn't indulge in any theorizing about what lay beyond the known world. "Our present object," he wrote in the *Geography*, "is to map our *oikoumene* as far as possible in proportion with the real [*oikoumene*]." The only true way of doing this—of creating an exact proportional correspondence between points on the earth and on a reduced likeness of it—

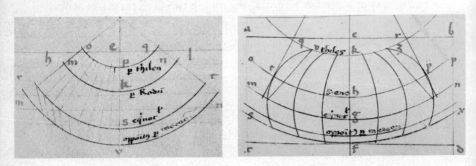

Figures 33, 34, and 35. Left: Ptolemy's first projection (1430). *Right:* Ptolemy's second projection (1430). *Far right:* Ptolemy's third projection (1525).

was to make a globe. But this, he noted, "does not conveniently allow for a size [of map] capable of containing most of the things that have to be inscribed on it, nor can it permit the sight to fix on [the map] in a way that grasps the whole shape all at once." A flat map offered a more practical alternative, he explained, but this required a cartographer to devise a projection: a mathematical formula allowing the three-dimensional coordinates of a globe to be plotted in only two dimensions, or, as he put it, on a "plane." A good map projection, he continued, should convey "a semblance of the spherical surface"—and in the *Geography* he laid out three different ways of doing this, the first being the easiest to construct but the least realistic, and the last being the hardest to construct but the most realistic (*Figures 33, 34, and 35*).

Having dispensed with definitions and technical explanations, Ptolemy moved on to what would make up the bulk of his *Geography*: raw data. On page after page he listed the names of provinces, towns, ports, islands, rivers, lakes, mountain ranges, seas, and more—some eight thousand names in all. No such list had ever been assembled, and in this sense alone the *Geography* was a remarkable achievement, representing precisely the kind of comprehensive but succinct guide to the geography of the ancient world that Petrarch, Boccaccio, and other early Italian humanists would begin to try to recreate for themselves more than a thousand years later.

But Ptolemy didn't just list the names of places. He also provided their

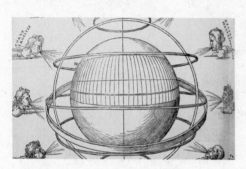

geographical coordinates—their latitude and longitude. Constructing a world map according to Ptolemy's specifications therefore became a sort of connect-the-dots exercise. You chose a projection, laid out a grid, plotted a set of coordinates, and a mathematically consistent and scalable picture of the world (*Figure 36*) would begin to emerge. That didn't mean that Ptolemy got everything right, however, as he himself was the first to admit.

Many parts of our *oikoumene* have not reached our knowledge, because its size has made them inaccessible, while other [parts] have been described falsely because of the carelessness of the people who undertook the researches; and some [parts] are themselves different now from what they were before because features have ceased to exist or have changed. Hence here [in world cartography], too, it is necessary to follow in general the latest reports that we possess, while being on guard for what is and is not plausible in both the exposition of current research and the criticism of earlier researches.

Mapmakers, in other words, would always have work to do.

*　　*　　*

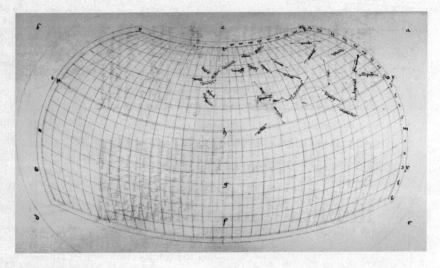

Figure 36. Plot of the mountain ranges of Asia on an otherwise blank grid representing the known world, using Ptolemy's second projection. From a manuscript of the German mathematician and astronomer Regiomontanus (fifteenth century).

No COPIES OF the *Geography* from antiquity survive, so it's impossible to answer one of the biggest questions about the work: Did it include maps?

Scholars have debated this question long and vigorously. On the one hand, it's hard to imagine that Ptolemy would have gone to the trouble of compiling all of his data, explaining all his terms, and devising his three different projections without also taking the final step of showing his readers what his world map would look like. On the other hand, Ptolemy describes so many places in his text, some of which are only separated by a twelfth of a degree, that only an exceedingly large world map could faithfully include all of the information he had collected—a map at least three feet tall by six feet wide. The largest known papyrus rolls used during Ptolemy's time, however, were only about two feet tall, and these were rare; Ptolemy probably wrote the *Geography* using rolls that were only a little more than a foot high. It may be, therefore, that Ptolemy didn't include a world map in his work but wrote it with large wall maps in mind—of the kind known to have formed part of grand imperial displays in ancient Rome. One such map, of unknown dimensions, is said to have adorned the wall of a portico in Rome during the reign of Augustus, in the first century A.D.

Even if Ptolemy did draw his own maps, they almost certainly didn't survive the many centuries during which the *Geography* was copied and recopied by scribes who had no training in cartography, mathematics, or even illustration. All it would have taken to lose the maps in the long transmission of the text from antiquity to the Middle Ages was one generation of scribes who decided not to reproduce them. Subsequent generations of copyists would then have merely copied the work as they received it—without maps. This phenomenon wasn't unique to Ptolemy. Not a single original world map survives from ancient Greece or Rome.

But just because Ptolemy's maps disappeared for centuries (if they ever existed), that didn't mean they were lost for good. What distinguished the *Geography* from the narrative descriptions of the world set out by Pliny, Mela, and others was that it contained those eight thousand precise coordinates. By gathering the latitudes and longitudes of so many places, and by explaining how to project them mathematically onto a plane, Ptolemy had, in effect, digitized a remarkably full picture of the ancient world. It didn't really matter if he had drawn maps or not. Nor did it matter whether the generations of scribes who had copied the *Geography* in the following centuries had managed to reproduce his maps. That's because, at least in theory, anybody who had access to the data provided in the *Geography* could use that information to re-create the world as Ptolemy had described it.

* * *

AFTER CASSIODORUS RECOMMENDED the *Geography* to his readers in the sixth century, the work faded from view in Europe. The Greek and Latin parts of the Roman Empire were drifting steadily apart.

The process had begun more than a century earlier, when the Roman Empire had officially been divided into a western and an eastern half. Each had its own emperor, its own capital (Rome, Constantinople), its own common language (Latin, Greek), its own religious leader (the pope, the patriarch), and its own brand of Christianity. Each also began to develop a sense of itself as the sole heir to the original Roman Empire, and of its own church as the sole heir to the original religious community founded by Christ. Inevitably, relations soured. When Rome fell to Europe's Germanic tribes, in the fifth century, the divide between the two empires grew: Byzantium (as the eastern half of the empire came to be known) flourished while Rome withered. This created no small measure of resentment among Latin Christians. A language barrier arose: the number of Greek speakers in the West quickly dwindled, as did the number of Latin-speakers in the East. Relations deteriorated particularly badly between the Latin and the Greek Churches, so much so that in 1054 they officially split into what would come to be known as the Roman Catholic and Eastern Orthodox churches—an acrimonious event now known as the Great Schism. Its legacy endures to this day.

The split launched a period of active political enmity between the two halves of the Christian world. In 1204, on their way to Jerusalem, the Latin soldiers of the Fourth Crusade sacked Constantinople, declared the city a subject of Rome, and occupied it mercilessly until the Byzantines managed to take it back, in 1261. Needless to say, the episode did nothing to improve relations between Rome and Byzantium, which continued to grow worse into the fourteenth century.

Such was the state of east-west relations within the Christian world when Petrarch began to agitate for the return of the papacy to Rome. But he didn't just want the western half of the Roman Empire back; he wanted Byzantium too. "I desire," he wrote to the pope in 1352, "to see that infamous empire, that seat of error, destroyed at your hands." Geography was destiny in this matter, he wrote; the imperial boot of Italy hovered over the territories of the "indolent and deceitful Greeklings," ready to crush them under its heel. Given his prejudices, Petrarch ranked the literary and cultural achievements of the ancient Greeks behind those of the ancient Romans, and he therefore focused his attention as a book hunter on rediscovering and reviving Latin texts. When it came to lost sources of information about ancient geography, Pomponius Mela's *Description of the World* was worth searching for, because it had been written in Latin; Ptolemy's *Geography*, written in Greek, was not.

The *Geography* had also dropped out of sight for centuries in Greek-speaking Byzantium, and it might never have been recovered at all had not a Byzantine monk named Maximos Planudes developed an obsession with finding it during the final years of the thirteenth century. Like Petrarch, Planudes was a learned scholar and poet who dedicated himself to the revival of lost classical texts. Unlike Petrarch, he was also a mathematician who spoke Greek. Exactly when and how Planudes managed to find a copy of the *Geography* is unclear, but find one he did. Chances are that he came across only the text of the work, but by following Ptolemy's instructions he managed to reconstruct its maps—and what he saw on the world map made him ecstatic (*Figure 37*). Here, mapped according to Ptolemy's detailed specifications, was the full extent of the *oikoumene* as it had been known in the second century, extending from the Fortunate Isles in the west, on the left, to the Chinese port city of Cattigara in the east, and from Thule in the north to Africa and the great unknown continent in the south.

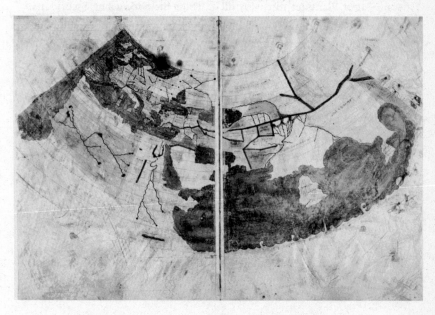

Figure 37. World map from one of the earliest surviving copies of Ptolemy's *Geography* (circa 1300), generally attributed to the Byzantine scholar Maximos Planudes.

Planudes immediately recognized the importance of what he had found. In about 1300 he celebrated his discovery in a poem, bearing the title "Heroic verses by the most wise monk Maximos Planudes on the *Geography* of Ptolemy, which had vanished for many years and then had been discovered by him through many toils."

A mighty marvel! Ptolemy the wise
Showed the great orb of the Earth's encircling shape
For all to see . . .
This work is such a one
As, rescued from oblivion after years
By one whose heart loves beauty, was revived
With all due speed into the light of day.

Several copies of the *Geography* appeared in Constantinople in the decades after Planudes' discovery, all of which included a single world map followed by twenty-six regional maps—and all of which have been traced back to Planudes' original copy. Had any one of those copies made it to the West in the early 1300s, the history of cartography and indeed the European Age of Discovery might well have unfolded very differently. But in the early 1300s the Latins and the Greeks weren't getting along. The *Geography* wouldn't make it back to the Latin-speaking world for another century.

• Italy and Greece •

CHAPTER EIGHT

THE FLORENTINE PERSPECTIVE

This great work [Ptolemy's Geography*] takes into full view the whole earth. It feeds not only military art but also philosophy, scripture, history, and poetry; the sweet life of agriculture, medicine, and art that animates the love of nature in the human breast. In sum, no greater need have our faculties than knowledge of the earth.*

—Francesco Berlinghieri (1482)

A̲ᴛ ᴀʙᴏᴜᴛ ᴛʜᴇ same time that Maximos Planudes announced the discovery of the *Geography*, Pope Boniface VIII announced a cosmological discovery of his own. He claimed to have discovered the existence of a new element in the material sphere—one every bit as pervasive as earth, water, air, and fire. It exerted such a powerful influence on sublunar affairs, he insisted, that it "feeds, clothes, and governs us," and "indeed appears to rule the world." Boniface made the discovery at Rome's first Grand Jubilee celebration, in July 1300. Greeting the many ambassadors who had come to pay tribute to him from all over Europe, he couldn't help noticing that no matter whom the ambassadors represented, they all originally hailed from

Florence. That was when it hit him. You Florentines, he told the ambassadors, must be "the fifth element of the universe."

At the beginning of the fourteenth century Florentines did seem to be everywhere. During the previous two centuries Florence had transformed itself from a provincial town, the home of cloth makers and small-time moneylenders, into one of the wealthiest and most economically powerful cities in the world. Bit by bit, the city's merchants had won control of the commerce of France and southern Italy, and they now managed a lively and highly profitable international trade in Italian corn, Sardinian cheeses, Tuscan oils, French wines, German copper, English wool, Cornish zinc, Persian furs, and Chinese silks. More and more money coming into the city meant more money to lend, and the city's bankers began making loans to Europe's noblemen, Church officials, and kings, sometimes at predatorily high rates of interest. Florence's bankers exerted unprecedented power over European finances and trade, and the city's merchants, forever seeking new markets, established significant presences outside of Italy: in Central Europe, the British Isles, the Iberian Peninsula, North Africa, Constantinople, and various trading outposts farther to the east. The effect, as one contemporary chronicler of the city put it, was that Florentines "have spread their wings over the world and have news and information from all corners."

Because of their rapidly expanding mercantile empire, Florentines had good reason to be interested in geography and exploration. So it's no surprise that when Europeans began to discover new Atlantic islands in the fourteenth century Florentine sailors would take part in the expeditions, Florentine merchants would spread the news about them, and Florentine humanists would help make sense of what had been discovered.

Humanism was the weak link in that chain. To a remarkable extent Petrarch himself *was* the humanist movement, and after his death, in 1374, his only heir apparent was Boccaccio—who promptly died the following year. Quite suddenly the future of the humanist program of revival became uncertain. Other similar classicizing movements had flickered to life in Europe in previous centuries, ensuring the survival of many of the ancient texts that made it to Petrarch's time, but each had gradually sputtered out. The odds seemed high that Petrarch's movement would suffer a similar fate.

But it didn't. Instead, during the next quarter century, a small group of Florentines would make their city into a center of classical studies and geographical learning and would transform Petrarch's idiosyncratic personal vision into a major force in European life. One figure more than any other was responsible for making this happen: a classics-obsessed notary and former papal secretary named Coluccio Salutati, whom Florentines chose in 1375 to be their city's chancellor. By keeping the humanist flame alive in Florence after the deaths of Petrarch and Boccaccio, Salutati helped ignite the astonishing explosion of literary, artistic, and scientific activity that today we call the Italian Renaissance. And as a part of that effort Salutati would engineer one of the pivotal events of the Renaissance: the return of Ptolemy's *Geography* to the West.

* * *

THE CHANCELLOR OF Florence was, in effect, the city's foreign minister. When Salutati took office it therefore became his job to represent his city to the world. He did so chiefly by writing letters. During his years as chancellor, which extended until his death, in 1406, Salutati wrote thousands of letters. The demands on his time as a correspondent were punishing. "Consider a little, my beloved son," he wrote wearily to a young friend in 1390, "the size of this great city, which, as it spreads out through almost the whole world, is compelled not only to fill the boundaries of Italy with letters but is forced to send letter after letter to all the princes of the world, wherever the Latin tongue and letters are known, both on account of public matters and because of the affairs of private individuals."

Salutati had immersed himself in classical studies long before being named chancellor, and as he began grappling with the challenges of his new job he quickly recognized that he had much in common with the statesmen and political philosophers of the ancient Roman Republic. Like him, they had lived in an Italian city-state that was rapidly expanding its political power and geographical reach. Like him, they had had to communicate and maintain diplomatic relations with a complex and ever-expanding world of Latin-speaking friends and enemies. Like him, they had lived in and officially represented a republic, not an imperial monarchy.

Salutati drew inspiration in his role as chancellor from the writings of

Cicero, the great orator, statesman, and champion of the Roman Republic. In particular he latched on to a concept that Cicero called *humanitas*. As Cicero used it, the word possessed many of the connotations we associate with *humanity* and *humaneness*, but it also possessed a more active and expansive connotation, as expressed in our idea of the *humanities*: that is, the pursuit of truth, knowledge, justice, and virtue through the study of the liberal arts. The essence of being human, Cicero argued, was the ability to study and learn, and to use this learning to cultivate virtue. *Humanitas* signified all of these things at once; it represented the full expression of what it was to be human. Taking Cicero's argument to heart, Salutati put *humanitas* at the heart of his classical studies and political philosophy—a move that would eventually lead to the terms *humanism* and *humanist*.

Humanitas was not a quality bestowed selectively by ancestry or religion. Everybody possessed it, and anybody could cultivate it. This was a philosophy guaranteed to catch on in a republic of self-made merchants. Those who applied themselves most virtuously to learning and the pursuit of virtue, as taught through the study of *humanitas*, could bring out the best in human nature—and it was they, not hereditary monarchs and noblemen, who were most suited to political leadership and cultural stewardship. History seemed to bear out its transformative possibilities: the ancient Romans had begun their rise to glory after rejecting the rule of kings and committing themselves to a republic based on the principles of *humanitas*, in the sixth century B.C.

After he became chancellor, Salutati began playing up the parallels between Florence and Rome. In particular, he made the case, based on his own careful study of Roman history, that Florence had originally been a Roman city, founded during the time of the republic. If this was true, and if Florentines could build a new republic in their city on a foundation of *humanitas*, then Salutati felt they could rightly lay claim to the cultural, political, and even imperial heritage of ancient Rome. One of his humanist disciples, Leonardo Bruni, who would succeed Salutati as chancellor in 1410, took this idea to its logical extreme. "To you, men of Florence," he wrote in 1402 in his *Panegyric to the City of Florence*, "belongs, by hereditary right, dominion over the entire world and possession of your parental legacy."

Salutati focused his attention on more prosaic matters. As he would put it proudly near the end of his life, he decided to make Florence "the home of the study of *humanitas*." In his role as chancellor he routinely mingled with Florence's most important and powerful citizens, and over time he won many of them over to the idea that a classical revival could not only advance their city's republican cause but also expand its political power and cultural influence. Gradually, an informal but tight-knit circle of scholars and patrons gathered around Salutati—and under his guidance they devoted themselves to the *studia humanitatis*, or study of the classical humanities, elevating it from the status of a private hobby to that of a formal educational curriculum. Humanism became a bona fide intellectual movement, and Salutati became its dean.

* * *

THE STANDARD PREHUMANIST approach to the study of the classics in medieval Europe was to focus on how the writings of the ancients had anticipated or could buttress Christian ideas. This was what early Christians had done in appropriating the classical three-part model of the world, and it was what the scholastics had done in appropriating Aristotle's theory of the cosmos. But because many of the pagan works of antiquity didn't mesh with Christian theology, Latin authorities often argued that such works should be ignored or suppressed. "If anything happens to be out of harmony and discordant," Cassiodorus had declared in his *Institutions*, "let us consider it something to be avoided." Pre-Christian history had no real chronological texture in this view: it amounted to a two-dimensional backdrop to the present, containing people and places and stories and myths and ideas that, drawn upon selectively, could help flesh out a Christian worldview. This was the historical perspective embodied in the medieval *mappaemundi*.

Salutati and his fellow humanists adopted a different approach. They wanted to understand ancient Roman culture on its own terms, not just as it appeared through a medieval Christian lens, and this meant trying to reconstitute and study the literature of antiquity as fully as possible. Lost manuscripts had to be found, variant editions of the same text had to be compared, the errors of copyists and interpolators had to be corrected, and theories had to be updated as new information became available. One had to read the poets to understand history, and the historians to understand the poets—and everything had to be situated in chronological context. Salutati made this happen, building the first methodical timeline of who had done and written what in Roman and Latin Christian history. The individual features of the past had to be charted methodically from point to point (as sailors charted the individual features of coastlines), and thousands of historical coordinates, gathered through the individual efforts of scholars working on different texts, had to be plotted onto a single relational grid (as Ptolemy had plotted his geographical coordinates). Only if history were mapped in these ways could its contours be seen clearly, its three-dimensional texture reproduced, its lessons absorbed.

In pioneering this newly methodical study of the classics, Salutati laid down many of the principles that would govern the humanist enterprise in the coming century. As he practiced it, humanism was a collective enterprise, a process of knowledge building in which scholars pursued their research individually but advanced their learning collectively. The job of the humanist was not only to revive classical learning as fully as possible but also to emend and build upon it for the modern age.

Salutati was a pioneering humanist in other ways, too. He made use of the prestige and powers of his office to raise the profile of his new movement. He wrote to correspondents all over Europe requesting help in tracking down classical manuscripts, and at great personal expense he amassed what would become the largest collection of manuscripts in the Latin-speaking world, some eight hundred in all. This collection, in turn, he made freely available to his friends, colleagues, and acolytes, effectively converting it into a precursor of the public library—an institution that had yet to be invented in Europe. His efforts to hunt down classical manuscripts also had a less tangible benefit: by cultivating relationships with like-minded men of letters all over Europe, Salutati helped create the rudiments of an international communications network that would greatly facilitate the spread of classical learning and ideals in the years to come. More actively than even Petrarch, he demonstrated how classical learning could be applied to the moral and political problems of his own time, and many of the men he nurtured and trained would go on to positions of power and authority in both Florence and Rome. He and his disciples would demonstrate to the world just how valuable humanists could be as bureaucrats, correspondents, diplomats, and statesmen, and by the beginning of the fifteenth century the Church itself had begun to turn to them for help in conducting its affairs.

It was a remarkable set of achievements. But as Salutati and his fellow humanists pursued their studies they began to understand themselves as significantly handicapped. They knew Latin well and could make their way through the works of the Romans with relative ease, but those works were full of references and debts to the writings of the ancient Greeks. To properly study the Romans, they realized, they would have to study the Greeks. But they had access to almost no Greek manuscripts, and even when they did they couldn't do much with them, because nobody knew the language.

In the 1390s, Salutati decided to address the problem. For help he looked in the obvious direction—toward Constantinople.

* * *

AT THE END of the fourteenth century the Byzantines were struggling for survival. In the preceding decades they had lost almost all of their eastern territories to the Ottoman Turks, a growing alliance of Muslim warrior peoples from Central Asia who, through an aggressive series of conquests, were fast superseding the Mongols as the dominant power in the Middle East. The Byzantine Empire now consisted of little more than Constantinople itself—and in 1395, eager to make the city the capital of their own empire, the Turks had laid siege to the city. The move threw the Byzantines into a panic. To defend their city they needed money and reinforcements, and they knew that in their weakened state they would have to turn to their Christian rivals in the West for help. Swallowing their pride, they dispatched a series of envoys to Europe.

It was a mission doomed to failure. Most Latins didn't care much about the fate of the "Greeklings," as Petrarch called the Byzantines, and many actively relished the idea of their fall. The Byzantine envoys therefore secured no concrete offers of assistance. In time they abandoned their efforts and returned home empty-handed, but not before one of those envoys, a scholar and diplomat named Manuel Chrysoloras, came to the attention of Coluccio Salutati.

During his visit to Europe, Chrysoloras had spent time in Venice, and while there he had briefly taught the rudiments of Greek to Roberto Rossi, one of Salutati's Florentine disciples. Rossi had returned home raving about Chrysoloras, and this had given Salutati an idea. Why not try to lure Chrysoloras to Florence as a teacher of Greek? Who better to bridge the gap between ancient Greece and Rome than a man engaged in trying to bring modern Constantinople and Rome together? So Salutati and his friends hatched a plan. They would send a thirty-five-year-old aspiring student of Greek named Jacopo Angeli da Scarperi to Constantinople to track Chrysoloras down.

It was a dangerous plan. The Turks were still laying siege to the city. Nevertheless, Angeli somehow managed to slip through the Turkish lines and locate Chrysoloras, and for the better part of a year he remained inside, writing regular letters back to Salutati that confirmed Roberto Rossi's high opinion of Chrysoloras as a teacher of Greek. Encouraged, Salutati himself began corresponding with Chrysoloras about coming to Florence to teach Greek—and Chrysoloras responded favorably. This was a rare opportunity, Chrysoloras realized; as a political envoy he had failed to convince the Latins to come to Constantinople's aid, but perhaps as a cultural envoy he might succeed.

In his letters Chrysoloras engaged with the Florentine humanists on their own terms. The study of Greek was a venerable Roman tradition, he wrote to Salutati in 1396: the ancient Romans had studied the Greek language in school, had prized Greek learning and literature, and had even sent their young men to Athens for training. It might even be said, he argued, that the best Roman writers owed much of their greatness to their knowledge of Greek.

Salutati needed little convincing. The idea that he might preside over the return of Greek learning to Florence made him ecstatic. "Tomorrow I shall reach my sixty-fifth year," he wrote to a Byzantine correspondent at the time, "but shall nevertheless someday see the origins whence is believed to have come whatever scholarship and learning Italy possesses!" To help himself realize this dream, Salutati turned to Palla Strozzi, a wealthy young businessman in his humanist circle, and secured the funding necessary to bring Chrysoloras to Florence. Salutati sent Chrysoloras a formal letter of invitation, and soon a deal was struck. The prospect of studying Greek with Chrysoloras reduced Salutati, the great statesman and man of letters, to a state of childlike giddiness. "Oh, how much patience my silliness will demand from you and Manuel!" he wrote to Angeli. "How much and in how many ways I will daily make you laugh aloud!"

Chrysoloras arrived in Florence in early 1397 and began teaching in February of that same year. With him he brought an assortment of Greek manuscripts to introduce to his students—including one that Palla Strozzi would later single out for special attention in his will. "I leave to my sons the *Cosmography* [*Geography*] in Greek," Strozzi wrote, "that is, the large illustrated one in vellum with a binding of black leather. They must preserve it and not sell it or dispose of it in any way, because this is the one that sir Manuel Chrysorolas, the Greek of Constantinople, brought with him to Florence when appointed to teach Greek in 1397. It was the first in these parts, and because he gave it to me I have looked after it. All those in Italy and some from outside derive from it."

Ptolemy's *Geography* had at last made it back to Europe.

* * *

THE GEOGRAPHY WAS one of only two Greek texts that Palla Strozzi mentioned in his will. The other was a deluxe edition of the Four Gospels, also brought to Florence by Chrysoloras. The juxtaposition of the two texts is significant, because it reveals just how highly Strozzi had come to value the *Geography* by the time of his death, in 1462. He wasn't alone. Not long before Strozzi's death, Tommaso Parentucelli—who later, as Pope Nicholas V, would go on to found the Vatican Library—had told Cosimo de' Medici that the *Geography* was one of the texts he imagined as the foundation of the ideal humanist library.

Salutati never valued the *Geography* quite that highly. Right up until his death, in 1406, he seems to have relied on it only as a catalog of ancient place-names. It was a research tool, a work to be consulted—alongside Pliny's *Natural History*, Mela's *Description of the World*, Solinus's *Gallery of Wonderful Things*, and Boccaccio's *On Mountains*—whenever geographical questions came up in his literary and historical studies. The only time he ever referred to the work in his voluminous correspondence, in 1405, he alluded to it simply to explain to a friend how he had discovered the ancient Roman name of a modern Italian town.

Salutati's lack of appreciation for the *Geography* is easy to explain.

Harried by official duties during his final years as chancellor, with his eyesight failing, he never actually managed to learn Greek. Chrysoloras himself only stayed in Florence until 1400—long enough to inspire a generation of humanists to study and teach the language, but not long enough to make Salutati proficient in it. But even if Salutati *had* developed a basic level of competence in Greek he would still have had trouble making sense of the *Geography*, which not only assumed a reading fluency in thousand-year-old Greek but also required a deep familiarity with the specialized vocabulary and principles of Greek mathematical cartography. It was all more than the old man had the time or energy for.

Florentines began to see the *Geography* in a new light only after it was translated into Latin. Chrysoloras seems to have started the job, and Jacopo Angeli completed it, sometime between 1406 and 1409. Angeli dedicated his translation to Pope Alexander V, and in his preface made it clear that he, at least, understood the *Geography*—which he had retitled the *Cosmography*, because it brought together celestial and terrestrial science—as far more than just a research tool for scholars. He began by reminding the pope of the intellectual and artistic achievements of the ancients, and of the humanists' achievements in reviving them. "Does not this present age of ours sparkle with genius," he wrote, "especially in our city of Florence, where the liberal arts have been awakened to great glory from their long sleep?" He then went on to describe how the ancient Romans and Greeks differed in their approaches to geography: how Pliny and other Romans had described the world anecdotally and unsystematically, whereas Ptolemy and the Greeks had mapped it with mathematical precision, focusing with care on questions of relative proportion, latitude and longitude, and "how our globe, which is spherical, can be described on a two-dimensional surface." These might be prosaic details, Angeli implied, but the power that the *Geography* conferred on its readers was anything but. It enabled them to see the true form of the earth from a previously inaccessible vantage point—the position occupied so often by Christ at the top of the *mappaemundi*.

It enabled a new kind of Christian imperial vision, in other words. "A kind of divine presentiment of your soon-to-be-realized empire," Angeli told the pope, "impelled you to desire the work so that you could learn clearly from it how ample would be the power you would soon hold over the entire world."

* * *

ONCE TRANSLATED INTO Latin, the *Geography* quickly took on a life of its own in Italy, especially in Florence. The earliest editions of the work contained no maps, which meant that their audience was limited primarily to classical scholars with a literary interest in the places of the ancient world. But soon manuscripts began to appear that contained not only Angeli's new translation but also copies of the maps originally made by Maximos Planudes and the Byzantines. Demand for lavish illustrated editions grew quickly among Florence's wealthy elite, and soon the modern world atlas as we know it today was born: a large, sturdily bound folio, designed for both reference and display, that contains an authoritative map of the world, a number of detailed regional maps, and an introductory essay that almost nobody reads.

Nothing like the Ptolemaic atlas had ever existed before, and Florentines reveled in the possibilities for armchair travel that it afforded. Later in the century the Italian poet Ludovico Ariosto would capture the appeal.

> Let him wander who desires to wander. Let him see
> England, Hungary, France, and Spain. I am content to live in
> my native land. I have seen Tuscany, Lombardy, and the
> Romagna, and the mountain range that divides Italy, and the
> one that locks her in, and both the seas that wash her. And
> that is quite enough for me. Without ever paying an
> innkeeper I will go exploring the rest of the earth with
> Ptolemy, whether the world be at peace or else at war.
> Without ever making vows when the heavens flash with
> lightning, I will go bounding over all the seas, more secure
> aboard my maps than aboard ships.

Even the idea of an instantly recognizable and easily replicable map of the world—*the* world map, an idea we now take for granted—was a novelty when the Ptolemaic atlas began to appear in the West. Before then, no single image came universally to mind at the mention of a world map. The phrase might conjure up one of any number of dramatically different ways of depicting the world: a schematic depiction of the cosmos, a simple T-O or zonal diagram, an elaborate *mappamundi*, a marine chart, or any number of idiosyncratic hybrids of the above. The world had no agreed-upon shape, design, or orientation; it might appear framed in a circle, a square, or a rectangle; it might show marvels and monsters or nothing but coastlines and names; and it might put any one of the cardinal directions at the top. But as the *Geography* spread, Ptolemy's vision of the world was reproduced again and again, in a form that is instantly recognizable to us today. For the first time people from different walks of life—scholars, merchants, statesmen, princes, clerics, lay readers—began to see the advantages of looking at the world from the same perspective.

* * *

FIFTEENTH-CENTURY FLORENCE was a place of shifting perspectives. As Florentines confronted the intricacies of their new international mercantile empire, as they contemplated the legacies of ancient Greece and Rome, and as they helped the Church develop its imperial ambitions, they developed a powerful need to see and understand the world as fully as possible—as it had been in the past, as it was in the present, and as it might be in the future. Ptolemy helped show them the way. He provided an authoritative picture of the world as the Romans had known it at the height of their empire, and he taught that world mapping was an ongoing, collaborative enterprise. His map of the world, he insisted, should be revised, corrected, and expanded according to new information, and the Florentines—who now regularly received reports of geographical discoveries from abroad, and who analyzed those modern reports in the context of ancient geographical learning—were uniquely suited to the job.

Ptolemy may also have helped Florentines develop another kind of perspective: linear perspective. The use of linear perspective, a drawing technique that allows an artist to convincingly mimic a three-dimensional subject on a two dimensional surface, is second nature to us today. You construct a geometrical grid that converges on a central vanishing point and then, in effect, you plot the coordinates of their picture along those lines.

The origins of the technique have been traced back to the thirteenth-century study of optics, but the architect Filippo Brunelleschi is generally credited with having perfected it in about 1415, in Florence—at about the time that the *Geography* began to circulate in the city. The claim that the *Geography* somehow triggered the use of linear perspective among Florentine artists is only suggestive, and has proved impossible to substantiate, but the idea of a possible connection between the two is intriguing.

Two views of Florence, one produced in 1350 and the other in 1493, reveal the changes in artistic technique brought about by the advent of linear perspective in the city (*Figures 38 and 39*). In the earlier view, selected features of the city appear crowded together haphazardly, with little concern for proportion; the image is nothing but foreground. In the later view, however, the city is depicted from a single imaginary vantage point and is laid out with a sense of proportion; it has a foreground and a background. This kind of shift is loosely analogous to the shift that took place in cartography during the same period: that is, the shift from the *mappamundi*, which contained features of the world crowded together haphazardly, to the Ptolemaic world map, which depicted the world from a single imaginary vantage point and laid it out with a sense of proportion. Ptolemy did *not* lay down the formal principles of linear perspective in the *Geography*, as some scholars have tried to argue, but he did discuss his cartographical method as a system of visual representation, and the terms he used would have appealed to artists grappling with the question of how to depict a three-dimensional subject on a two-dimensional surface. By using his grid, he wrote, they could draw a "portrait"—and this portrait, he went on to explain, would depict "the known world as a single and continuous entity," would focus on "the proportionality of distances for all things," and would preserve "a semblance of the spherical surface."

There's no doubt that the *Geography* provided a general kind of inspiration to many of the great figures of the Italian Renaissance. Notable among them was Leon Battista Alberti. By turns an architect, an artist, an art theorist, a cartographer, a cryptographer, a grammarian, a satirist, a sculptor, and a surveyor (among other things), Alberti clearly spent time studying the *Geography*. In a work titled *Praise of the Fly*, for example, he mentioned Ptolemy by name and described his "picture of the world" as being

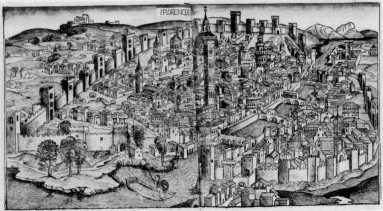

Figures 38 and 39. *Top:* Florence before the advent of linear perspective
(circa 1352–58) *Bottom:* Florence after the advent of linear perspective (1492).

organized by "intersecting parallels drawn perpendicular to one another."
He also seems to have had the *Geography* in mind when he composed his
highly influential 1435 treatise *On Painting*—the first work to codify the
principles of linear perspective and put them into writing. Artists, Alberti
wrote, could organize a picture of a room by using a tiled floor to create
"parallels," and these parallels in turn could be used to plot the "latitude"
and "longitude" of the objects in the room. Elsewhere in the treatise Alberti
used similar terms to describe a "veil" he had designed to help artists trace
the outlines of what they wanted to draw.

Nothing more convenient can be found than the veil, which among my friends I call the intersection, and whose usage I was the first to discover. It is like this: a veil loosely woven of fine thread, dyed whatever color you please, divided up by thicker threads into as many parallel square sections as you like, and stretched on a frame. I set this up between the eye and the object to be represented . . . the position of the outlines and the boundaries of the surfaces can easily be established accurately on the painting panel; for just as you see the forehead in one parallel, the nose in the next, the cheeks in another, the chin in the one below, and everything else in its particular place, so you can situate precisely all the features on the panel or wall which you have similarly divided into appropriate parallels. Lastly, this veil affords the greatest assistance in executing your picture, since you can see any object that is round and in relief represented on the flat surface of the veil.

Alberti's focus on the organizing power of the parallel, his claim to be able to represent a round object on a flat surface, and his emphasis on properly putting each part of a portrait in "its particular place"—all bring to mind ideas set forth in the *Geography*. Even the specific example Alberti uses in explaining the function of his veil recalls the analogy that Ptolemy used to introduce his mapping technique: the idea that world cartography doesn't just reproduce "an ear or an eye" but rather recreates "a general view, analogous to making a portrait of the whole head."

Alberti even produced one small treatise on mapping itself: the *Description of the City of Rome*, written sometime in the 1430s or 1440s. The work introduced a new coordinate-based system for the mapping of a city and then showed how it could be used to create a map of Rome. The *Description* differed greatly from the *Geography*—Alberti taught how to map a city, not the world, and he didn't rely on latitude and longitude—but it did bear a resemblance to it in several respects. Like Ptolemy, Alberti opened his treatise with a technical introduction—a guide to the methods and tools necessary for the new kind of mapping he had in mind. Also like Ptolemy, he proposed to determine the locations of various sites of interest by plotting them as the intersection of two different sets of coordinates: in Alberti's case, the distance and direction from Rome's Capitol, at the city center.

Whether Alberti actually drew a map of Rome to accompany the *Description* is the subject of debate, just as it is in the case of the *Geography*; if he did, it hasn't survived. But because Alberti appended detailed coordinate tables to his introduction, again just like Ptolemy, it's easy to imagine what his map would have shown: Rome as though seen directly from above.

* * *

THIS IDEA—THAT THE world could be seen from above, and that the individual human mind, with the help of mathematics, could grasp its measure—would be one of Ptolemy's great gifts to the artists and thinkers of the Italian Renaissance. For some Florentines, in fact, especially in the latter half of the fifteenth century, the *Geography* became a symbol for the entire humanist enterprise. The scholar Francesco Berlinghieri expressed this view in a verse adaptation of the *Geography* that he published in 1482. Ptolemy, he wrote, not only shows his readers how to map the world but also "raises us above the limits of an earth obscured by clouds" and demonstrates "how, with true discipline, we can leap up within ourselves, without the aid of wings." The study of geography as Ptolemy had taught it, in other words, was a metaphor for the *studia humanitas* as a whole; carefully pursued, both could expose the divinely established order of things. Everything from the anatomy of the tiniest insect to the complex workings of the entire cosmos could be studied, surveyed, drawn, and mapped, and the effort would help modern Christians understand God's plan for the world better than ever before. This was the true gift of the humanist movement: it taught Christians that God's vision of the world—its geography and its history—was fully accessible to them. Cicero captured the ethos in his *Dream of Scipio*, in an utterance that the Christian humanists of fifteenth-century Florence would take to heart. "Understand," Scipio is told as he gazes down at the earth from space, "that you are god."

Leonardo da Vinci, the archetypal Florentine Renaissance Man, certainly felt that way. "The deity that the painter has," he wrote near the end of the fifteenth century, "causes the mind of the painter to transmute itself into a similitude of the divine mind." Da Vinci captured this idea in visual form in one of his many memorable notebook doodlings (*Figure 40*). The picture shows an artist, possibly da Vinci himself, using a "perspectograph"—

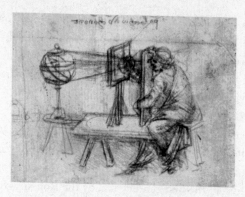

Figure 40. The divine perspective: an artist uses a "perspectograph" to draw the cosmos. Leonardo da Vinci (circa 1478–80).

a device presumably not unlike the one proposed by Alberti—to draw an armillary sphere. The symbolism is obvious and powerful: the artist can see and reproduce the world and the cosmos as God created them.

Like Alberti, da Vinci knew the *Geography* well. He owned a copy of the work and referred to it repeatedly in his writing—especially when explaining the workings of the human anatomy, which, he felt, had been designed according to the same principles as the earth itself. "By the ancients man was termed a lesser world," he would write in 1509,

> and certainly the use of this name is well bestowed, because, in that man is composed of water, earth and fire, his body is an analogue for the world; just as man has in himself bones, the supports and armature of the flesh, the world has the rocks; just as man has within himself the lake of the blood, in which the lungs increase and decrease in breathing, so the body of the earth has its oceanic seas which likewise increase and decrease every six hours with the breathing of the world; just as in that lake of blood the vessels originate, which make ramifications throughout the human body, so the oceanic sea fills the body of the earth with infinite vessels of water.

Not surprisingly, when da Vinci decided to embark on a systematic description of human anatomy, he chose to follow the model laid out by

Ptolemy in the *Geography* (which he referred to by Jacopo Angeli's title, the *Cosmography*). "In fifteen entire figures," da Vinci explained, "you will have set before you the microcosm on the same plan as, before me, was adapted by Ptolemy in his *Cosmography*; and so I shall afterwards divide them into limbs as he divided the whole world into provinces."

Many of Leonardo's ideas about the human body as a geographical microcosm come together in one of his most iconic images: Vitruvian Man (*Figure 41*). The drawing, a careful study of the human body and its proportions, is based on information contained in *On Architecture*, a wide-ranging treatise produced in the first century B.C. by the Roman writer Vitruvius. The original work contained no illustrations, it seems, but at one point Vitruvius did describe in precise detail an ideal set of human proportions. The palm of the hand, he wrote, is equal to the width of four fingers; the foot is the width of four palms; and so on. Leonardo seems to have been the first Renaissance artist to reconstruct this ideal figure based on the measure-

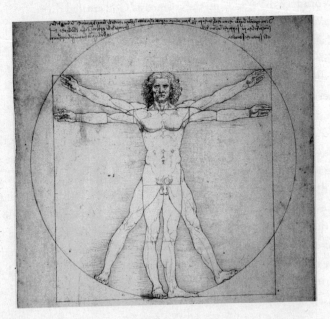

Figure 41. Leonardo's Vitruvian Man (circa 1490): a map of the microcosm and the macrocosm. Note the schematic similarity to the Lambeth Palace Map (*Figure 13*) and other *mappaemundi*.

ments provided in *On Architecture*—and it's easy to imagine that in doing so he had the *Geography* in mind. Reconstructing a proportionally harmonious picture of a man based on the data recorded by Vitruvius, after all, was not unlike the job of reconstructing a proportionally harmonious picture of the world based on the data recorded by Ptolemy.

Leonardo may also have drawn Vitruvian Man with a *mappamundi* in mind. Consider the visual similarities, for example, between Leonardo's Vitruvian Man and Christ as he appears on the Lambeth Palace Map (*Figure 13*). In each case a stand-in for humanity as a whole is inscribed within a kind of cosmic circle, with his arms and legs at its circumference, and his navel at its center. The macrocosm and the microcosm, in other words, come together as one. This was the new perspective that Ptolemy's *Geography* helped the humanists of Florence obtain—on the world and on themselves. As Cicero put it in the *Dream of Scipio*, "You have a god's capacity of aliveness and sensation and memory and foresight; a god's power to rule and govern and direct the body that is your servant, in the same way as God himself, who reigns over us, directs the entire universe."

Understand that you are god.

· Unknown Land ·

CHAPTER NINE

TERRAE INCOGNITAE

Our oikoumene *is bounded to the east by the unknown land that is situated next to the eastern peoples of Great Asia . . . to the south likewise by the unknown land that encloses the Sea of India and surrounds Aithiopia . . . to the west by both the unknown land surrounding the Aithiopian Bay of Libye and the adjacent Western Ocean; and to the north by the continuation of the Ocean . . . and by the unknown land that is situated next to the most northern countries of Great Asia.*

—Claudius Ptolemy, *Geography* (circa A.D. 130–150)

WHEN JACOPO ANGELI dedicated his translation of the *Geography* to Alexander V in 1410, the Church didn't have just one pope. It had three.

The problem was some thirty years old. Pope Gregory XI had moved the papacy back from Avignon to Rome in the late 1370s, but after his death in 1378, Church officials had disagreed about who his successor should be, and rival camps in the two cities ended up electing two different popes, setting in motion the brief but turbulent period in European history known as the Papal Schism. In 1409, hoping to bring an end to the schism, a congregation of cardinals at the Council of Pisa had elected Alexander V to replace both popes, but to no avail. Neither of the two sitting pontiffs had agreed to step down.

Three popes! It was an intolerable situation, and to address it Europe's religious and political leaders convened another council, in the fall of 1414, this time in the southern German town of Constance. The council would turn out to be one of the largest and most culturally significant gatherings in the history of medieval Europe—and would help spread a new kind of geographical thinking throughout Europe.

Estimates of the size of the Council of Constance vary dramatically, but all sources agree that it was huge. Between 40,000 and 150,000 people are thought to have converged on the city for the event, which dragged on for more than three years. Some seven hundred high-ranking Church officials from all over Europe and Byzantium attended the council, among them one pope, three patriarchs, twenty-nine cardinals, and an extended assortment of important archbishops, bishops, and abbots. Accompanying them came a sprawling retinue of close to eighteen thousand personal assistants, secretaries, copyists, scholars, legal advisers, servants, and others. The archbishop of Mainz alone arrived with an escort of five hundred people. Almost every city and feudal state in Europe was represented at the council, either by its ruler or by a high-level delegation. Eighteen dukes and archdukes, eighty-three counts, seventy-one barons, and 1,500 knights turned up for the event, and they, too, brought thousands along with them. Europe's major universities and schools sent representatives, and poets, artisans, laborers, and merchants all poured into the city to serve the needs of the gathering—as did Florentine bankers. Cosimo de' Medici himself set up shop in Constance. The town had to absorb more than just people, or course. By one estimate, some thirty thousand horses had to be fed and stabled during the council.

The humanists arrived in good number. Some came for their own reasons, as was the case with Cosimo de' Medici, but most came as part of ecclesiastical or political delegations, from Italy, France, Spain, England, Germany, and elsewhere. Many sought out Manuel Chrysoloras, by now a revered figure in the West, who arrived at the outset of the council to represent the interests of the Byzantine emperor. Chrysoloras's prestige had grown so dramatically in Europe that some at the council considered him a serious candidate for the papacy—a unifying figure whose election as pope might bring an end not only to the Papal Schism among the Latins but also to the Great Schism between the Latins and the Greeks. (He died not long after arriving at the council, however.) Several of Chrysoloras's former students, classical scholars who had been part of Coluccio Salutati's circle in Florence, also made appearances—among them a thirty-four-year-old Florentine humanist named Poggio Bracciolini, who traveled to Constance from Rome as a papal secretary.

An avid correspondent, Poggio wrote often to friends and colleagues back in Italy, and his letters—among them one in which Poggio admits to "doing nothing in Constance"—shed much light on council life. Few of those at the council, in fact, had much to do while high-ranking prelates solemnly debated the schism and other matters of official Church business. As the council ground tortuously on, Poggio and some of his humanist colleagues, who cared a great deal more about recovering ancient texts than they did about reunifying the papacy, couldn't help observing the proceedings with increasingly wry detachment. "Salvation itself," Poggio observed at one point, "could scarcely save this gathering."

Enforced idleness had unexpected benefits. Not since antiquity had so many learned figures from so many regions of the world gathered in one place. Men of like interests who had not known one another before, or who had only corresponded from a distance, were now thrust alongside one another in meeting halls and churches and eateries and lodging houses, with plenty of time to kill. This led to a vigorous exchange of ideas—and manuscripts.

The council rapidly evolved into a sort of unofficial international book fair. Many of the humanists and scholars who had come to the city had brought all sorts of manuscripts with them, for business and pleasure, and before long they began to share them. Constance was also full of idle scribes who had been brought to the city as part of the many delegations attending the council, and these scribes began privately copying manuscripts on demand. The geographical setting also enlivened the manuscript trade in Constance. Several important monasteries, abbeys, and cathedrals dotted the nearby regions of Germany, France, and Switzerland, and word soon got around among the humanists that these institutions possessed rich but neglected collections of ancient manuscripts. In 1415, inspired by Petrarch's example, Poggio and several of his colleagues began to explore the countryside around Constance in search of lost classical texts—and met with spectacular success. Their haul attracted attention all over Europe. "Few things have been accomplished at Constance that would take preference in importance," one of Poggio's correspondents told him, "over the discoveries of books that have been made." Most of the Latin classics that have come down to us today owe their survival to the burst of manuscript hunting begun by Poggio and other humanists during their stay in Constance— although they also owe their survival, of course, to the many generations of anonymous and uncelebrated scribes who kept the works alive long enough for them to be rediscovered.

Manuscripts copied in Constance soon made their way all over Europe, dispersed by the many travelers who made their way to and from the city during the council. Many of the continent's oldest libraries, even those far removed from southern Germany, today still contain evidence of the flurry of copying, buying, and selling that took place. "This book," one notation found in several manuscripts in a Swedish monastery reads, "was purchased in Constance at the time of the General Council." A work brought to the Univeristy of Krakow, in Poland, contains a similar notation: "Purchased by Paul Vladimir, magistrate, in Constance." Other examples abound.

Geographical texts figured prominently among the works circulating in Constance. Jacopo Angeli's new translation of Ptolemy's *Geography*, finished less than a decade earlier and still barely known outside of Italy, attracted special attention—and not just among midlevel humanists and papal secretaries with too much time on their hands. Even the French cardinal Guillaume Fillastre, one of the busiest and most important Church officials at the council, took notice.

Fillastre was one of the first non-Italians to develop an interest in Ptolemy's *Geography*, which he seems to have learned about not long after Jacopo Angeli translated the work into Latin. But only in Constance did Fillastre finally manage to procure a copy for himself, probably through the connections he had made at the council to Manuel Chrysoloras and the Florentine humanists. Whatever the exact source of his copy, Fillastre treasured it. "This book, which I sought for many years," he wrote proudly in the manuscript, "I obtained from Florence and had copied here, as a gift for the library of the Reims Church. I beg that it be well taken care of. I believe it to be the first in France."

*   *   *

THE GEOGRAPHY'S ARRIVAL in Constance, much more so than its arrival in Florence, marks the point at which many Europeans began to reconsider their fundamental preconceptions about geography—and about the very function of a world map. For all of their enthusiasm about Ptolemy, most of the early Italian humanists didn't grasp that his mapping system might be more valuable than his maps. This seems like a curious oversight. For some two centuries, after all, starting with their overland forays into Asia and continuing with their ocean voyages out into the Atlantic, Europeans had been expanding their geographical awareness, and by the early fifteenth century they had learned enough about the world to recognize that Ptolemy had gotten a lot wrong. But throughout the fifteenth century, edition after edition of the *Geography* appeared containing maps that faithfully reproduced the world as Ptolemy had described it. Many modern commentators, vexed by the early humanists' apparent lack of interest in updating Ptolemy's maps to reflect modern reality, have therefore described the fifteenth-century embrace of the *Geography* as a step backward in the history of cartography.

That's a mistake. Most of the early humanists—those living in the early decades of the fifteenth century—had their eyes fixed on the past, not the present. Few of them realized that the *Geography* might make possible a new perspective on the modern world. That would come later, in the second half of the century. The early humanists thought of the *Geography* in a far more limited way: as a way of reconstructing a picture of the ancient world, not developing a new one of their own.

Poring over his new copy of the *Geography* in Constance, Cardinal Fillastre began to think differently. He was well versed in the full range of traditional Christian geographical teachings. A student of Isidore of Seville and other medieval authorities, he had combed through their writings and jotted down specifically geographical excerpts from them in a private manuscript of his own. Maps themselves appealed to him greatly: on the walls of his home church in Reims were two large custom-made *mappaemundi*, one in his private quarters and the other in his library. Those maps no longer survive, but according to Fillastre they looked much like a small world map that appeared on the opening page of another manuscript that he seems to have acquired in Constance: Pomponius Mela's *Description of the World*.

At first glance, Fillastre's little map seems unremarkable (*Figure 42*). It's a brightly colored *mappamundi* based on the standard T-O scheme. But at its margins it exudes something unconventional for its time: a sense of uncertainty. Fillastre wrote *terra incognita* ("unknown land") in three different places at his map's northern and southern limits, and those two words changed everything. Instead of limiting himself to a depiction of what was known, as so many medieval mapmakers had done before him, Fillastre, inspired by Ptolemy's example, decided to include the unknown. The study of geography, his map announced, demanded inquiry, revision, and debate.

Fillastre studied and compared his new copies of the *Description* and the *Geography* during the Council of Constance and sent a commentary on the two works back to his colleagues in Reims. It's a revealing document. Unlike other early humanists, Fillastre didn't approach Ptolemy and Mela primarily for help in reconstructing the ancient world. He had more practical—

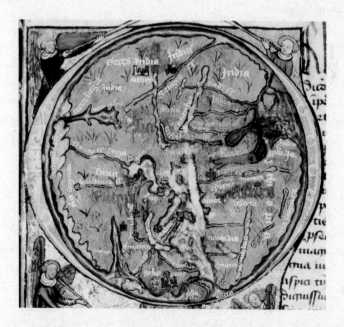

Figure 42. Cardinal Fillastre's world map (circa 1418). East is at the top.
Note the words *terra incognita* at the northern and southern limits of the world,
and the abbreviation *Ind. Prb. Jo.* ("India of Prester John") at the far right,
where East Africa meets the Indian Ocean.

and more contemporary—concerns. How did what Ptolemy and Mela had to say about the world's places correspond to what appeared on T-O maps, zonal maps, and the *mappaemundi*? Did the world consist of only three parts, all exposed together to the air on one side of the globe? Or did a separate Antipodal continent exist apart from them, on the other side of the torrid zone? How could traditional geographical teachings be reconciled with the reports of modern travelers? And, of course, the big question: What did the world—the whole world—really look like?

Mela took on that question at very outset of his book. "Let me untangle what the shape of the whole is, what its greatest parts are, what the condition of its parts taken one at a time is, and how they are inhabited; then, back to the borders and coasts of all lands, as they exist to the interior and on the sea coast." But before launching into his guided tour he stood back and surveyed the earth as though from above, just as Cicero had done in the *Dream of Scipio*.

> The uplifted earth . . . is encircled on all sides by the ocean. In the same way, the earth also is divided from east to west into two halves, which they term hemispheres, and it is differentiated by five horizontal zones. Heat makes the middle zone unlivable, and cold does so to the outermost ones. The remaining two habitable zones have the same annual seasons but not at the same time. . . . The former zone is unknown, because of the heat of the intervening expanse.

Fillastre didn't like this. It contained two ideas he considered fundamentally incompatible. An earth uplifted in one part of the globe and surrounded by water seemed to correspond to the T-O model of the world, on which the *mappaemundi* were based. But that model, and Mela's description, implied that the whole extent of the known world was habitable. So how could Mela then go on, in the same breath, to say that the uplifted and habitable earth actually spread across the uninhabitable torrid zone? "From the beginning of the book," he wrote, "[Mela] himself imposes a great difficulty on his readers, and on those who compare the division into five zones with a spherical map. The author's description of all of the parts along the shores of the ocean implies that the land inside the circle of the ocean is almost entirely habitable, whereas the division into zones says that only two are habitable."

Complicating matters was the theory of the off-center earth that had been developed in the previous two centuries by the scholastics to explain why the earth wasn't completely submerged at the center of the watery sphere. By the early fifteenth century the theory was being widely taught at European universities, and Fillastre knew it well. Just a few years earlier, in fact, his close colleague and friend Cardinal Pierre d'Ailly had summed the theory up in a cosmographical treatise titled *Imago mundi*, or *The Image of the World*. One of France's leading theologians and a professor at the University of Paris, d'Ailly, along with Cardinal Fillastre, helped run the Council of Constance. "Water does not surround the whole earth," d'Ailly had written in his treatise, "but it leaves a part of it uncovered for the habitation of animals. Since one part of the earth is less heavy and weighty than another, it is therefore higher and more elevated from the center of the world."

D'Ailly included a simple diagram of the cosmos in his manuscript that illustrated the concept very clearly (*Figure 43*). To anybody who looked at the diagram and accepted its model of the cosmos, one thing was obvious: the earth couldn't emerge from two diametrically opposed sides of the watery sphere. This meant that no separate continent could possibly exist in the ocean on the other side of the world. Land in the southern hemisphere might still exist—but if it did, logic dicated that it had to form a part of the

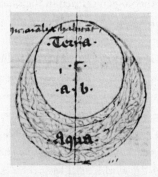

Figure 43. The earth emerging from one side of the sphere of water, from *The Image of the World*, by Pierre d'Ailly (early fifteenth century). One side of the earth is deeply submerged in the water, making it impossible for land to be exposed on the opposite side of the globe.

same single landmass as Europe, Africa, and Asia. The entire continental world, even its unknown Antipodal regions, therefore *had* to be contiguous. Hence Fillastre's decision to write *terra incognita* at his map's northern and southern limits. By adding those two words Fillastre implied, as Ptolemy had, that not just the known world but the whole world could be mapped. The unknown reaches of the world had begun to beckon.

Fillastre drew another important conclusion. If the whole world was indeed a single contiguous landmass, and if the known world occupied almost all of the earth's northern habitable zone, then the unknown parts of the world had to extend into the frigid and torrid zones—and those regions might well be habitable. Ptolemy and Mela had both suggested as much, by locating many different cities and peoples in the equatorial regions of Africa and India, and both ancient authorities and modern travelers had reported finding inhabited lands in the far north. Fillastre needed no more evidence than that. "It seems necessary to affirm," D'Ailly wrote, "that no part of the earth is uninhabitable."

* * *

PIERRE D'AILLY, THE author of *The Image of the World*, felt the same way, and he and Cardinal Fillastre almost certainly discussed the matter during their off hours at the Council of Constance.

D'Ailly approached geography from a more traditional perspective than Fillastre. D'Ailly was a theologian trained in the scholastic tradition, not a humanist, and in *The Image of the World* he distilled what he and his colleagues at the University of Paris would have taught their students about the nature of the world and its place in the cosmos: the theory of the spherical cosmos, the principles of astronomy, the movements of the heavens, the size of the earth and its various parts, the climate zones, the debates about the Antipodes and the habitable extent of the world, and more. D'Ailly explored these subjects from a religious perspective, very much in the way that Roger Bacon had in his *Opus majus*. "It seems," he wrote in his opening lines, "that the image of the world, or at least the description of the world that one can make by representing it as though in a mirror, is not without use in elucidating the Scriptures, which so often mention its diverse parts, especially those of the habitable world."

D'Ailly, in fact, drew heavily on Bacon in composing *The Image of the World*, often paraphrasing much of what Bacon had written, and at times even copying him word for word, especially in a chapter titled "On the Size of the Habitable Earth." Again and again, d'Ailly echoed or copied verbatim statements that he found in the *Opus majus*: "India is near Spain"; "the beginnings of the Orient and of the Occident are close"; "from the end of the Occident to the end of the Orient by land is a tremendous distance"; "water runs from pole to pole between the end of Spain and the beginning of India." (That last statement, of course, calls to mind Bacon's odd diagram of the ocean channel separating Spain and India [*Figure 27*].) The east-west extent of the ocean was something of an obsession for d'Ailly, and his thoughts about it led him to believe, as Cardinal Fillastre did, that the traditional image of the world might have to be rethought. "It is apparent," he wrote, "that the habitable earth is not round like a circle."

D'Ailly included an unusual world map in *The Image of the World* (*Figure 44*). The map is all text—nothing but a jumble of place-names superimposed on a standard climate zone map, roughly according to their geographical location. In the northern hemisphere, which appears at the top, the names (if stared at for a moment) resolve themselves into a typical T-O configuration: Europe, Africa, and Asia make up the known world, and Spain and India appear at its western and eastern limits. Instead of making Jerusalem the center of the world, as a medieval theologian might have done, d'Ailly, following Bacon, makes it Aryne—the city Arab astronomers used to designate the astronomical midpoint of the world. These similarities have led some modern scholars to argue that d'Ailly's map was based on the map Bacon sent to the pope with his *Opus majus*.

The most interesting and novel part of d'Ailly's map is what it shows in the south. At first glance this would appear to be precisely nothing: the map's southern hemisphere is almost entirely blank. But there is one small legend, written vertically, that creeps down along the map's southwest side, and what it has to say is crucial. The legend refers to the parts of the world "before the *climata*"—that is, those parts of the world that exist to the south of the known world. "Toward the equator and beyond it are many habitations," the text reads. "As it is learnt from authoritative histories."

This innocuous-seeming legend is every bit as laden with significance

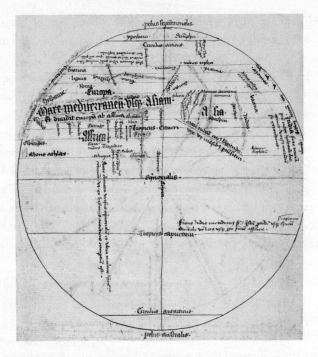

Figure 44. Zonal world map, from a manuscript version of *The Image of the World,* by Pierre d'Ailly (early fifteenth century). North is at the top. Europe, Africa, and Asia occupy the northern habitable zone. The legend on the left that extends vertically into the southern hemisphere suggests, contrary to traditional teachings, that crossing into the southern hemisphere is possible.

as Fillastre's use of *terra incognita*. Its message is clear: the torrid zone and much of the world to its south are in fact habitable, even accessible, and d'Ailly slyly reinforces this idea by forcing his readers' eyes to travel from above the equator, where the legend starts, almost all the way to the Antarctic Circle, where it ends. The legend draws the imagination south, in other words—beyond the limits of the known world and into a realm of new geographical possibilities. This appears to be a cartographic first: no earlier map is known to suggest that crossing over the equator and into the southern hemisphere is possible.

In the years after the Council of Constance, Cardinal Fillastre himself would come across evidence that traveling unknown lands in the other

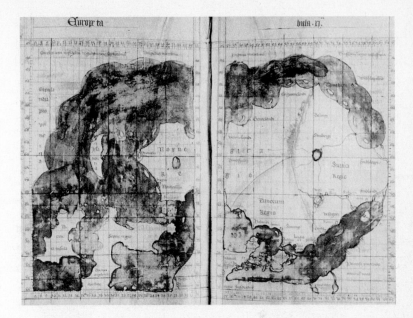

Figure 45. Map of northern Europe, by Claudius Clavus (1424). The western side of Greenland is at the top left, flush with the edge of the map. Iceland is the crescent-shaped island to its right. Below Iceland are Ireland and Scotland, and to their right are Denmark and the Scandinavian Peninsula.

direction was also possible. In or about 1424 a Danish traveler and church-man known as Claudius Clavus visited the Vatican, where he befriended Poggio Bracciolini and other humanists. Claudius had traveled extensively above the Arctic Circle, in Scandinavian territories, and even claimed to have had access to descriptions of Greenland. At the instigation of the humanists, it seems, Clavus drew a map of the places he had visited, and in so doing, as Fillastre would later put it, made Ptolemy's picture of Europe's northern regions "more complete" (*Figure 45*).

The Clavus map occupies a notable place in the history of cartography. It's the earliest known "modern" Ptolemaic map; it's the first detailed map to show territories well above the Arctic Circle, up to 75 degrees north; and it's the first map to situate Greenland in something like its proper location—to the north and west of Iceland. It's thus sometimes optimistically described as the earliest surviving map to show evidence of North America.

Cardinal Fillastre, who lived in Rome in the 1420s, somehow got his hands on Claudius's map. Fillastre had by now acquired the standard set of maps to accompany his *Geography* (a single world and twenty-six regional maps), and in 1427 he decided to have Claudius's bound in alongside them. The decision marks yet another milestone: Fillastre's manuscript survives today as the first edition of the *Geography* to include a modern map—Clavus's—constructed according to Ptolemaic principles.

Fillastre knew that Clavus's map broke new ground. "Ptolemy," he wrote, "was perhaps unaware of these regions."

* * *

UNKNOWN LANDS, UNCHARTED routes of travel, new kinds of world maps: these subjects dominated the off-hours geographical conversations of the humanists and scholars at the Council of Constance. The *Geography* no doubt provoked the most discussion (d'Ailly himself seems to have first encountered the text during the council), and much of the debate surely veered in abstract scholarly directions. But the conversations were by no means all theoretical. The nature of the council itself, teeming as it was with churchmen and diplomats and merchants from all over Europe and beyond, invited a practical exchange of information and ideas. People shared travel stories, read letters about new discoveries abroad, planned business ventures, hatched missionary schemes, and perhaps—the evidence isn't clear—studied marine charts that depicted islands newly discovered in the western ocean. They had *terra incognita* on their minds—and they therefore surely listened with rapt attention to the announcement made before the council's general congregation on June 5, 1416, by a certain Egidio Martinez, one of six envoys sent to the council, well after it had begun, by King João of Portugal.

Martinez first apologized for his delegation's tardiness in getting to Constance. But he had a good excuse. His king, he said, had recently sent a naval expedition into the Mediterranean and had successfully conquered Ceuta—an important port on the North African coast.

This was breathtaking news, on many levels. One was geographical: Ceuta (pronounced "Soota") was located on the stretch of coastline where Africa juts up and almost touches Spain. Another was religious: Ceuta had long belonged to the Muslims of North Africa, the Moors. Yet another was economic: Ceuta was a thriving commercial entrepôt that sat at the very northern end of an ancient trans-Saharan caravan route—one that brought gold, ivory, and slaves from sub-Saharan Africa to the coast. (More than half of Europe's supply of gold at the time came by camel across the Sahara, after being mined in distant regions of West Africa unknown to Europeans.)

But the victory had a profound strategic and symbolic dimension too. Ceuta sat at the foot of a mountain known to Christians as Monte Almina, considered by many to be one of the Pillars of Hercules—the two legendary rock formations (one being Monte Almina and the other perhaps Gibraltar) that since ancient times had signified the boundary between the Mediterranean and the Atlantic. The Moors had built an elaborate fortress on top of the Monte Almina, from which they could survey the Straits of Gibraltar, defend their territory, and control shipping lanes in and out of the straits. When the Portuguese captured Ceuta they captured its fortress—and this, Martinez declared triumphantly to the general congregation at the council, meant that Portugal now was in possession of "the port and key of all Africa."

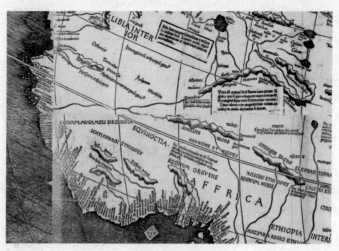

+ Northwest Africa +

CHAPTER TEN

INTO AFRICAN CLIMES

It is found that all that coast goeth to the south, with many
promontories, according to what this our Prince had added to the
navigating chart.

—Gomes Eannes de Azurara, *The Chronicle of*
the Discovery and Conquest of Guinea (1453)

T HE PORTUGUESE HAD won Ceuta from the Moors in a single day of
fighting. The news must have stunned the audience at the Council of Constance. Against all odds, a little European kingdom at the distant southwestern edge of the Iberian Peninsula—a place most Europeans considered
little more than a source of salt cod, sardines, cork, and port wine—had
somehow decisively defeated the Moors on their own soil. With the Holy
Land under Muslim control, with the Turks threatening Byzantium, and
with the Church in deep schism, this at last was some good news for Christendom. For the first time in more than a century a Christian army had

established a foothold in the Islamic world. There was another cause for celebration, at least among the theoretical geographers at the council who had studied Africa on the *mappaemundi*. The Latins, it seemed, might finally be back on the trail of Prester John.

The search for the great priest-king had reached a dead end in Asia by the middle of the previous century. "Not one hundredth part is true of what is told of him," wrote Friar Odoric of Pordenone, a Franciscan who in the early decades of the fourteenth century had spent years in India and China, and had been on a constant lookout for signs of Prester John. But this didn't mean that the search was over. In his letter Prester John had described ruling over three Indias, but in the fourteenth century Europeans had only explored two of them: Nearer India and Farther India. Thwarted by the Muslim presence in North Africa and the Near East, they had yet to explore the third—Middle India, the hazily understood African territories south of Egypt and east of the Nile. If Prester John wasn't in Asia, then perhaps he was in Africa.

That's exactly where a Dominican missionary named Friar Jordanus of Sévérac put him in 1324. In a geographical work known as the *Book of Wonders*, Friar Jordanus, recently back from extensive travels in both Africa and Asia, identified the kingdoms of Ethiopia as the "Third India" and estimated the region's Christian population to be three times that of all Western Christendom. And, he identified the emperor of the Ethiopians as none other than Prester John, a ruler "more potent than any man in the world, and richer in gold and silver and in precious stones . . . [who] ruleth over all his neighbors towards the south and the west." Such reports helped spread the idea of an African Prester John in the decades that followed. By the early fourteenth century the idea had taken hold among Europe's scholarly geographical authorities—which is why, for example, the eastern portion of Ethiopia on Cardinal Fillastre's *mappamundi* (*Figure 42*) bears the label "*Ind. Prb. Jo.*," the India of Prester John.

An African Prester John became a feature on some marine charts in the fourteenth century, too—the lavishly illustrated charts of the so-called Catalan School, produced by the mapmakers of Majorca. The earliest known map to locate Prester John in Africa is a product of the Catalan School: the Dulcert chart of 1339 (*Figure 28*), which makes reference in a legend to "the Christians of Nubia and Ethiopia, who are under the dominion of the Negro Christian Prester John." Other Catalan chart makers followed suit, including Cresques Abraham, maker of the grand Catalan Atlas (*Plate 6*), and by the early fifteenth century an African Prester John had become a fixture on most Catalan marine charts. The one drawn by Mecia de Viladestes in 1413, for example, shows Prester John seated in full regalia in an area south of Egypt, just below the point where the two branches of the Nile converge (*Figure 46*).

Figure 46. Africa and the Atlantic as the Portuguese knew them on the eve of the invasion of Ceuta: a detail from the Viladestes chart (1413). At the bottom right is Prester John. To his immediate left is the River of Gold, flowing west across the continent. Offshore, from south to north, are islands both real and imaginary: the Canaries, Madeira, the Azores, Brasil, and Saint Brendan's Isle.

As it appears on the Catalan charts, Africa contains several instantly recognizable features, and all of them appear on the Viladestes chart, drawn on the eve of the invasion of Ceuta. King João of Portugal and his advisers almost certainly pored over a map very much like the Viladestes chart as they developed their plan of attack—and as they began to imagine what else they might find and conquer in Africa. What surely attracted their attention most, especially once they had taken Ceuta, was what they saw on their charts as they looked to the south of the Mediterranean coast, past the Atlas Mountains and the Sahara: a series of portraits of wealthy and powerful African kings. These kings ruled over a vast territory that was cut through from east to west by the River of Gold, presumed to be the source of the gold brought north to the Mediterranean coast by desert traders. Represented as a branch of the Nile, it forks south of Egypt on the map and flows across the continent until it reaches the Atlantic at an uncharted part of Africa's northwest coast, approximately at the latitude of the Canary Islands. This was, as one early Catalan chart put it, "the end of Africa and the western land."

Military ventures inevitably produce unexpected results. So it was with the capture of Ceuta. Once in possession of the city the Portuguese found it hard to defeat the Moors elsewhere in North Africa. Nor were they able from their base in Ceuta ever to wrest control of the immensely profitable trans-Saharan trade in gold and other riches. But ultimately these failures didn't matter, because by turning their attention to Africa the Portuguese set in motion a new phase in the European history of discovery. Recognizing that they had no hope of dislodging the Moors in North Africa, or of reaching the great sub-Saharan kingdoms by land, the Portuguese began to raid—that is, explore—Africa's northwest coast. They set their sights, in particular, on the mysterious, uncharted region not far south of the Canaries, where the Catalan charts suggested that the River of Gold flowed out into the sea. If the Portuguese couldn't reach the kings and riches of Middle India by land, then perhaps they could reach them by sea.

*   *   *

THE MAN WIDELY credited with launching the European exploration of Africa is Prince Henry of Portugal, King João's third son. In the first half of the fifteenth century Henry—often referred to in the history books as Prince Henry the Navigator—helped direct a program of discovery that would take the Portuguese south along the coast of West Africa to regions never visited by Europeans before. Henry's reputation derives largely from a single source: a chronicle of Portugal's early African exploration written in

1453 by the court historian Gomes Eannes de Azurara, titled *The Chronicle of the Discovery and Conquest of Guinea*. Azurara's *Chronicle* is an invaluable source of information about the first Portuguese voyages of discovery, but it has to be read with caution; it profiles Henry and his role in the Portuguese exploration of Africa in a style that can only be described as hagiographical.

What drew Henry to Africa? Azurara described several specific lures. One was Henry's status as a noble, which required him, Azurara wrote, "to carry out very great deeds"—precisely the sort of chivalrous mindset that Miguel Cervantes would later spoof so gloriously in *Don Quixote*. Many of the earliest encounters that Azurara describes between the Portuguese and the inhabitants of the West African coast do indeed have a *Quixote*-like quality to them—although for the Africans the encounters were far more tragic than comic. At various intervals in Azurara's account dashing knights and squires ("shouting out 'St. James!' 'St. George!' 'Portugal!' ") leap ashore brandishing their swords and do battle with the African infidel. It makes for a good tale, but the enemy they took on is now known to have consisted of the small communities of fishermen and nomadic tribesmen who eked out a forlorn existence along the hot and dusty Saharan coast at the time. But no matter. Henry wanted to make a name for himself as a Crusader doing battle with the Moors, and by sending his men against the Africans of the Saharan coast, and by describing his battles with them to the rest of Europe as a succession of triumphant conquests against the Moors, he could do just that.

Crusading wasn't the only reason for Henry's interest in Africa. Azurara also credited him with a genuine desire to learn more about the continent and its commercial prospects. "If there chanced to be in those lands some population of Christians," Azurara wrote, "or some havens into which it would be possible to sail without peril, many kinds of merchandise might be brought to this realm . . . which traffic would bring great profit to our countrymen." Henry also had very specific interests. "Because every wise man is obliged by natural prudence to wish for a knowledge of the power of his enemy," Azurara continued, "[Henry] exerted himself to cause this to be fully discovered, and to make it known determinately how far the power of those infidels extended." Henry also sought "to know if there were in those parts any Christian princes in whom the charity and love of Christ was so ingrained that they would aid him against those enemies of the faith." This in turn was related to a sense of evangelical mission—namely, Henry's "great desire to make increase in the faith of our Lord Jesus Christ and to bring to him all the souls that should be saved."

After prefacing his *Chronicle* with these descriptions of Henry's motivations, Azurara opened the work itself by celebrating a famous first: the rounding of the legendarily treacherous Cape Bojador. The cape—today identified with Cape Juby, which lies near the point where southern Morocco meets the northern end of the western Sahara, near the latitude of the Canaries—is the one identified on that early Catalan chart as "the end of Africa and the western land." Bojador is indeed a difficult cape for mariners to negotiate. Powerful ocean currents in the area run perpendicular to the coast over shallow reefs, making inshore navigation a dangerous business, and trade winds blowing out of the northeast make the return journey up the coast to Portugal tedious, especially for ships, like those used by the Portuguese, unable to advance into the wind. Rumor had it that before Henry's time the cape was impassable—a story in all likelihood peddled by North African mariners eager to discourage Europeans from venturing south into the waters of the Atlantic, known to some as Green Sea of Gloom. "No one knows what exists beyond this sea," the Muslim geographer Muhammad al-Idrisi, of Ceuta, had written in the twelfth century. "No one has been able to learn for certain, because of the dangers to navigation caused by the impenetrable darkness, the great waves, the frequent storms and violent winds, and the multitude of sea monsters."

Azurara made the most of the Bojador legend. "Up to [Prince Henry's] time," he wrote, "neither by writings nor by the memory of man was known with any certainty the nature of the land beyond that Cape. Some said indeed that Saint Brendan had passed that way; and there was another tale of two galleys rounding the cape, which never returned"—a reference to the Vivaldi expedition of 1291. Until Henry arrived to prove them wrong, Azurara continued, everybody agreed that the cape simply could not be passed.

For, said the mariners, this much is clear, that beyond this Cape there is no race of men nor place of inhabitants: nor is the land less sandy than the deserts of Libya, where there is no water, no tree, no green herb—and the sea so shallow that a whole league from land it is only a fathom deep, while the currents are so terrible that no ship having once passed the Cape will ever be able to return. Therefore our forefathers never attempted to pass it: and of a surety their knowledge of the lands beyond was not a little dark, as they knew not how to set them down on the charts, by which man controls all the seas that can be navigated.

Such tales supposedly only spurred the brave and noble Henry on. Starting in the early 1420s, as Azurara tells it, the prince dispatched a series of expeditions south along the African coast, each with the express mission of rounding the famous cape. On board were Portugal's most experienced sailors and fiercest warriors—"yet," Azurara noted, "there was not one who dared to pass that Cape of Bojador." Even a personal squire of Prince Henry's named Gil Eanes, given the job of rounding the cape in 1433, found himself "touched by the self-same terror" that had afflicted his predecessors, and "only went as far as the Canary Islands, where he took some captives and returned to the Kingdom."

Henry would not be bowed. Henry sent Eanes straight back to Bojador in 1434, urging him to "strain every nerve to pass that Cape"—and this time Eanes pulled it off. He and his men sailed around the cape and made a landing on a beach some distance beyond it, where they found sweet-smelling herbs but no signs of human habitation. They didn't linger. Their mission accomplished, they turned around and headed home to report the good news.

Overjoyed, Henry once again ordered Eanes back into African waters, this time with instructions to sail as far beyond Bojador as he could. Eanes rounded the cape successfully a second time and sailed some fifty leagues beyond it to the south. Once again he and his men made a landing on a desolate and uninhabited stretch of coast. But this time, according to Azurara, they came across something new—the "footmarks of men and camels."

* * *

EUROPEAN SAILORS HAD actually been exploring this part of the African coast for decades, since at least 1401. A few accounts of those journeys, deliberately overlooked by Azurara in his *Chronicle*, circulated in Portugual during Henry's lifetime, and they may well be the reason that Henry believed that Cape Bojador could indeed be passed.

One account that Henry probably came across was *The Canarian*. Written by two priests living on the island of Lanzarote, in 1402, when Henry was just eight years old, the book chronicles a madcap attempt by two Frenchmen, Jean de Béthencourt and Gadifer de La Salle, to take control of the Canary Islands. By this time, Europeans had been haphazardly reconnoitering the Canaries and other Atlantic islands for more than half a century, at least since Lancelotto Malocello's famous voyage. A number of different charts and maps from the late fourteenth century make it clear that by then mariners and geographers knew of the existence of the Madeira island group, to the north of the Canaries, and of the Azores, a group even farther out into the Atlantic to the northwest.

Béthencourt and La Salle wanted in on the goods that these expeditions were bringing home—especially the slaves. They set sail in May of 1402, and within a few months had managed to establish a precarious hold on a few of the Canary Islands. They focused most of their attention on the islands, but they also let their imaginations drift to the conquest of territories on the African mainland, where there was more money to be made. Only a year before their arrival in the Canaries, they knew, fifteen European sailors had sailed from Lanzarote to Bojador, and had returned with a haul of black slaves.

The authors of *The Canarian* refer to this raid in a passing aside, and the very nonchalance of the reference is revealing: it's a matter-of-fact description of what appears to have been a routine excursion from the Canaries to Cape Bojador and back, made more than thirty years before Gil Eanes supposedly became the first European to visit the cape. The authors' complete silence on the subject of the cape's legendary dangers also suggests that by the early fifteenth century at least some Europeans already knew this part of Africa far better than Azurara would have had his readers believe—and both the Catalan Atlas (*Plate 6*) and the Viladestes chart (*Figure 46*) bear this out. Just below Cape Bojador, accompanied by a picture of a European ship equipped with oars and a sail, both maps make reference to an expedition sent to the region in 1346 in search of the River of Gold.

According to *The Canarian*, Béthencourt and La Salle developed plans to explore the African mainland for a considerable distance south beyond Cape Bojador. The section of *The Canarian* that describes these plans reads suspiciously like an appeal for funds, written with a rich European prince in mind. Getting to the region was easy, the two men wrote, and so was finding "pilots who knew the harbors and these countries." The Canaries themselves were perfectly placed as a staging ground for attacks on the West African mainland, and unlike the redoubtable Moors of North Africa, the inhabitants of the Saharan coast would fall easily to Christian armies, because they "have no armor nor any knowledge of warfare." Even the land itself was surprisingly inviting: "flat, wide, and broad; and supplied with all good things, with fine rivers and large towns."

It's not hard to imagine the powerful effect that this appeal would have had on the young Prince Henry as he dreamed of making a name for himself. Here were two authors with firsthand experience of Africa who were practically begging a prince like Henry to sponsor an expedition to the region beyond Cape Bojador, and who were promising that the venture would be easy, profitable, and strategically advantageous. Nobody had yet responded to the call, so who better to seize the day than the young and valiant new governor of Ceuta?

The authors of *The Canarian* didn't limit themselves to describing Africa's northwest coast. They also described the marvels and riches to be found in the African interior. They hadn't been there themselves, they admitted, but had read the work of somebody who had: "a book by a mendicant friar" who had traveled "through all the countries, Christian, Pagan, and Saracen, of those parts."

Four fifteenth-century manuscripts of this book survive. Titled *The Book of Knowledge of All Kingdoms*, and written by an anonymous Castilian who may or may not have actually been a friar, the work first appeared sometime in the latter half of the fourteenth century, during the years that Prince Henry and the Portuguese were contemplating the exploration of Africa. The author provided a succinct, blow-by-blow account of what he claimed were his extensive travels throughout the known world. Much of the book is obviously fictitious; in some places the author covers great distances at impossible speeds, and in others he describes countries and peoples now known not to exist: Gog and Magog in the east, the dog-headed race of Cynocephali in the north. But, as with Mandeville's *Travels*, *The Book of Knowledge* straddles the line between fact and fiction.

Africa in *The Book of Knowledge* is an often inchoate blend of the real and the imaginary, the verbal equivalent of what appears on the Catalan charts of the time. Parts of the author's description of the Bojador region have at least the ring of truth. "I traveled along the coast a very long way," he writes, "and crossed all the sandy beaches that are not inhabited by men, and I arrived at a land of black people, at a Cape they call Buyder, which belongs to the King of Guynea, near the sea. And there I found Moors and Jews. And know that from Cape Buyder to the River of Gold there are 860 miles." When the author moves inland, though, his itinerary quickly becomes a mixture of hearsay, legend, and stereotypes. He visits the Atlas Mountains and the desert communities of the Sahara. He travels east up the River of Gold toward the Nile, stopping along the way to visit some of the powerful kings who live along its banks. After much traveling he reaches "Nubia and Etiopia," a distant region of East Africa that he claims forms "one fourth of the entire face of the earth," and there finds himself in Christian lands, among men who are "as black as pitch . . . [and] burn themselves with fire on their foreheads with the sign of the cross." The men tell him stories of the lost Vivaldi brothers, whose expedition, they say, foundered on a river in the area, and a relative of whose, "a Genoese man that they called Sorleonis," later passed through looking for them. (One of the Vivaldi brothers indeed did have a son named Sor Leone.) From the African Christians he meets he learns about the location of the Earthly Paradise, which he is told can be found in southern Africa, on the other side of the equator, where the waters of Paradise descend from "some very high mountains that border on the circle [sphere] of the Moon . . . and make such a great noise that the sound can be heard two-days' journey away." These are the legendary Mountains of the Moon, which geographers from ancient times through the Middle Ages had located in Africa, and whose descending waters may derive from some early description of Victoria Falls.

It's at this point in his travels, among the black Christians of East Africa, in a land irrigated by all four rivers of Paradise, that the anonymous Castilian locates the continent's most powerful Christian king—a ruler who, he claims, "governs many great lands and many cities of Christians." The identity of that king should come as no surprise. He is, the author writes, "Prester John, the Patriarch of Nubia and Etiopia."

Reaching these parts of East Africa might have been possible for a mendicant friar on foot. But what about Portuguese sailors? The author

Figure 47. The Catalan-Estense world map (circa 1450). At the bottom left is the supposed Ethiopian Gulf, promising easy access to the Middle Indian kingdoms of East Africa, the Indian Ocean, and the Spice Islands.

of *The Book of Knowledge* gave no hints about whether the River of Gold was navigable so far inland. But he did describe another route that seemed to offer Portuguese ships a way of reaching East Africa. Not too far south of the River of Gold, he claimed, a huge Atlantic gulf cut into West Africa and extended almost all the way in to Ethiopia—a gulf identified on some fifteenth-century maps and charts as the Sinus Aethiopicus, or Ethiopian Gulf. One such map—a Catalan-style world map drawn in about 1450 and known as the Catalan-Estense world map—depicts the gulf almost exactly as the author of *The Book of Knowledge* describes it (*Figure 47*), and it's fair to assume that both this map and the book convey an idea of what Prince Henry envisioned his sailors might find if they were to follow the African coast south. Both made it seem impossible, or at least impractical, for European sailors to reach East Africa by rounding the continent, as it extended too far into the southern hemisphere. But both works made reaching East Africa by sea seem easy. All you had to do, according to *The Book of Knowledge*, was turn east into the Ethiopian Gulf and sail for fifteen days.

* * *

DREAMS OF GOLD and Prester John lured the Portuguese down the African coast—and so did the search for slaves.

Portuguese sailors brought home little of value in the years immediately after the rounding of Cape Bojador in 1433. The most promising discovery came in 1436, when, some 120 leagues south of the Cape, Henry's sailors came across an inland waterway that they assumed was the mouth of the River of Gold. Further investigation proved this theory wrong—but it also revealed that the waterway was a refuge for thousands of basking, blubbery sea lions. The animals offered the sailors at least some hope of a profit, Azurara wrote, in the form of skins and oils that could be sold at home, so the sailors carried out a "great slaughter" and returned home with their bounty. Prince Henry, the owner of a monopoly on the production of soap in Portugal, professed himself greatly pleased.

But he wouldn't send ships back to West Africa for years. Instead, the following year he devoted his time and energy to a Portuguese assault on the North African city of Tangier, a venture that would prove to be a costly failure. Only in 1441, when the Tangier campaign had ended, did Henry once again turn his attention to the coast of West Africa—and when he did, according to Azurara, he resumed his efforts modestly, by dispatching a single ship, manned by twenty-one men, in search of more sea lions. The mission surely seemed of little consequence to the prince, who was still nursing his wounded pride in the aftermath of the Tangier debacle. But the voyage represents a landmark event in history: the moment when the official Portuguese slave trade in Africa can be said to have begun.

Henry's sailors successfully rounded Cape Bojador, located and slaughtered more sea lions, and loaded up their vessel with skins and oil. But then their captain gathered them on deck for a speech. They could return home now, he told them, knowing that they had done what had been asked of them. But would it not be shameful to arrive back in Portugal, he continued, "having done such small service"? Why not bring honor to Prince Henry and themselves by carrying out some loftier deed? "How fair a thing it would be," he declared, "if we who have come to this land for a cargo of such petty merchandise were to meet with the good luck to bring the first captives before the face of our Prince." Roused by the speech, the sailors greeted the captain's proposal with approval and sailed off in search of Africans to capture.

Later that year, after the first small cargo of African slaves had arrived in Portugal (ten of them, probably Berbers), Henry recognized that his sailors had opened the door to something big. Slave trading was by no means a Portuguese invention, of course. It was an ancient practice, taken for granted in most parts of the world. For centuries, traders had brought slaves from the African interior north to the Mediterranean, and by the late medieval period Genoese merchants had virtually monopolized the European side of the business. (They were equal-opportunity slavers, however, who also ran a profitable trade in the opposite direction, rounding up Orthodox Christians, Eastern European pagans, Mongols, and various prisoners of war in Europe and Central Asia for transport to the Middle East and North Africa.)

As Portuguese sailors began bringing slaves back to Europe, Henry realized that he stood to make great profits by eliminating the Arab and Genoese middlemen who had for so long dominated the North African slave trade with Europe. Azurara claimed that Henry also had a loftier goal in mind as he began to oversee the capture and enslavement of more and more Africans: "salvation for the lost souls of the heathen." Henry was doing his captives a favor. "For though their bodies were now brought into some subjection," Azurara explained, "that was a small matter in comparison of their souls, which would now possess true freedom for evermore."

A flurry of Portuguese slave raids ensued in the 1440s. Shiploads of what Azurara identified only as "Moorish captives" were captured at points successively farther south of Bojador, some of whom were plucked right out of dugout canoes as they paddled out to greet or trade with Portuguese ships. Skeptics in Portugal who had previously complained about the great expense of Henry's African ventures developed a sudden change of heart when they noticed "the houses of others full to overflowing of male and female slaves"; overcome with envy, Azurara wrote, they had to "turn their blame into public praise." As the leader of a decades-long program of discovery and conquest, Henry was at last acknowledged by his people for what he really was: another Alexander the Great.

Although Azurara described the prospect of this new slave trade as a "delight," he also evinced a rare sympathy for the captives. He did so most memorably in a passage describing a large cargo of slaves unloaded in the port town of Lagos, on Portugal's south coast. The passage—one of the earliest surviving descriptions of the slave trade to take the slaves' perspective into account—recreates a heartrending scene that is surely as old as the institution of slavery itself.

The unloading of the slaves took place on August 8, 1444. The Portuguese ships had anchored in the harbor two days earlier, and word of their cargo had spread quickly. Early in the morning, as the sailors readied to unload their slaves, a large crowd of townspeople gathered in a nearby field, many of them having stopped work for the day—"for the sole purpose," Azurara wrote, "of beholding this novelty." Once brought ashore, the slaves were herded together into the field. It was a "marvelous sight," Azurara reported, but then he shifted his perspective.

What heart could be so hard as not to be pierced with piteous feeling to see that company? For some kept their heads low and their faces bathed in tears, looking one upon another; others stood groaning very dolorously, looking up to the height of heaven . . . others struck their faces with the palms of their hands, throwing themselves at length upon the ground; others made their lamentations in the manner of a dirge. . . . But to increase their sufferings still more, there now arrived those who had charge of the division of the captives, and who began to separate one from another . . . and then it was needful to part fathers from sons, husbands from wives, brothers from brothers. . . . As often as they had placed them in one part the sons, seeing their fathers in another, rose with great energy and rushed over to them; the mothers clasped their other children in their arms, and threw themselves flat on the ground with them.

Among the onlookers was Prince Henry, who seems to have staged the entire event in order to publicize the newfound profitability of his expensive Africa ventures. Mounted on a powerful steed and surrounded by a royal retinue, Henry observed the proceedings with satisfaction. "He reflected with great pleasure," Azurara wrote, "upon the salvation of those souls that previously had been lost." But profits, too, were on his mind. Just the year before, in 1443, his older brother Pedro, now the regent of Portugal, had granted him exclusive rights of navigation and commerce south of Cape Bojador. From that point forward, anybody wishing to sail into those waters had to first obtain Henry's approval, and upon their return had to turn over a fifth of their cargo to him—the so-called Royal Fifth.

After receiving Pedro's grant, Henry reduced the number of expeditions he himself officially organized to explore the West African coast. Once again he threw himself into devising and implementing a variety of foolhardy schemes for the conquest of North Africa, something that would preoccupy him for the rest of his life. But the exploration of the coast didn't stop. Noblemen and merchants seeking fame and profit in West Africa now began approaching Henry for permission to launch their own private expeditions, and he freely granted them licenses.

The easy pickings along the Saharan coast had now disappeared. Africans living along the Saharan coast knew to flee inland at the sight of Portuguese ships. Slavers and profit seekers sponsored by Henry therefore had to press farther and farther south in search of unsuspecting victims, and the result was inevitable. In 1444, a Portuguese squire named Diniz Diaz put out to sea and, according to Azurara, "never lowered sail till he had passed the land of the Moors and arrived in the land of the blacks."

The land of the blacks. Here, at last, was the rich sub-Saharan region so vividly displayed on the Catalan charts. The land was green and lush, a startling sight after the barrenness of the Saharan coast. Opening out into the sea at several points along the coast were great rivers, any one of which, the sailors started to imagine, might be the River of Gold. Diaz and his men took four black captives—"the first to be taken by Christians"—and then continued sailing south until they came across an even more remarkable sight: a volcanic peninsula on the coast of present-day Senegal that they named Cape Verde (today known as Cap Vert).

Although they didn't know it, Diaz and his men had reached Africa's westernmost point, about 15 degrees above the equator. What they *did* know, and what must have thrilled them as they saw it, was that beyond this cape to the south, the coast began a gradual turn to the east. Anybody who knew the Catalan charts of the time would have understood what this meant: the Portuguese had reached Ethiopian Gulf. Azurara summed up the euphoria of the moment by putting words into the mouth of a sailor who reached Cape Verde the following year. "Let us press on by all means," the sailor tells his captain, after they have reached the Cape, "be it even to the Terrestrial Paradise."

*　　*　　*

WORD THAT THE Portuguese had reached Guinea, as the land of the blacks was often called, traveled fast in Europe. Not surprisingly, the Italians were some of the first to act on the news. Increasing numbers of merchants from Genoa, Venice, and Florence began to establish themselves in Spain and Portugal. Gradually, they made voyages farther and farther along the Guinea coast in search of goods to trade, and soon enough they realized what the Crusade-minded Henry and his squires hadn't: that it was easier, safer, and more profitable to barter with the locals for slaves than to try to seize captives themselves. To Azurara, the arrival of the merchants and their practical approach to the exploration of West Africa represented the end of an era. The heroic phase of Portuguese discovery was over, he felt—and so he decided to end his *Chronicle* in 1448. "After this year," he explained, "the affairs of these parts were henceforth treated more by trafficking and bargaining of merchants than by bravery and toil in arms."

But it wasn't just merchants and sailors who had noticed Portugal's African discoveries. For decades, ever since the invasion of Ceuta, the Portuguese had worked hard—and successfully—to convince the Church that their raiding trips along the West African coast were part of an organized Crusading campaign against the Moors. Prince Henry was instrumental in this effort, and indeed nobody in Portugal had better credentials for the job. From 1419 until the end of his life, in 1460, he led Portugal's branch of the Order of Christ (a successor organization to the Templars), and in that capacity he wrote to the Church in Rome repeatedly with reports of his valiant efforts to wrest West Africa away from the infidel.

The news from Portugal pleased the Church. During the first half of the fifteenth century, a succession of popes issued a series of official decrees, or bulls, giving religious sanction to the Portuguese conquest of all African territories not already in Christian hands. The most famous and influential of these bulls, the wording of which scholars have shown derives almost directly from the Portuguese themselves, appeared on March 13, 1456. Formally titled *Inter caetera*, the bull has often been described as "the charter of Portuguese imperialism," and for good reason. Noting that Prince Henry and the king of Portugal had "conquered for the Christian religion" many "islands, lands, harbors, and places situated in the ocean toward the southern shore in Guinea," the bull granted Portugal's Order of Christ, with Henry at its head, spiritual jurisdiction and power over the lands extending "from Capes Bojador and Nam as far as through all Guinea, and past that southern shore all the way to the Indians"—a likely reference to the inhabitants of East Africa, not Asia.

The pope himself didn't handle correspondence with the Portuguese about their African voyages, nor did he draft the bulls sanctioning them. That was the job of his secretaries—and by the 1440s, when the Portuguese reached Guinea, many of these secretaries were Italian humanists. Among them was Poggio Bracciolini, the Florentine who had made such a name for himself during the Council of Constance as a hunter of books. Poggio had already drafted at least one bull concerning Portuguese exploration in the Atlantic, and now, in the 1440s, with privileged access to the reports that Prince Henry was spoon-feeding the Church, he and his colleagues followed the southward course of Portuguese discovery with mounting excitement. The Portuguese, they realized, were opening up a startling new chapter in the history of exploration—one that might settle once and for all those unresolved questions about lands in the southern hemisphere and life in the torrid zone.

Eager to learn as much as they could, Poggio and his colleagues plied their Portuguese friends in the curia with questions about Henry's African discoveries. Poggio even wrote to Prince Henry himself, in 1448, in a letter that wonderfully captures the spirit in which he and other humanists had begun to interpret Portugal's African discoveries. He compared Henry's accomplishments to those of Alexander the Great and Julius Caesar, and then went on to transform Henry into a symbol of the humanist enterprise—a heroic modern figure whose discoveries first built on and then exceeded those of the ancients. "When I asked my many Portuguese friends about your deeds," he wrote to Henry, "I was told that, moved by a certain greatness of soul and incited by virtue's goal, you sailed with several triremes along the most distant shores of the Ocean Sea and went where none of the ancients, neither emperor nor king, ever went, as far as we have heard or read. For they say that you went to the south ends of Africa and even reached Ethiopia."

The Portuguese, for their part, recognized that the humanists had much to offer them in return: continuing papal support of their African ventures, an extensive ecclesiastical and political network within which to publicize their efforts, and—thanks to the humanists' growing knowledge in classical geography—access to what the ancient Greeks and Romans had taught about Africa.

It was a relationship that would only become stronger in the coming decades, and as it did, the Italians and the Portuguese began to think along similar lines. With the notable exception of Ptolemy, almost all of the Latin and Greek authorities who had discussed Africa had described it as being surrounded by water. That's what the traditional *mappaemundi* showed, too. So when reports began to trickle back from Portuguese sailors that the African coast had begun to turn to the east below Cape Verde, people began to wonder: What if the Portuguese weren't actually sailing into the Ethiopian Gulf? What if they had found a passage right under the continent itself?

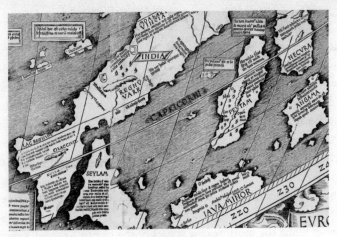

• India and Spice Islands •

CHAPTER ELEVEN

THE LEARNED MEN

We know very well that Ptolemy was ignorant of many things.

—Flavio Biondo (1453)

SOMETIME BETWEEN 1415 and 1420—as the Church held its council in Constance and the Portuguese established their presence in Ceuta—a young Venetian merchant named Niccolò Conti left Italy to do business in Damascus. For the next two decades he would travel through the Near East to India and beyond, before finally returning to Italy some twenty-five years later. What he had to report about his travels would prompt Europeans to take a new look at a part of the world that had been off-limits to them for decades: the Far East. Many of the stories that Marco Polo had recounted about the region, Niccolò reported, were true.

Niccolò's account echoed Marco's in many respects. "Heretics who are called Nestorians are scattered all over India," he claimed, "as the Jews are among us." Sailing east from India, he had reached a huge island known to its inhabitants as Sumatra; it measured some six thousand miles in circum-

198

ference, produced an abundance of gems, precious metals, spices, and fruits, and was home to cannibals and idolaters. He had sailed east beyond Sumatra to two more huge islands, Greater and Lesser Java, and there—"towards the extreme confines of the world"—he had picked up stories of islands even farther to the east, whose dark-skinned inhabitants traded spices and parrots with the Javanese. This seemed to be the same vast Indian Ocean archipelago that Marco Polo had described sailing through—and Niccolò confirmed that it extended below the equator. "The natives of India," he reported, "steer their vessels for the most part by the stars of the southern hemisphere, as they rarely see those of the north. They are not acquainted with the use of the compass, but measure their courses and the distances of places by the elevation and depression of the pole."

Niccolò also traveled overland in Asia, through parts of Burma, Thailand, and Vietnam. He didn't reach Cathay, but he did relate stories he had heard about it. It was the most distant of the three Indias: the one that "excels the others in wealth, humanity, and refinement, and is equal to our own country in the style of life and in civilization," where "the men are extremely humane," "the merchants very rich," and "the houses and palaces and other ornaments are similar to those in Italy."

When Niccolò at last decided to return home, he sailed back to India, around the southern tip of Arabia, and into the Red Sea. Married now and with four children in tow, he stopped briefly at ports along the coast of Ethiopia and then made for Mecca, in Arabia, on the other side of the Red Sea. From Mecca he planned to make the final leg of his journey home overland, but he received a rude welcome among the Arabs. Suspicious of his intentions, they forced him and his family to convert to Islam, and only after the deed was done was he allowed to join a trade caravan to Cairo, and from there to return to Italy.

One of Niccolò's first acts upon arriving home was to seek atonement from Pope Eugenius IV for having converted to Islam. But the pope wasn't in Rome, he learned. He had traveled to Florence, where another Church council had just begun. Great flocks of international delegates had descended on the city for the event, just as they had in Constance, and this time they came from all over not just Europe but the whole of the Christian world. With the Papal Schism resolved, the delegates had launched an effort to resolve the differences between the eastern and western branches of Christendom. Like the Council of Constance before it, this new council would exert a lasting influence on European geographical thought—in no small part because of the stories that Niccolò Conti would tell there.

*   *   *

THE GREEK DELEGATION to the Council arrived in February 1439. Led by John VIII Palaeologus, the emperor of Constantinople, it consisted of two hundred Byzantine notables and some seven hundred men in all—a dignified assemblage of clerics and scholars, the sight of whose bejeweled hats, shaggy hair, long beards, lavishly embroidered silk robes, and painted eyebrows at once impressed and amused the Florentines. The delegation itself arrived under duress. Their capital was once again under threat. The Turks, who earlier in the century had had to abandon their siege of Constantinople in order to fight the Mongol warlord Tamerlane in the East, had resumed their campaign to take the city. The Byzantines recognized that their only hope of resisting the Turkish advance was to try to forge a military alliance with the Christian powers of Europe. This meant the need for a reunion, however distasteful, with the Latin Church, and in the late 1430s the emperor had convinced many of his leading citizens to swallow their pride and join him on a new mission to Italy and the West.

Pope Eugenius invited representatives from other Eastern Christian churches to attend the council as well. He hoped, as he wrote in one letter, "that the Catholic church scattered over the world should be one, should feel and think the same, and that there should be one fold under one shepherd." The response was impressive. Delegations of Armenians, Chaldeans, Copts, Ethiopians, Georgians, Greeks, Maronites, Nestorians, and Russians all made their way to Florence, and for more than two years, from 1439 to 1442, the city offered Florentines a preview, in microcosm, of what a new global Christian empire under Roman rule might look like. In their magnificent new Renaissance churches and public plazas, on their picturesque medieval stone bridges, along their narrow cobbled streets, in their bustling markets and inns—everywhere Florentines looked they encountered an exotic assortment of peoples and cultures and languages, the likes of which hadn't been brought together in Europe since ancient Roman times.

The Byzantines attracted the most attention, and not just for religious reasons. Their presence in the city thrilled the humanists. If Coluccio Salutati had been able to engineer a flowering of Greek studies in Italy by bringing just one man, Manuel Chrysoloras, to Florence, then what might happen with hundreds of Byzantium's best and brightest in the city? The Council of Constance had dramatically energized the study of classical Latin language and literature two decades before, and the Council of Florence might now do the same for Greek. The two long-divided worlds of classical learning, like the long-divided churches of East and West, might finally be brought back together. The Latin humanists and clerics shared a political agenda, too; they intended to bring about union on *their* terms. As the new Romans, they would return the schismatic Greeklings to the subsidiary status they had occupied at the height of Rome's power, and would regain control of what was rightfully theirs: the political and cultural legacy of antiquity.

All eyes were on the East, then, when the penitent Niccolò Conti arrived in Florence seeking atonement for his conversion to Islam. Conti managed to obtain an audience with the pope, who found himself so captivated by what Conti had to say that he asked one of his papal secretaries to write his stories down. The man he turned to was Poggio Bracciolini.

Poggio eagerly took on the job. The official business of the council was turning out to be just as slow-moving and dull for him as it had been in Constance. For months on end, high-ranking representatives of the Latin and Greek churches had been negotiating about arcane matters of theological contention. Chief among these was the wording of the so-called Filioque ("And the Son") Clause in the Nicene Creed. "We believe in the Holy Spirit," the Greeks said, "who proceeds from the Father." "We believe in the Holy Spirit," the Latins said, "who proceeds from the Father *and the Son*." Back and forth they went. And back and forth they went.

As the council dragged on, the humanists of Florence recognized that they might enliven the event for themselves by convening informal scholarly symposia and inviting visitors to the city to take part. These symposia—gatherings that Poggio would refer to as "the meetings of learned men"—soon took on a life of their own. Ducking out of official meetings during the day, and convening again in private homes at night, the humanists and their guests met to share ideas, debate theories, compare notes, and exchange rare manuscripts. What had happened two decades before in Constance was happening again—and once more the talk turned often to geography. Poggio, for one, was enthralled by the idea that these symposia might expand his knowledge of the world. "I was seized with desire," he wrote, "to learn those things which appear to have been unknown to ancient writers, philosophers, and also to Ptolemy."

* * *

WHAT SORTS OF geographical conversations took place during the council? Details are hard to come by, but momentary flashes of illumination do appear in the historical record—and with a strobelike quality they reveal the learned men of Florence, at different times and in different places, scrambling to gather as much information about the world as they can from the host of travelers and scholars who have suddenly descended on their city. One glimpses the physician and mathematician Paolo dal Pozzo Toscanelli in conversation with an Ethiopian monk about the geography of Africa. A venerable Byzantine scholar named George Gemistos Plethon discusses the theories of Ptolemy and other Greek geographers with the humanists. Toscanelli and Plethon together pore over the Ptolemaic map of northern Europe made by Claudius Clavus the Dane. Members of the Orthodox delegation from Kiev respond to questions about Russian and other northern lands. Poggio and his colleague Flavio Biondo grill a delegation of Ethiopians about the size of their country, the names of their cities and towns, the sources of the Nile, and the stars of their night sky. Through a poor interpreter, Poggio talks to a Nestorian from Central Asia and understands him to say that the Great Khan still rules much of the East. Although disparate and fragmentary, these images nonetheless reveal the learned men of Florence engaging in what would become one of the defining enterprises of fifteenth-century humanist geography: the attempt to build a new picture of the world by comparing and contrasting ancient and modern geographical accounts.

Nothing offered the humanists a better chance to further this effort than the serendipitous arrival of Niccolò Conti. Based on the stories that Niccolò told Poggio in private, and others that he told at the gatherings of the learned men, Poggio produced a written account of Niccolò's travels.

The account, which would become Book IV of a larger work by Poggio titled *On the Vicissitude of Fortune*, is often assumed to be a straightforward report of what Niccolò told Poggio. But in fact it's a record only of what struck Poggio as noteworthy about Niccolò's travels. For that reason it reveals as much about the interests of midcentury Florentine geographers as it does about Niccolò's experiences themselves.

Poggio left no doubt about what he felt to be the most noteworthy thing Niccolò had accomplished in his travels. "So far as our records inform us," he wrote, "he went farther than any former traveler ever went, for he crossed the Ganges and traveled far beyond the island of Taprobane, a point which no European had previously reached, with the exception of a commander of the fleet of Alexander the Great and of a Roman citizen in the time of Tiberius Claudius Caesar, both of whom were driven there by tempests."

That one remark speaks volumes. Poggio is trying to understand a modern journey in an ancient geographical context. Niccolò almost certainly didn't make the connection that Poggio implies between Sumatra and Taprobane; he was a merchant, not a scholar. In all likelihood, he recounted as much about his travels as he could recall, and then the humanists took over, trying to figure out where Niccolò had gone by thumbing through their copies of Pliny and Mela and Ptolemy, and by studying the Far East as it appeared on the *mappaemundi*, the marine charts, and Ptolemy's maps.

Poggio's remark also reveals something else of importance. It shows him to be a believer in an emerging humanist article of faith: that at last modern Europeans were poised to exceed the reach and power that the Greeks and Romans had obtained at the height of their empires. Understanding the extent of this empire-to-be was every bit as important to the humanists—and to the Church that so many of them served—as was understanding the empire-as-it-had-been. As they struggled to cope with the flood of new geographical information that was becoming available in Europe in the mid-1400s, the humanists began to see that Ptolemy offered them a way to map not just the world of the past but also the world of the present. A new phase in the reception of the *Geography* had begun, in other words; the humanists began to imagine extending Ptolemy's world in the east and the south, just as Claudius Clavus the Dane had done in the north.

* * *

THE MOST CAPTIVATING speaker to attend the gatherings of the learned men during the Council of Florence was the Byzantine scholar George Gemisthos Plethon. To humanists who heard him speak, he was a prophet, "another Plato." An octogenarian neopagan with flowing silver-white hair, Plethon had traveled to Florence as a senior member of the official Greek delegation, but he cared more about philosophy than religion. During the official proceedings his manner was sullen, but at the humanist symposia he came alive, especially in a famous series of lectures he delivered about Plato, whose works, he claimed, had been unfairly overshadowed in the West by Aristotle's. Plethon's lectures so moved Cosimo de' Medici that after the council he sponsored the creation of a special Florentine academy devoted to the translation and study of Plato.

In addition to championing Plato, Plethon introduced the humanists to his favorite Greek geographer: the historian Strabo, who in the early decades of the first century had written a giant work also known variously as the *Geography* or the *Geographical Sketches*. The book, not yet translated into Latin, was a rambling narrative summary of Greek geographical learning and lore that Strabo intended for "statesmen and commanders," and Plethon insisted that it had much to recommend it.

Strabo differed sharply from Ptolemy on a few important geographical questions. Most notably, Ptolemy described the Indian Ocean as enclosed—bounded on three sides by the known world, and on a fourth, in the south, by an unknown land. This made Taprobane and the other islands of the Indian Ocean inaccessible to mariners sailing from Europe; the only way for European merchants to reach them, therefore, appeared to be by traveling overland through Muslim territories in the Holy Land and the Near East. But Strabo didn't agree. "Homer declares that the inhabited world is washed on all sides by Oceanus," he wrote, "and this is true." The Indian Ocean was not enclosed, in other words, and those dazzling archipelagoes described by Marco Polo and Niccolò Conti might be accessible to European mariners after all.

Europeans had good reason to hope that this was the case. The Turks were rapidly taking control of the Near East and were keeping more of the profits from the spice trade for themselves. The idea of sailing directly to the Far East therefore began to develop a new appeal in Europe, notably among the Italians, who dominated the European end of the trade. Newly interested in the region, thanks to the accounts of Ptolemy, Strabo, and Niccolò Conti, merchants and scholars alike in Italy sought out more information about the region, and inevitably this led them back to Marco Polo's *Description of the World*. Niccolò's stories in particular, made Marco's suddenly seem much more credible. "I, Jacomo Barbarigo," one Venetian wrote in a copy of Marco's Book made at the time of the Council of Florence, "have read this present book of Marco Polo, and I have found many things that he says to be true, and this I testify through the revelation of Ser Niccolò Conti, who has been a long time in that part of India, and similarly through many Moorish merchants with whom I have spoken."

For more than a century after Marco Polo had written his Book, the geographers who tried to make sense of his account had had no uniform mental image of the world onto which they could try to project the places he described. Even the picture of the world that Ptolemy had provided in the *Geography* wasn't of much help, because Marco, and Niccolò Conti after him, had clearly traveled beyond the limits of Ptolemy's *oikoumene*. But as the humanists began to focus on developing an expanded vision of the East with the help of both ancient and modern accounts, they realized that in fact Ptolemy could help them. Expandable in every direction and designed according to the principles of mathematical consistency, his mapping system would for the first time allow geographers all over Europe to consult—and update—the very same map.

*   *   *

How might Europeans actually reach the Far East by sea? Strabo provided some clues. "The Ethiopians," he wrote, "live at the ends of the earth, on the banks of Oceanus." Africa, in other words, was habitable right up to its southern limits, and at those limits it bordered the ocean—which meant that sailing around it should be possible. The Greek historian Herodotus in fact claimed the feat had already been accomplished. "As for Libya [Africa]," he had written in the fifth-century B.C.,

we know it to be washed on all sides by the sea, except where it is attached to Asia. This discovery was first made by Necos, the Egyptian king, who on desisting from the canal which he had begun between the

Nile and the Arabian Gulf, sent to sea a number of ships manned by Phoenicians, with orders to make for the Pillars of Hercules and return to Egypt through them. . . . The Phoenicians took their departure from Egypt by way of the Erythraean [Red] Sea, and so sailed into the southern ocean. When autumn came, they went ashore, wherever they might happen to be, and having sown a tract of land with corn, waited until the grain was fit to cut. Having reaped it, they again set sail; and thus it came to pass that two whole years went by, and it was not till the third year that they doubled the Pillars of Hercules, and made good their voyage home.

The humanists had read similar claims in the works of their own Latin authorities. Pliny the Elder spoke of a fleet of vessels that had set out from the great ancient city of Carthage, in North Africa, and had successfully explored "the circuit of Africa." Pomponius Mela referred to Africa as being "terminated to the east by the Nile and everywhere else by the sea," and went on to describe the shape of the continent so confidently that he seemed to be relying on firsthand accounts. "[Africa] is actually longer than it is wide," he wrote, "and is widest where it abuts the Nile." In the south, he added, it "gently hones itself to a point."

Europeans of the fifteenth century are generally portrayed as having had no idea about the size, the shape, and the circumnavigability of Africa. To a great degree that's true: the Portuguese really didn't know what they would find as they sailed south. But at least a few Europeans had an idea of what to expect.

Even during times of great conflict between Christians and Muslims, Italian merchants maintained active contacts with Arab and Ottoman merchants, and from them they learned a great deal about areas of the world that they themselves couldn't reach. Muslim traders were regularly sailing across the Indian Ocean and traveling in the Far East in the fifteenth century. Their understanding of the area's geography was far superior to most Europeans', and it's quite possible that during their time in the East some Muslim travelers came across maps like the famous Kangnido Map—a mid- to late-fifteenth-century copy of a map made in Korea in 1402, and the earliest surviving world map in the Chinese cartographical tradition (*Figure 48*).

The world as it appears in the Kangnido Map, and as Muslim traders may have described it to counterparts in the West, would have come as a welcome sight to fifteenth-century Europeans newly fixated on the Far East. On the eastern side of the map, south of China, is a multitudinous island realm: just the sort of region that Marco Polo and Niccolò Conti had described. To the west is India, a giant peninsula extending deep into the Indian Ocean in the south. The Indian Ocean itself is open, and on its western side lies a clearly recognizable—and circumnavigable—Africa.

During the early 1400s a few maps appeared in Europe that also suggested an awareness of Africa's true shape. A case in point is the world map of Albertin de Virga, made in Venice between 1411 and 1415 (*Figure 49*). Drawn before the Portuguese had even captured Ceuta, and probably incorporating knowledge obtained from Muslim or Chinese merchants, the map confidently portrays the continent as bulging out to the west in the north, and then honing itself gently to a point in the south.

Such maps may have helped inspire the Portuguese exploration of Africa—as a famous story about Prince Henry's older brother Pedro sug-

 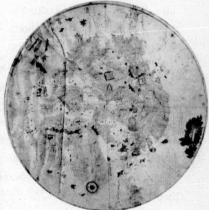

Figures 48 and 49. Left: the Kangnido map. A fifteenth-century Korean copy of a 1402 Chinese original, the map reveals that the shape of southern Africa was known decades before Europeans first reached the continent's tip, in 1487. Africa is on the far left; India and China are on the right. *Right:* the world map of Albertin de Virga (circa 1411–15), showing that some Europeans knew the real shape of Africa by the early fifteenth century. Africa is at the bottom left.

gests. By all accounts less of a Crusader and more of a scholar than Henry, Pedro is said to have visited Venice and Florence in the 1420s, as part of a grand European tour. During his stay in Italy, arranged in part by Poggio Bracciolini, he discussed Africa and the Indies with some of the region's leading humanist geographers. According to the sixteenth-century Portuguese historian Antonio Galvão, the Venetians sent Pedro home with a copy of Marco Polo's Book, accompanied by "a map of the world that had all the parts of the world and earth described." This map, Galvão went on to explain, "set forth all the navigation of the East Indies, with the Cape of Good Hope, as our later maps have described it; whereby it appeared that in ancient times there was as much or more discovered than now there is." Upon returning to Portugal, Pedro shared the map with his brother—an act by which, Galvão wrote, "Don Henry, the King's third son, was much helped and furthered in his discoveries." Pedro, it would seem, may have played a more important role in setting the Portuguese exploration of Africa in motion than Prince Henry's hagiographer, Azurara, chose to reveal.

*   *   *

THE PORTUGUESE DELEGATES to the Council of Florence didn't have much real news to report about Africa when they arrived in Florence in 1439. Gil Eanes had officially rounded Cape Bojador only five years earlier, and Portuguese sailors had yet to sail beyond the parts of northwest Africa that had appeared for at least a century on the Catalan charts. When it came to geography the Portuguese arrived at the council as students, eager, like Prince Pedro, who was now the country's regent, to learn what the city's renowned scholars might teach them about Africa. They also arrived eager to learn from Africans themselves—especially from the Ethiopians who had traveled to Florence for the event. Many in the city, in fact, assumed them to have been sent by Prester John himself.

That assumption must have befuddled the Ethiopians. Who was this king that people kept badgering them about? They did indeed serve a powerful and rich Christian king, but they only had two names for him: his given name, Zare'a Ya'qob, and his regal title, Constantine, Emperor of Ethiopia. Undaunted, the Latins at the council persevered in calling him Prester John, and Pope Eugenius even dispatched a representative to Africa to try to find him. That move may well have prompted the Portuguese to do the same; it's only after the Council of Florence that the Portuguese began officially to describe their voyages of discovery in Africa as motivated by a desire to locate Prester John.

The Portuguese did arrive in Florence with some interesting geographical news to report. In the two decades since the capture of Ceuta they had expanded their knowledge of the islands of the Atlantic beyond the Canary Islands, and by 1420 they had begun settling Madeira and Porto Santo, two lushly wooded and previously uninhabited islands to the north of the Canaries and to the west of Morocco. A little more than a decade later they had also laid claim to the Azores—a chain of small islands, also uninhabited, that lay almost a thousand miles due west of Portugal. European sailors had sporadically visited these island groups throughout the fourteenth century, as part of their travel to and from the Canaries, but the Portuguese nevertheless considered themselves to be embarking on a new enterprise when they formally took possession of them. They were settling a new island world—and they were doing it, they knew full well, in the very waters where the ancients had located the Fortunate Isles, and where Saint Brendan had discovered the Promised Land of the Saints. When the first two children were born on Madeira, they were given fitting names: Adam and Eve.

Two other legendary islands began to appear regularly on charts of the Atlantic during the late fourteenth and early fifteenth centuries. The first, located somewhere vaguely to the west of Ireland, was known by such names as Brasil or Hy-Brasil—names that may derive from the Gaelic Uí Breasail, or "descendants of Breasal," an ancient Irish clan. At the time the Portuguese were rapidly expanding their seagoing trade with northern Europe, and it was in the British Isles, it would seem, that Portuguese merchants and mariners first began to bring home stories about the Isle of Brasil. They also learned about two verifiably real islands off to the northwest of Ireland—Iceland and Greenland, both of which had long been settled by the Norse. If those two islands were real, as plenty of northern sailors who had visited them on trading voyages could attest, then why not Brasil, Saint Brendan's Isle, Marco Polo's Cipangu, and other islands too? Perhaps the Norse had actually found some of these islands already. Centuries before,

Leif Eriksson and others had sailed off into the fog to the west and south of Greenland, where they had discovered long pine-studded stretches of coastline—parts of Labrador, Newfoundland, and the northeastern coast of the United States. "South from Greenland is Helluland," one fifteenth-century Icelandic text recorded about these discoveries, "next to it is Markland, thence it is not far to Wineland the Good, which some men think is connected with Africa." Although the Portuguese are unlikely to have had access to this sort of text, their royals, diplomats, merchants, and fishermen had all established contacts with Scandinavian peoples by the middle decades of the 1400s, and they may well have heard stories of the Norse discoveries through any one of these channels.

Another legendary Atlantic island began to make its way onto maps in the early fifteenth century: Antilia, also known to the Portuguese as the Island of Seven Cities. Located south of Brasil, the island first appears on a map in 1424, drawn on a marine chart made by the Venetian Zuane Pizzigano (*Figure 50*). Depicted on the chart as a huge red rectangle lying due west of the Azores, the island is impossible to miss. The chart provides no information about Antilia, but a later source, the 1492 globe of Martin Behaim, describes its story in detail.

> In the year 734, when the whole of Spain had been won by the heathen of Africa [the Moors], the above island Antilia, called *Septe citade*, was inhabited by an archbishop from Porto in Portugal, with six other bishops, and other Christians, men and women, who had fled thither from Spain, by ship, together with their cattle, belongings, and goods. [In]1414 a ship from Spain got near to it without becoming endangered.

The presence of Antilia on many fifteenth-century nautical charts—often accompanied, as on the Pizzigano Chart, by another rectangular island called Satanezes, just to its north—has prompted an endless churn of speculation. Are these two islands supposed to represent two of the big islands of the Caribbean? Might they even demonstrate an awareness of North and South America? Are they Java and Sumatra? Vast webs of theory have been spun out of these thin threads of speculation, but nothing can be said with any certainty. What these charts make clear, however, is that by

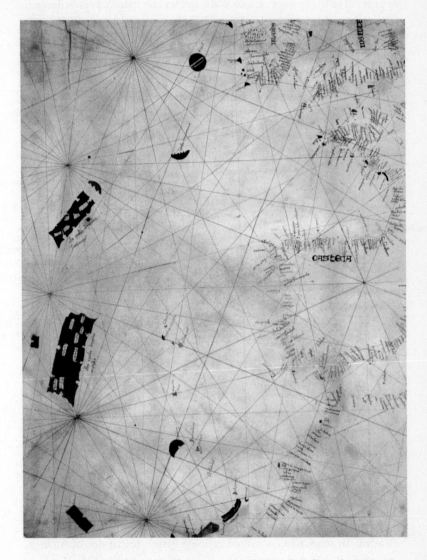

Figure 50. Detail from the Pizzigano chart (1424). On the left are the mythical Atlantic islands of Antilia (bottom) and Satanezes (top). Opposite them on the right are the coasts of Africa and Europe. The little circle just west of Ireland is the Isle of Brasil.

the early fifteenth century at least some sailors and mapmakers believed in the existence of two large islands far off in the ocean to the west of Europe, and to the south of Greenland and Iceland.

Island talk had a special appeal to the classical scholars at the Council of Florence. They, too, were discovering legendary islands in the Atlantic—in their books. The Greeks in particular had had much to say on the subject. As Plethon no doubt reminded the humanists in his lectures, Plato had written at length about Atlantis, the mother of all Atlantic islands. "The island," he claimed, "was larger than Libya and Asia together; and it was possible for the travelers of that time to cross from it to the other islands, and from the islands to the whole of the continent over against them which encompasses that veritable ocean"—a possible reference to the Antipodes. Atlantis and its fabled civilization, Plato wrote, had sunk under the sea in a single day centuries before, but there were plenty of other worlds still to explore. "I believe that the earth is very vast," he made Socrates say, "and that we who dwell [around the Mediterranean] inhabit a small portion only about the sea, like ants or frogs about a marsh, and that there are other inhabitants of many other like places."

Other authorities followed suit. Plethon's favorite geographer, Strabo, fused Atlantis with Ogygia—a distant island described in Homer's *Odyssey* where Calypso kept the storm-tossed Ulysses in exile for seven years. Several episodes of the *Odyssey*, Strabo claimed, had taken place not in the Mediterranean but in the Atlantic somewhere out beyond the Pillars of Hercules. Coupled with well-developed Greek theories about the Antipodes, this idea led Strabo to suggest that "two inhabited worlds, or even more" might exist in the northern hemisphere, and that they probably lay "in the proximity of the parallel through Athens that is drawn across the Atlantic Sea." Writing about a century later, Plutarch got even more specific. Ogygia, he claimed, lay "far out at sea, a run of five days off from Britain as you sail westward." Beyond it, he continued, echoing Plato's description of Atlantis, came other islands, and then "the great mainland, by which the great ocean is encircled."

The Roman knew similar stories. All self-respecting humanists knew the famous passage from the *Aeneid*, for example, in which Virgil had prophesied that the empire of Rome would eventually extend across the ocean to the vast new southern land that lay "beyond the stars, beyond the paths of the year and the sun." Seneca had made a similar prophecy in the play *Medea*, a retelling of the Greek myth of Jason and the Argonauts. "An age will come after many years," he wrote,

> when the ocean will loose the chains of things, and huge land lie revealed; when Tethys [a Greek sea goddess] will disclose new worlds, and Thule no more be the ultimate.

*   *   *

THE LATINS AND the Greeks signed a decree of union in Florence in 1439, largely on Rome's terms, and church bells soon rang out in celebration throughout Western Europe. But in Byzantium the mood was somber, even hostile. Feeling betrayed by their own emperor and responding to an outpouring of popular discontent, most of Constantinople's religious and civic leaders flatly refused to recognize the validity of the decree. No grand military alliance between the Latins and the Greeks would materialize to repel the Ottoman advance, and soon enough, in 1453, the inevitable came to pass: Constantinople fell. Even then some opponents of the union with Rome remained defiant. "Better the turban of the Turk," one prominent Byzantine reportedly insisted, "than the tiara of the Pope." To this day the Roman Catholic and Greek Orthodox churches remain divided.

The Council of Florence *did* help create unity in one important area— geographical theory. Led by the systematic investigations of the humanists, Europeans from all walks of life began to seriously consider the idea that the whole world, not just the known world, might be susceptible to understanding and exploration. Why not? Virgil, after all, had predicted that the Roman Empire would eventually span the entire globe, and modern travelers, urged on by Europe's new Alexanders and Caesars, were already pressing outward beyond the limits of the known world. Thule in the north, the Fortunate Isles in the west, Cape Bojador in the south, Taprobane in the east: none was ultimate anymore. The ocean was unloosing its chains, and new worlds were coming into view.

A newly global conception of the earth emerged at the Council of Florence, in fact, nudged into being by an important shift in geographical perspective. If the torrid zone really was habitable, and the equator pass-able, as so many ancient authorities and modern travelers seemed to sug-gest, then in principle the entire southern hemisphere was accessible to Europeans. Here was a solution to the growing problem of how to reach the Indies from the West. Instead of trying to travel overland *across* the known world to the East, why not sail *under* it in the south? The ocean didn't have to be a barrier that kept the different regions of the world apart. It could become a giant global waterway that brought them all together.

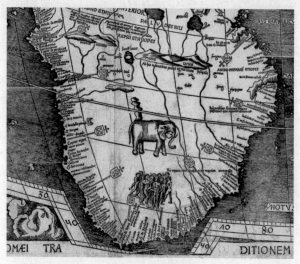

• Southern Africa •

CHAPTER TWELVE

CAPE OF STORMS

Let none doubt the simple arrangement of the world, and that every part may be reached in ships, as is here to be seen.

—The globe of Martin Behaim (1492)

*I*N THE YEARS after the Council of Florence, the Portuguese turned often to their new Italian friends for help in making sense of their African discoveries. One of these friends was Poggio Bracciolini, who in the 1450s became chancellor of Florence, and who during that decade not only corresponded directly with Prince Henry but also seems to have entertained an offer to write his official biography. The Portuguese also turned for advice to another of Florence's learned men: Paolo Toscanelli, the physician and mathematician who during the council discussed the geography of Africa with Ethiopian monks, and the geography of northern Europe with George Gemisthos Plethon. Portuguese ambassadors in Florence sought out Toscanelli in 1459 to discuss geographical questions, and for the meeting

Toscanelli borrowed—according to the friend who lent it to him—"a large *mappamundi* with legends, and complete in everything."

That map, unfortunately, is lost. But just a few years earlier the Portuguese had asked the Venetian monk Fra Mauro to produce for them a grand *mappamundi* (*Figure 51*) showing as full a picture of the world as Italian learning could provide—and this map, delivered to the Portuguese in about 1459, still survives. Mauro's and Toscanelli's maps, although no doubt different in their particulars, are likely to have demonstrated the same basic conception of the world, one that makes clear just how large Africa and the Far East had begun to loom in the minds of Italian geographers and Portuguese explorers.

Figure 51. The Fra Mauro world map (circa 1459). South is at the top, in the Islamic style, focusing attention on southern Africa, the Indian Ocean, and the Far East.

Fra Mauro's map is a magnificent but tortured hybrid: a giant circular world map, almost six feet in diameter, that's sometimes described as the last great medieval *mappamundi*. But it's much more than that. Oriented with south at the top, in the Islamic style, the map reproduces the traditional medieval Christian vision of the world but overlays it with a wealth of information drawn from other sources: Portuguese and Italian marine charts, the writings of Ptolemy and Strabo, accounts of Arab traders and merchants, the stories of Marco Polo and Niccolò Conti, and more. It's gorgeous to look at, and its southern orientation gives pride of place to the very parts of the world that had begun to obsess the Italians and the Portuguese at the time: southern Africa, the Indian Ocean, and the Spice Islands. On the left side of the map the Far East appears just as Marco Polo had described it: a land of unsurpassed civilization and wonders and geographical extent, presided over by the Great Khan. At the top left, in the Sea of Cathay and the Indian Ocean, a mesmerizing and exotic island world beckons. And at the top right, surrounded by water, is an Africa that, once again, in Pomponius Mela's phrase, gently hones itself to a point. Mauro left no doubt about whether sailing around Africa was possible. "Some authorities write that the Indian Ocean is closed like a lake," he declared, "but Solinus says that the Indian Ocean is navigable from the southern part to the southwest, and I affirm that some ships have gone out and come back by that route. Pliny also confirms this . . . [as do] those who have experienced this journey, men of great prudence, in agreement with those authorities." Who exactly those men of great prudence were, he didn't say.

The Italians had more than an academic interest in this route. Merchants from Venice, Genoa, and Florence had long controlled the European end of the spice trade, but the rapid rise of the Turks in the Near East posed a growing threat to business as usual. When Constantinople finally fell, in 1453, many Italian merchants realized that they would have to look for a new way of obtaining spices and other goods from the Far East—and they realized that the Portuguese, by sailing down the coast of Africa, might be leading the way. So they began to drift to the Iberian Peninsula, drawn by the promise of profitable ocean journeys to new African regions and beyond.

Prince Henry died in 1460, and with his death the Portuguese advance along the African coast came to a halt for close to a decade. It resumed again in earnest in 1469, when a prominent Lisbon merchant named Fernão

Gomes was granted a five-year trading monopoly in exchange for a promise to explore at least one hundred leagues (approximately three hundred miles) of new coastline each year—a rate of exploration far more ambitious than anything the Portuguese had previously attempted. During the next few years Gomes's sailors would sail east for nearly 1,500 miles along the underside of the West African bulge, at about 5 degrees north of the equator, and would discover extensive new populated regions—the Slave and Gold Coasts—that would bring Portugal great profits. By the early 1470s they had sailed so far to the east that they—and the European geographers who followed their progress—began to imagine that they had indeed reached the underside of Africa, as it appeared on the maps of Fra Mauro and others. The farther east Gomes's men sailed, the more plausible this idea began to seem. In 1474, deciding that the course of African discovery had become too important to leave in the hands of a private merchant, King Afonso V of Portugal granted lifetime control over trade and exploration in the region to his nineteen-year-old son and heir, Prince João.

João embraced his new job with zeal, and right from the start he seems to have decided that he would pioneer a route around Africa to the Indies. He had good reason to think that his sailors could make better progress than their predecessors. The Portuguese had recently begun using a new kind of ship known as the caravel: a light, maneuverable vessel with two or three masts, based on the model of an Arab fishing vessel known as the *qarib*, that could move quickly over open water but could also sail into the wind and navigate the ins and outs of shallow coastal waters. But just as João took control of African trade and exploration and began making grand plans to sail to the Indies, two of Fernão Gomes's sailors returned to Portugal with some disheartening news. After traveling along the bottom of the West African bulge for some 1,500 miles, they reported, they had arrived at a point (in present-day Cameroon) where the coast of Africa once again turned sharply to the south—and stretched out in that direction as far as the eye could see.

* * *

King Afonso and Prince João evidently greeted this discovery with consternation. But the news prompted one of their advisors, a Lisbon canon named Fernão Martins, to recall a conversation he had had some years earlier with his friend Paolo Toscanelli. The best way to sail to the Far East, Toscanelli had proposed to Martins, might not be to sail east at all, but rather to sail west.

Theoretically, the idea made good sense. Why try to reach the Indies by sailing south into the torrid zone, across the equator, and around a continent of unknown extent when the evidence was mounting that the western ocean might be navigable? The Portuguese had already established a presence in the Azores, after all, a thousand miles to the west of Europe, and according to the ancients, the *mappaemundi,* and the marine charts there were lots of other islands out there still to be found: Antilia, Atlantis, Brasil, Saint Brendan's Isle, Ogygia, Satanezes, and others. Simple geometry made the idea seem reasonable, too. If the known world stretched 180 degrees from west to east, as Ptolemy had taught; if Cathay extended far beyond that to the east, as Marco Polo had claimed; and if, indeed, the island of Cipangu lay some 1,500 miles out into the Sea of Cathay—then Seneca and Roger Bacon and Pierre d'Ailly must all have been right in asserting that only a relatively short distance separated Spain from India.

During the 1460s and early 1470s the Portuguese had little reason to ponder such ideas. They were making impressively long and profitable journeys east along the Slave and Gold Coasts, and the Indies didn't seem far off. But when the news came that Africa's west coast had turned again to the south, suddenly the idea of sailing around Africa became more daunting— and Fernão Martins therefore decided to mention Toscanelli's suggestion to King Afonso.

The king was intrigued. He instructed Martins to write to Toscanelli, now a grand old man approaching the age of eighty, requesting a written summary of his theory, accompanied by a map. Martins duly carried out his orders, and in the summer of 1474, Toscanelli sent him what was destined to become one of the most famous letters of all time. The original letter and map no longer survive, but three very similar early versions of it do, including the version that follows, transcribed early in the sixteenth century by the Spanish historian Bartolomé de Las Casas.

Paolo, physician, to Fernão, greetings.

I was delighted to hear of the intimacy and friendship you enjoy with your most illustrious and noble sovereign. Although I have often spoken of the short distance that lies between here and the Indies, where the spices grow, and the sea route that is shorter than the one you pursue by way of Guinea, you tell me his majesty now desires me to provide him with some statement to that effect, or some kind of demonstration that would make that route easier to comprehend and follow. I know that I am in a position to furnish such proof in the form of a sphere representing the earth, but I believe I can do so more clearly and convincingly by providing a sea-chart of the kind used by mariners. I accordingly submit such a chart to his majesty, devised by myself and drawn by my own hand, on which is laid out all the western part of the world, from the coast of Ireland southwards as far as the very end of Guinea, together with all the islands that lie in between and, directly opposite them, the east coast of the Indies with all the islands and places along the route, together with advice about where and how far a traveler must depart from the line of the equator, and an indication of the distances one must travel to reach these places, which are most fertile and abundant in all manner of spices and jewels and precious stones. Do not marvel at my characterizing the region where the spices grow as "the West" although these are commonly known as "the East," for any man who sails westward will always find these lands to the west, just as he who sets out overland to the east will find them in the east. The straight lines that are drawn lengthwise on this chart show the distances from east to west, while those that are drawn across the chart show the distances from north to south. I have also depicted on this chart many places in India where one may seek refuge in the case of storms or contrary winds and any other unforeseen eventuality, and also that you may be as fully informed as you seek to be about the whole region.

You must know that the only people who live and trade throughout all these islands are merchants, and that there are as many ships, seamen, and merchants in the area as in any other part of the world, especially in one particularly noble part known as Zaiton, where every year a hundred large ships load and unload their cargoes of peppers, not to mention the many others that carry spices of other kinds. The land hereabouts is densely populated with a multitude of provinces, kingdoms, and cities without number, all under the rule of a sovereign known as the Great

Khan, a name which in our own tongue means King of Kings, who resides most of the time in the province of Cathay. His ancestors greatly desired to have contact and dealings with Christians, and some two hundred years ago sent an embassy to the Holy Father asking him to send them a large number of learned and wise men who might instruct them in our faith, but those who were sent were forced to return home because of difficulties they encountered along the way. In the time of Pope Eugenius IV there came an ambassador who told of their great friendship for the Christians, and I had many conversations with him, in which he told me of many things: of the great size of their royal palaces, the wondrous breadth and length of their rivers, the vast number of cities on the banks of these rivers, and of how on one river alone there are two hundred such cities, with broad bridges of great length made entirely of marble and adorned with marble pillars. This is as noble a country as any known to man, for not only can one find there many things of great value but there is also gold and silver and precious stones as well as every manner of spice in great quantity, many of which have never been brought to our shores. The governance of this magnificent province and the conduct of its wars is in the hands of wise and learned men: philosophers and astrologers and others of great skill.

Given in this city of Florence on the twenty-fifth day of June, in the year one thousand four hundred and seventy four.

The language that follows in Las Casas's copy of the letter seems to have been lifted from Toscanelli's map itself, and has prompted scholars and amateurs alike to try to reconstruct that map (*Figure 52*).

From the city of Lisbon, in a line directly to the west, there are twenty-six spaces marked on the map, each representing two hundred and fifty miles, before you arrive at the great and most noble city of Kinsai, which is a hundred miles in circumference, the equivalent of twenty-five leagues and has ten bridges made of marble. The name of the city, translated into our language, is City of Heaven, and marvelous things are related of its great buildings and magnificent monuments and vast wealth. It [these twenty-six spaces] occupies about one-third of the sphere [that is, of the circumference of the globe]. This city lies in the province of Mangi, near the city of Cathay, where the king spends the greater part of his time. And between the island of Antilia, known to you as the Island of the Seven Cities, and the most noble island of Cipangu, are ten spaces—that is, two thousand five hundred miles, or two hundred and twenty-five leagues. This island is most rich in gold and pearls and precious stones, and you should know that the temples and royal palaces are covered in solid gold. Yet because the route is unknown, all these things are hidden from us, even though one can voyage there without danger or difficulty.

Nobody knows how seriously King Afonso considered Toscanelli's proposal. But within a year he had more pressing matters to attend to, because Portugal went to war against Spain. In open violation of the papal bull of 1456 that had granted Portugal jurisdiction over all territories "from Capes Bojador and Nam as far as through all Guinea," Spain had begun sending its own merchant ships on missions to Africa, something the Portuguese decided had to be stopped.

The fighting would last five years, until 1479, when the two countries signed a peace accord, known as the Treaty of Alcáçovas, that allowed both

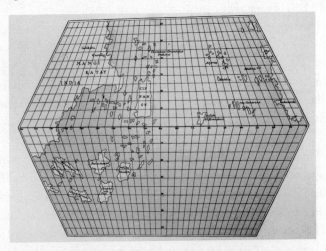

Figure 52. A nineteenth-century reconstruction of the lost map of Paolo Toscanelli, showing the ocean separating Europe and Africa (on the right) from Cipangu, Cathay, and the Indies.

sides to claim victory. For Portugal, this meant a new assertion of authority over Atlantic and African territories. In exchange for concessions to Spain on the Iberian Peninsula, Portugal received control of the Azores, the Cape Verde Islands, the Madeiras, and

> all the trade, lands, and barter in Guinea, with its gold mines, or in any other islands, coasts, or lands, discovered or to be discovered, found or to be found ... or in all the islands hitherto discovered and to be discovered, or in all other islands which shall be found and or acquired by conquest from the Canary Islands beyond toward Guinea ... excepting only the Canary Islands [listed by name], and all the other Canary Islands, acquired or to be acquired, which remain to the kingdom of Castile.

All things considered, this was a very good deal indeed for Portugal— especially since the treaty also affirmed Portugal's exclusive right to license trade and exploration along the Guinea coast. The Portuguese were now free, as the *Inter caetera* bull of 1456 had put it, to continue their explorations "past that southern shore [of Guinea] all the way to the Indians."

* * *

KING AFONSO DIED in 1481 and was succeeded by Prince João. One of João's first moves as king was to commission the building of a fortress on the Guinea coast, on the underside of the West African bulge, to protect Portuguese interests there. His sailors chose a southward-looking site on the coast of present-day Ghana, and by 1482 the fortress, Saint George of the Mine, was in place. The fortress and the town that grew up around it would soon become known as Elmina, or "the Mine," and would serve as Portugal's headquarters for African trade and exploration for decades afterward.

As soon as the fortress at Elmina was ready, João decided to resume exploration of the African coast to the south. Summoning an experienced Portuguese sea captain named Diogo Cão, he instructed him to sail as far as he could below the equator.

Cão left Lisbon in 1482. With him he brought a number of carved stone columns topped by a cross, known as *padrões*, which he planned to erect along newly discovered stretches of the coast, as a way of marking his progress and claiming the territories for Portugal. Other Portuguese sailors would continue the practice, and soon chart makers began to draw *padrões* on their maps: iconic records of Portuguese discovery.

Cão sailed south as far as present-day Angola—and there, for reasons unknown, he decided that he and his men were at last about to round the tip of Africa. A marine chart survives that provides an idea of what he thought he had discovered (*Figure 53*). Copied in Venice in 1489 from a lost original made during Cão's voyage, it marks the southernmost point Cão reached on his journey with a *padrão*, and then shows the coast turning east, forming a southern cape that Cão believed was the tip of the continent.

Figure 53. A detail from the "Ginea Portogalexe" chart (circa 1489), showing the point along the coast of modern-day Angola at which Diogo Cão believed he had discovered the route around the southern tip of Africa.

No such cape exists along the coast of Angola. But nobody in Portugal knew that, and when Cão returned home, in 1484, and reported the news of his discovery, he was greeted as a hero. King João granted him a royal pension and made him a nobleman, and soon Portugal went public with its news. "Last year," a Portuguese ambassador announced in Rome in 1485, "our men, having completed the greater part of the circuit of Africa, reached almost to the Prassum Promontorium."

Anybody familiar with Ptolemy's *Geography* would have understood the significance of that remark. The Prassum Promontorium—clearly visible in a popular edition of the *Geography* that had been printed in Rome in 1478—was a prominent cape that Ptolemy had located at the very bottom of his Africa, on its eastern side (*Figures 54 and 55*). If the Portuguese had almost reached the cape, in other words, they had almost reached the Indian Ocean.

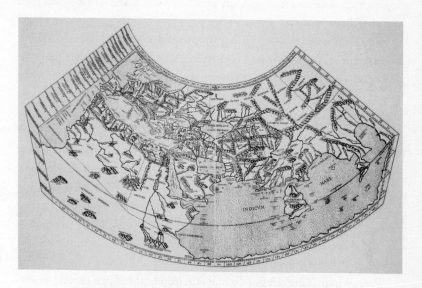

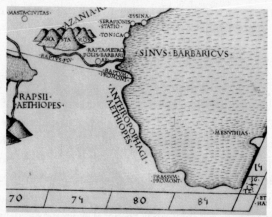

Figures 54 and 55. Top: world map from the Rome edition of Ptolemy's *Geography* (1478). Africa is on the left. The Prassum Promontorium is at the bottom of its east coast, where it meets the map's border. *Bottom:* the Prassum Promontorium, at the southern limits of Ptolemy's known world. From the Rome *Geography*'s map of southern Africa.

Cão set sail again in 1485, but his promised cape failed to materialize. He probed south for hundreds of miles beyond the southern limit of his previous journey, but to no avail: the arid and uninviting desert coast of Namibia stretched out indefinitely to the south. He planted his last *padrão* at about 22 degrees south, at a site now known as Cape Cross—and then vanished from history.

But good news reached Lisbon from another part of Africa at about the same time. While Cão had been sailing south, the Portuguese had sent scouts into the West African interior from their base in Elmina. Exploring present-day Nigeria, they got word that some 250 leagues farther inland there lived a great ruler—"a King the most powerful of all in those parts," the historian João de Barros would write in the following century. All of West Africa's kings professed an allegiance to this king, the Portuguese were told, just as the kings of Europe professed an allegiance to the pope. This all had a very familiar ring to it—"wherefore," de Barros wrote, "the king and his cosmographers, taking into consideration Ptolemy's general map of Africa, the *padrões* on the coast that had been set up by his discoverers, and also the distance of two hundred and fifty leagues to the east . . . concluded that he must be Prester John."

Prester John! Could it be? To find out, João sent out two expeditions— "ships by sea and men by land," according to João de Barros, "in order to get to the root of this matter."

* * *

THE MAN JOÃO CHOSE for the land-based hunt was Pêro da Covilhã, an Arabic-speaking squire who had long worked in the service of the Portuguese royal family and who could pass as a Muslim. For the better part of the next decade Covilhã would travel south and east in search of Prester John and the Indies—first to Alexandria and Cairo, then across the Red Sea to Aden, and then across the Indian Ocean to India's west coast. He explored the coast with care, learning all he could from Hindu and Muslim merchants about the spice trade, the local geography and customs, and the currents and weather systems of the Indian Ocean. He found himself particularly impressed by the port city of Calicut (today Kozhikode), on India's southwest coast. The city, he would report in a letter to King João, was a bustling entrepôt in the spice trade, rich in "cinnamon and pepper [and] cloves that came from beyond." If the Portuguese ever made it around Africa and into the Indian Ocean, he implied, *this* was where they should head.

Covilhã then sailed back to East Africa, where he picked up some very promising information: traders on the Red Sea coast told him that Africa did indeed come to an end in the south. If the Portuguese kept following the west coast to the south they would eventually reach the continent's tip, and just on the other side they would find a giant island that the Arabs called the Island of the Moon—Madagascar. From there, Covilhã reported with excitement, the king's ships "could easily . . . reach the coast of Calicut, for it was sea all the way."

Covilhã went on to explore the interior of Ethiopia for years and finally managed to locate a minor Christian king named Eskender. He never did find Prester John, however—and with the failure of his mission the long hunt for Prester John in Africa effectively came to an end. But the dream of finding him *somewhere* still lived on. "Prester John," the Portuguese printer Valentim Fernandes wrote in 1502, in an introduction to a new edition of Marco Polo's Book, "is out there in Cathay."

* * *

To LEAD HIS ocean-going search for Prester John and the Indies, King João turned to Bartolomeu Dias, a respected Portuguese sea captain. Given orders to travel as far as he could down the African coast, Dias set sail from Lisbon in August of 1487.

Dias had better luck at sea than Covilhã had on land. He and his crew successfully located Diogo Cão's last *padrão*, on the Namibian coast, and then continued sailing south. Northerly currents and prevailing winds that blew inshore made their progress slow and difficult. When heavy weather blew in on December 4, 1487, Dias ordered his ships out to sea to ride out the storm. Running behind gale-force winds, with their sails only at half-mast, the ships sailed south for thirteen days out of sight of land. When the skies finally cleared Dias ordered a course set to the east, assuming that he would eventually return to the coast. But his ships sailed for several days without managing to find land. On a hunch he ordered them to tack to the north. Not long after, he and his men finally caught sight of the coast—and discovered that it now traveled east, not south.

Dias and his men made a brief landing, took on water and supplies, had a skirmish with cattle herders who had pelted them with stones, and then continued on their way. Several days of easterly sailing later, the coast began a long, gradual bend to the north—suggesting that Dias had rounded the continent's tip, and "giving great hope," as one Portuguese historian would put it in 1508, "of the discovery of India."

At this point Dias's crew—short of supplies, sick with scurvy, struggling to make progress against coastal currents—decided they had gone far enough. Dias convinced them to continue on for a few more days, long enough to confirm that the coast continued to run on to the northeast, but in March 1488 he finally turned his ships around and began the long journey home. This time he stayed closer to the coast, and he and his crew now saw what they had missed on the outward passage: a high and majestic cape jutting out into the Atlantic near the very bottom of Africa's west coast. Dias erected a *padrão* at the site and named it the Cape of Storms, as a tribute to the harsh weather that had carried his ships around it. King João would later change the name to something more uplifting: the Cape of Good Hope.

When Dias returned home, in December of 1488, he received nothing like the hero's welcome that Diogo Cão had received after his first voyage. King João granted Dias only a modest pension for his efforts, in fact, and no record exists of any sort of public celebration after his expedition's return—a curiously indifferent reaction, it would seem, to one of history's great feats of exploration. But the explanation is a simple one: nobody knew for sure if Dias had actually discovered anything of significance. At least one orator in Portugal believed that despite Dia's discovery, the Portuguese had yet to obtain their objective. "Each day," he declared in March of 1489, well after Dias's return, "we strive to reach the capes Prassum and Raptum [a cape Ptolemy locates to the north of Prassum] . . . and from there to the approaches of the Indian Ocean."

The Portuguese would only recognize the significance of Dias's voyage a decade later, after they had finally managed to sail to India and back. That's the point in history at which they began to tell the story of Dias's voyage as it's still told today: as a epic journey in which Dias and his men ride out the famous storm and eventually discover—as João de Barros would put it—"that great and notable Cape, hidden for so many hundreds of years . . . which, when it is seen, not only reveals itself but opens up an entire new world of countries." Uncertainty has no place in this heroic version of the narrative; Dias here comes across as a bold and visionary prophet. "Dias had been chosen like Joshua," Valentim Fernandes wrote in his Marco Polo introduction, "to enter into that New World, which we can indeed call the Promised Land."

That's easy to say in retrospect. But the news Dias brought back to King João must actually have been quite dispiriting. He had sailed more than a thousand miles beyond Diogo Cão's final *padrão* and yet he still couldn't say with certainty that he had reached the Indian Ocean. And even if he had actually pioneered the ocean route around Africa, he had revealed its coastline to be far longer than anybody had expected. According to João de Barros, stories about the fierceness of the storm that had blown Dias far south gave rise among Portuguese mariners to "a new myth of the dangers [at the Cape of Good Hope] as there once had been concerning Cape Bojador."

In short, the news that Dias brought back to Portugal was that sailing around Africa to the Indies would be longer, more fraught with peril, and less practical than anybody had expected. Small wonder, then, that nobody celebrated Dias as a hero.

* * *

WHAT DIAS DID come home with was a map. Like all good Portuguese mariners, he had carefully noted his progress on a marine chart, and upon his return he brought it ashore and laid it out for the king to see. The chart is now lost, but copies evidently were made and began to circulate—and one of them seems to have made it into the hands of a German mapmaker named Henricus Martellus.

Frustratingly little is known about Martellus. All that can be said about him is that he worked in Italy during the 1480s and 1490s, in partnership with a well-known Florentine mapmaker named Francesco Rosselli; that he made a number of maps on his own; and that he was one of the very first mapmakers in Europe to extend Ptolemy's Africa to the south to include Dias's discoveries.

Martellus's most famous map, which dates from 1489 or 1490, presents a strikingly new vision of the known world (*Plate 9*)—one that would powerfully lodge itself in the imaginations of European explorers and geographers for years to come. Martellus designed his map according to Ptolemaic principles, but wherever possible he updated and expanded Ptolemy's original world map to include modern discoveries. Nowhere is this more obvious than in the case of Africa. The continent is depicted in the style of the marine charts, with neatly spaced place-names running along its north and west coasts, and right down around the Cape of Good Hope to the point at which Dias and his crew had turned back. To focus attention on what Dias had discovered—and to signal that in showing an open Indian Ocean he was breaking with long-established Ptolemaic tradition—Martellus came up with a memorable visual conceit: he rammed the southern tip of Africa right through the bottom of his map's frame.

Martellus also introduced something new on the eastern side of the Indian Ocean. Relying on Marco Polo, just as Fra Mauro and others had done before him, he extended the known world all the way to the coast of Cathay, and there he added a long, imaginary north-south peninsula that gradually bends to the west as it makes its way down into the Indian Ocean. This peninsula, possibly a Martellus innovation, is sometimes known as the Dragon's Tail, and it would soon become a regular feature on many world maps.

At a glance the Dragon's Tail looks quite a lot like South America, and its sudden appearance at the end of the 1480s—at a time when the Portuguese are known to have already sponsored several voyages of discovery out to the west of the Azores—has prompted some scholars to argue that that's exactly what it is. But what the Dragon's Tail actually represents is most certainly nothing more than a vestige of the giant southern land that stretched between Asia and Africa under the Indian Ocean on Ptolemy's world map.

Martellus produced another world map at about the same time—a painted wall map, approximately four feet by six feet, now owned by Yale University (*Figure 56*). In many respects the Yale map corresponds to its smaller cousin, but there are a few important differences. The Yale map, for example, shows the south coast of Africa extending far to the east beyond the Cape of Good Hope, suggesting that a journey of intimidating length would be required to sail around the continent and into the Indian Ocean.

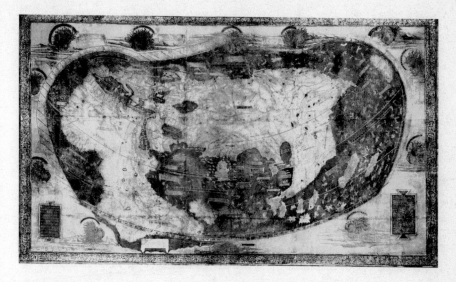

Figure 56. The wall map of Henricus Martellus (circa 1491–92). Combining geographical ideas from Ptolemy's *Geography*, Portuguese marine charts, and the Book of Marco Polo, the map shows the recently discovered route around the southern tip of Africa, and a maze of islands in the Far East, dominated at the map's edge by the giant island of Cipangu.

At the eastern edge of the Yale map, Martellus also extends his picture of the world out beyond the end of Asia and into the Sea of Cathay, which appears very much as Marco Polo had described it: an island-studded ocean world dominated in the north by Cipangu and in south by the two Javas. This was the vision of the Far East, more than any other, that European sailors would carry with them in their imaginations when, just a few years later, they began making their way west across the Atlantic.

Martellus got a lot of his distances wrong, it turns out: he portrayed Africa as extending to about 45, instead of 35, degrees south—a mistake he may well have picked up from Dias himself. This error made the voyage south around Africa and east to the Indian subcontinent appear some one thousand miles longer than it actually is. Martellus also dramatically exaggerated the width of Europe and Asia, estimating it to be 230, instead of 130, degrees, an error that added some seven thousand miles to the distance between the west coast of Portugal and the east coast of Cathay.

These inaccuracies derived from Ptolemy, who had famously overestimated the width of Eurasia, but Martellus compounded them by extending his Far East east some 50 degrees beyond where Ptolemy's Asia had ended, at 180 degrees—and on his Yale map Martellus extended his world another 40 degrees to the east, in order to include Cipangu. He had mapped 270 degrees, or three-quarters of the globe, in other words, leaving only 130 degrees of unmapped ocean between Europe and Cathay, and only 90 degrees between Europe and Japan. As it happens, these distances correspond almost exactly to those laid out by Toscanelli.

The world was coming together in bits and pieces. Toscanelli had mapped the uncharted western ocean; Martellus had mapped the whole of the known world, including parts of Africa and Asia not known to Ptolemy; and in 1492 a mapmaker named Martin Behaim did the inevitable: he constructed a globe that made those two partial visions of the world into one.

* * *

BEHAIM IS A slippery character. Born in Nuremberg in 1459, he seems to have begun his professional life as a mediocre and dissolute man of business. But in the early 1480s he moved to Portugal, got married, and moved a thousand miles out into the western ocean, to the Portuguese island of Fayal, in the Azores—the perfect place from which to contemplate the geographical notions of Martellus and Toscanelli.

As a trader living in Portugal and the Azores, Behaim had ample opportunity to join Portuguese voyages to Guinea, and on his globe he makes the very unlikely claim that he captained one of the ships that took part in Diogo Cão's first voyage. He does seem to have worked his way into King João's inner circle in Portugal, however, and for a time, in the 1480s, he served as an advisor to the court on matters relating to celestial navigation and geography. During that period he may have come across Toscanelli's famous letter and chart—and they in turn may have given him the idea of making his globe.

Behaim's globe is not the work of an expert geographer (*Figure 57*). It's a muddle, full of mistakes, misperceptions, and outright inventions. But it also neatly sums up a number of the geographical ideas and traditions that had begun to converge in the popular imagination in Europe by the early

1490s. Behaim's world is the world as many Europeans knew it on the eve of the discovery of the New World.

Behaim crammed everything he could onto his globe, and explained his major sources in a legend.

Be it known that on this Apple [globe] here present is laid out the whole world, according to its length and breadth, in accordance with the art of geometry—namely, the one part as described by Ptolemy . . . and the remainder from what the Knight Marco Polo of Venice caused to be written down in 1250 [sic]. The worthy Doctor and Knight John Mandeville likewise left a book in 1322 [sic] that brought to the light of day the countries of the East, unknown to Ptolemy, whence we receive spices, pearls, and precious stones, but the Serene King John of Portugal has caused to be visited in his vessels that part of the south not yet known to Ptolemy. . . . Towards the west the Ocean Sea has likewise been navigated further than what is described by Ptolemy, beyond the columns of Hercules as far as . . . the Azores.

Figure 57. The globe of Martin Behaim, showing the Far East as European geographers imagined it in 1492. Cathay and Mangi (China) are on the left; Cipangu (Japan) is at the bottom right, surrounded by the islands of the Indies.

Behaim borrowed from plenty of other sources, too, including the maps of Martellus, or at least maps very like them. The eastern-trending tip of Africa, the Dragon's Tail, and the islands of the Sea of Cathay: all appear on Behaim's globe just as they do on Martellus's maps. Behaim's version of the oceanic space between Cipangu and Europe corresponds well, in terms of size and content, to how Toscanelli described it; Antilia, for example, lies precisely where Toscanelli had suggested it could be found.

More than any map before it, Behaim's globe made sailing west to the Indies seem feasible—and Behaim seems to have decided that he himself was the man for the job. The evidence for this comes in the form of a long and remarkable letter sent by the prominent Nuremberg physician Hieronymus Müntzer to King João in 1493.

> To the Most Serene and Invincible João, King of Portugal, the Algarves, and Maritime Mauritania, first discoverer of the Islands of the Canaries, Madeira, and the Azores—Hieronymus Müntzer, a German doctor of medicine, most humbly commends himself. Because heretofore you have inherited from the most serene Prince Henry, your uncle, the glory of sparing neither effort nor expense in extending the bounds of the world, and by your diligence you have subjected to your rule the Seas of Ethiopia and Guinea, and the maritime nations as far as the Tropic of Capricorn, together with their products, such as gold, grains of Paradise [a peppery African spice], pepper, slaves, and other things, you have by this display of talents won renown and immortal fame, and great profit for yourself besides. . . . Wherefore Maximillian, the most invincible King of the Romans, who is Portuguese on his mother's side, desired to have Your Highness invited to seek the very rich shore of Cathay in the Orient. . . .
>
> Be not disturbed by [those] who lack experience and who have said that only one quarter of the earth is not covered by the sea. . . . For in matters pertaining to the habitableness of the earth more confidence is to be placed in experience and trustworthy accounts than in fantastic imaginations. For you know that many official astronomers denied there was any habitable land below the tropics and the equatorial regions. By your own experience you have found reasoning like this to be vain and false. . . .

O, what glory you will gain if you make the habitable Orient known to your Occident! Likewise, what profits would its commerce give you, for you will make those Isles of the Orient tributaries, and their kings, amazed, will quietly submit to your sovereignty. . . . If you carry out this mission you will be exalted like a god, or a second Hercules; and, if you wish you will take along on this journey as a companion Martin Behaim, deputed by our King Maximilian especially for this purpose. . . .

Farewell! From Nuremberg, a city in Upper Germany, 14 July A.D. 1493.

As a former court advisor in Lisbon, Behaim had good reason to expect that King João would consider this plan seriously and perhaps seek him out. But the letter came just a little too late.

A few months earlier, on March 4, 1493, a battered caravel seeking refuge from heavy seas had nosed its way into Portugal's Tagus River estuary. The vessel, which flew a tattered Spanish flag, had come to anchor some four miles south of Lisbon harbor, not far from where a Portuguese man-of-war lay at anchor. In command of the man-of-war was none other than Bartolomeu Dias, who, concerned about the unannounced arrival of a Spanish vessel in Portuguese waters, boarded the caravel with a small party of armed men and demanded an explanation. They soon got one. A tall Genoese mariner, forty-two years old, with blue eyes, a fair, ruddy complexion, and a haughty bearing, announced himself as the vessel's captain. His name was Cristoforo Colombo, he said—and he had just sailed west to the Indies and back.

• Portugal, Spain, North Africa, and the Atlantic •

CHAPTER THIRTEEN

COLOMBO

*Just as one thing leads to another and starts a train of thought,
while he was in Portugal [my father] began to speculate that just as
the Portuguese had sailed so far south, it should be possible to sail as
far west, and to find land in that direction.*

—Ferdinand Columbus (circa 1538)

THE PORTUGUESE KNEW Colombo well. He had been based in Portugal in the late 1470s and early 1480s, and near the end of his time there, in 1483 or 1484, he had even met with King João himself, to request official Portuguese backing for a voyage of discovery across the Atlantic. João had listened with interest as Colombo laid out how he intended to sail to Cipangu, Cathay, and the Indies, but in the end the king had declined to support the venture. Diogo Cão had just returned to Portugal from his first African voyage with the news that he had found a sea passage around Africa to the Indies, and João saw no reason to start hunting for one to the west. He had turned Colombo down for another reason, too. According to João de Barros, he had found Colombo to be "a big talker and boastful in setting forth his accomplishments, and full of fancy and imagination with his Isle Cipangu."

By the spring of 1493, João had good reason to rue that assessment. Diogo Cão had disappeared on his second voyage of discovery, in 1485, without ever having found the Prassum Promontorium. Bartolomeu Dias had reached the Cape of Good Hope four years later, but his voyage had made clear that the route around Africa and into the Indian Ocean would be far longer and more treacherous than anybody had expected—if such a route even existed. And now this upstart Colombo, soon to be celebrated across Europe as Christophorus Columbus, the Latin form of his name, had sailed west to the Indies—and had claimed them for Spain.

* * *

BORN IN GENOA in 1451, Columbus had first appeared in Lisbon in 1476, a young sailor with an unremarkable background in the wool trade. For much of the following decade, until he left for Spain in 1485, he had based himself in the city, living and working among its growing community of Genoese, Venetian, and Florentine expatriates, and during that period he accumulated experiences as a sailor that he had every right to boast about. Even before he arrived in Portugal he had already spent years crisscrossing the Mediterranean; he had sailed to Tunis and Marseille in the early 1470s, in the service of King René of Anjou, and at least once had made the voyage to the Greek island of Chios, at the time a trading colony controlled by the Genoese. He had also fought in a naval battle and survived a ship-wreck—which, as legend has it, is how he first arrived in Portugal. In May of 1476, the story goes, he joined a convoy of Genoese ships on a trading mission from Chios to Portugal, England, and Flanders. On August 13, not long after it had sailed past Ceuta and through the Straits of Gibraltar, the convoy was attacked by French and Portuguese warships, not far from the south coast of Portugal. A daylong skirmish ensued, and several of the Genoese ships were sunk, including Columbus's own. Wounded and cast adrift, Columbus caught hold of a floating oar and kicked his way some six miles to shore, near Lagos—the Portuguese port where, several decades earlier, Prince Henry had proudly supervised the unloading of some of the first West African slaves.

Penniless and in need of work, Columbus made his way to Lisbon, and there the city's Genoese community took him in. He couldn't have arrived at a better time. Unlike Genoa, whose fortunes were in decline, Lisbon was a city on the rise. Its port bustled with merchants, sailors, and ships on their way to and from every imaginable destination: the well-established ports of the Mediterranean and North Atlantic; the newly settled islands of the African Atlantic; and the ever-expanding coastline of Guinea. The talk was of gold, spices, and slaves—exotic new sources of wealth guaranteed to spark the imagination of a young sailor eager to make a name and a fortune for himself. Even the setting was inspiring. Continental Europe came to an end just a few miles to the west of Lisbon harbor, and beyond it lay the open ocean.

Columbus decided to stay and soon was sailing again. But now he saw Atlantic, not Mediterranean, horizons. By his own account, he sailed north to Iceland in 1477 and visited Ireland at about the same time. In 1478 he sailed south to Madeira, as a business agent for a Genoese sugar merchant, and a few years later he married into a Portuguese family that owned land on Porto Santo, Madeira's sister island. He and his wife lived on Porto Santo for a time, where he seems to have prospered as a merchant, and in the early 1480s, probably from his base there, he sailed on Portuguese ships to Guinea, and, even made a stop at the brand-new Portuguese fort and trading colony at Elmina. "I have sailed in every part of the world now known to man," he would later tell the Spanish Sovereigns, and no doubt he made a similar claim to King João when proposing to sail west across the Atlantic.

The claim had merit. Columbus had, in fact, sailed along almost every stretch of coastline depicted on the marine charts of his time, and he had visited regions of the earth that had long been declared off-limits by scholarly geographers: the frigid zone in the north and the torrid zone in the south. During his many voyages up and down the length of the Atlantic seaboard he had learned the tricks of his trade from some of the best sailors and navigators in the business; he had developed a nuanced understanding of the Atlantic's currents and wind systems; and, significantly, he had picked up stories about the islands that still awaited discovery to the west.

When Columbus had first arrived in Portugal, the search for new Atlantic islands was intensifying. For more than a century, sailors setting out from the Iberian Peninsula had been finding, raiding, settling, and exploiting an ever-growing number of islands: the Canaries, Madeira and

Porto Santo, the Cape Verde Islands, the Azores. From their vantage point a thousand miles to the west of Europe, Portuguese settlers in the Azores had for decades been reporting sightings of land farther out to sea, and had described how pieces of cane and driftwood, unlike anything found in Europe, regularly washed ashore onto their shores. The stories tantalized some adventurous souls, who, eager to discover new islands for themselves, began to venture out beyond the Azores to the west. Some requested official backing from the Portuguese court. A few documents recording those requests survive, and they spell out the names of the islands to be found, among them Antilia, the Island of the Seven Cities, and Saint Brendan's Isle.

Columbus presumably knew of these expeditions—and of others being launched farther north by the English. Especially active was a group of merchants and fishermen based in Bristol, in the south of England. These "men of Bristol," as they are known today, had long been making regular trading visits to the colonies on Iceland and Greenland, and Columbus may well have made his own journey to Iceland in their company. (He makes explicit mention of men from Bristol when describing his journey.) In addition to sailing to Iceland and Greenland, the men of Bristol probed the cold and stormy waters of the North Atlantic in search of cod, and on some unrecorded voyage during the fifteenth century they caught sight of *something* to the west of Greenland. Whatever this place was, real or imaginary, and whenever it was that they saw it, by the 1470s they had developed a firm belief in its existence. If Columbus sailed with any of them, to Iceland or on any of his other voyages, he would surely have heard about it.

The men from Bristol did more than just talk about this new place in the west. So convinced were they of its existence—and perhaps of the great schools of cod that swam in its vicinity—that on July 15, 1480, some of them boarded a ship in Bristol harbor and set out to find it. The commander of the vessel was John Lloyd, identified in a chronicle of the time as "the [most] knowledgeable seaman of the whole of England"—and his mission was to find none other than "the Island of Brasylle," a name clearly borrowed from the Brasil of Irish legend. Lloyd and his sailors searched unsuccessfully for the island for nine months before being forced back to Ireland by bad weather, but they soon put out to sea again. On July 6, 1481, according to Bristol customs records, "Thomas Croft of Bristol, armiger, customer of the said lord the King . . . laded, shipped, and placed forty bushels of salt . . . and not with the intention of trading but of examining and finding a certain island called the Isle of Brasil." Nothing else is known about this expedition or its fate, although the amount of salt it carried suggests that finding fish was indeed the primary objective.

During his itinerant years in the 1470s and 1480s, as he sailed in Atlantic waters off the coasts of Europe and Africa, Columbus heard stories about the lands and islands that lay out in the Atlantic to the west, and at some point—nobody knows quite when—he decided that they were worth studying. "Consequently," his son Ferdinand would later recall, "he wrote down all the indications that he heard from people and sailors, in case they might be helpful to him. [And he] was able to make such good use of these things that he began to believe beyond any doubt there were many lands to the west of the Canary Island and the Cape Verde Islands, and that it was possible to sail to them and discover them."

* * *

INITIALLY, COLUMBUS IMAGINED finding just a few new islands out in the Atlantic—a discovery that might win him the honor of becoming a latter-day Lanzarotto Malocello, the Genoese sailor who had rediscovered the Canary Islands more than a century earlier. But he soon began thinking in grander terms. Perhaps he took inspiration from the story of the Vivaldi brothers, also from Genoa, who in 1291 had sailed out beyond the Pillars of Hercules in search of India, never to return; their story, complete with hazy reports of sightings of the brothers in Africa, was still circulating in the fifteenth century. The growing Portuguese obsession with finding a sea route to the Indies also influenced his thinking. But one thing in particular seems to have helped him develop the idea of sailing west to Cipangu and Cathay: the letter that Paolo Toscanelli had written in 1474 to Fernão Martins.

The story of Columbus and the Toscanelli letter is the stuff of epic scholarly controversy. As the story is usually told, Columbus came across the letter some time before making his proposal to King João, in the late 1470s or early 1480s. Thrilled by what he read, and by the fact that the letter came from a learned Italian scholar with close ties to the Portuguese, he wrote a letter of his own to Toscanelli in which he laid out his ideas for a voyage west to the Indies and asked for Toscanelli's opinion of such an enterprise.

By this time Toscanelli was a weary old man with only a year or two left to live. He had neither the time nor the energy to engage in lengthy correspondence with an uneducated Genoese stranger living hundreds of miles away. And so, at least according to Bartolomé de Las Casas and Ferdinand Columbus, who are the sources of the story, he simply sent Columbus copies of the letter and the map he had previously sent to Martins. "Noting your magnificent and great desire to voyage to the land where the spices grow," he wrote,

> I send, in reply to your letter, a copy of a letter I wrote some time ago, before the war with Castile, to a friend of mine, a gentleman in the household of his Most Serene Highness the King of Portugal, in reply to a letter he had written to me on behalf of His Majesty about this whole matter, and I enclose a mariner's chart identical to the one I sent him, and which should provide answers to the questions you ask.

Some time later Toscanelli reportedly responded to a second letter from Columbus and briefly reiterated his endorsement of the eager young sailor's plans. "The said voyage is not only possible," he wrote, "but it is sure and certain and will bring honor, inestimable gain, and the widest renown among all Christians."

The authenticity of this correspondence is hotly contested, and many scholars have decided that it was forged—by Columbus himself, perhaps, or by one of his early biographers, in order to suggest that Columbus was not just an accomplished Genoese mariner but a man of letters guided by one of the most learned authorities of his age. The debate shows no sign of ending. But there's no doubt that Columbus did see Toscanelli's original letter to Martins, because a copy survives in Columbus's own hand.

Reading the letter gave Columbus a boost of confidence. It was one thing for sailors to approach King João with plans to find a new Atlantic island or two; the king routinely considered such proposals. But Columbus intended to suggest to King João that Portugal's entire program of discovery—the result of decades of investment, exploration, and study—was misguided. No matter how accomplished and widely traveled he might be, Columbus knew that as a simple sailor he would never be able to convince the king to change course. But here was a trusted geographical authority who had proposed just that. Coming across Toscanelli's letter seems finally to have emboldened Columbus to make his case to the king.

No firsthand account of the conversation survives. Secondhand reports suggest that Columbus boasted of his accomplishments as a sailor; proposed to sail west in search of Cipangu and Cathay; requested more than one ship to carry out his mission; and promised to return bearing gold. Columbus probably presented Toscanelli's letter as the blueprint for his proposal, supplemented with many of the island stories he had gathered over the years. No doubt Columbus also relied on a variety of maps to make his case: marine charts showing the coastline of Africa and the islands of the Atlantic; *mappaemundi* showing Asia as medieval cosmographers had long depicted it; a copy of the world map from the Rome edition of the *Geography*, printed in 1478, showing Ptolemy's known world extending halfway around the globe; a copy of a Martellus-like map of the world that expanded Ptolemy's Asia far out beyond the end of Ptolemy's east to include the Far East of Marco Polo; a copy of Toscanelli's chart; and perhaps even a Behaim-like globe that laid out the previously unmapped expanse of ocean that Columbus was proposing to cross.

Columbus evidently so importuned João that the king couldn't say no—at least not on the spot. Instead, he turned the matter over to his advisors for review, ordering, according to João de Barros, "that [Columbus] confer with D. Diogo Ortiz, bishop of Ceuta, and Master Rodrigo and Master José, to whom the king had committed these matters of cosmography and discovery." Months went by, during which time Columbus used all of his powers of persuasion to try to convince the king's advisors to endorse his plan, but to no avail. "They all considered the words of Christovão Colom vain," de Barros would write, "simply founded on imagination, or things like that Isle Cipangu of Marco Polo." Their opinion confirmed João's own—and so, in 1485, the king officially turned Columbus down and sent Diogo Cão back to southern Africa.

* * *

ONE OF THE most enduring myths about Columbus is that the cosmographers who reviewed his plan in Portugal—and, later, in Spain—rejected it because they refused to accept the idea that the world was round. But King João and his advisors in fact had a far more sophisticated understanding of geography than Columbus, and they rejected his proposal for perfectly legitimate reasons. Consider Cipangu. The only European ever to have described the island in any detail was Marco Polo. Launching an expensive new voyage of discovery into uncharted waters in search of an island that might or might not exist, with nothing more to go on than the testimony of a questionably truthful Venetian merchant who had lived some two hundred years earlier, seemed like a dubious proposition at best. So did estimating the circumference of the world based on the vague distances that Polo had recorded in his Book. "When the traveler leaves Kan-chau," a typical passage read, "he journeys eastward for five days through a country haunted by spirits, whom he often hears talking in the night, till he reaches a kingdom called Erguiul. This is subject to the Great Khan and forms part of the great province of Tangut, which comprises many kingdoms." What could a careful cartographer make of *that*?

The Portuguese had more reliable information at their disposal. In the 1470s and 1480s they had developed a newly precise estimate of the size of a geographical degree, based on the wealth of new data about latitudes that they were bringing home from Africa, and their estimate led them to believe—correctly, as it turns out—that the earth's circumference was far larger than Columbus believed.

Measuring east-west distances had always been a problem for geographers. The rough estimates made by Marco Polo and other travelers were of little practical use. The only reliable way of determining east-west distances was to measure longitude, but this was so difficult that only learned astrologers even tried it—and they focused their gaze on European, not Asian, skies. The result was that all estimates of east-west distance in the fifteenth century were untrustworthy, especially when it came to distant places. King João's advisors knew this well, and when Columbus came to them with a proposal that depended entirely on such estimates, they naturally considered it suspect. They preferred to think about distance in terms of latitude, not longitude.

Latitude had always been far easier to determine than longitude, because it corresponded to the height of the Pole Star above the horizon. This meant that even simple travelers early in the Middle Ages had a rough way of estimating how far they were from home. "By Jordan," an Icelandic priest wrote after visiting the Holy Land in 1150, "if a man lies flat on the ground, raises his knee, places his fist upon it, and then raises his thumb from his fist, he sees the Pole Star just so high and no higher." Because most medieval Europeans spent their time in a tiny belt of northern latitudes, where destinations and routes of travel were well-known, they had little need to measure latitude any more precisely than that. Even as late as the 1450s the Italian merchant and explorer Alvise Cadamosto, on a voyage to West Africa in the service of Prince Henry, recorded his latitude in only the most basic of terms. At anchor in the mouth of the River Gambia, he noted that the Pole Star appeared "very low down over the sea . . . about a third of a lance above the horizon."

As the Portuguese began to press farther south, they realized that if they could devise a more exact way of determining latitude, they could spend more time sailing offshore, out of sight of land. Coastal sailing was a slow and tedious business for the Portuguese in Africa: to avoid running aground, they had to advance cautiously during the day, methodically dropping and raising a lead line to measure the depth of the waters they were passing through, and they always had to be on the lookout for a safe place to anchor at night. The daily ritual varied little. "At dawn we made sail," Cadamosto wrote, "always stationing one man aloft and two in the bows of the caravel to watch for breakers that would disclose the presence of shoals."

Open-water sailing offered a chance to make much swifter progress. If the Portuguese could accurately determine the latitude of a given place— a river mouth, or a notable cape, or the site of a *padrão* planted by a previous explorer—then they didn't have to waste a lot of time hugging the coast looking for it. Instead they could put out to sea, set a course south, and then sail for days at a time, until the height of the Pole Star told them they had reached approximately the right latitude. At that point they could turn east and head straight for land, keeping the Pole Star at a constant height, in a practice known as "running down the latitude."

For help in measuring latitude, the Portuguese turned to their astronomers. Since at least the thirteenth century, astronomers in Europe had been using the quadrant, a simple point-and-shoot device, to measure the height of the Pole Star (*Figure 38*). At some point during the middle of the fifteenth century the Portuguese recognized that the quadrant could help them at sea, and so on their African voyages they began to bring along astronomers, who gradually amassed data on the latitudes of prominent features along the African coast. The method was far from perfect. Getting an accurate bearing of the Pole Star at sea was impossible, and even on land errors were the norm, especially when uneducated sailors rather than learned astronomers took the sightings. But using a quadrant was more effective than sailing with only a compass and a chart, as comes across clearly in the first known reference to a quadrant's being used at sea. "I had a quadrant when I went to these parts," a Portuguese squire named Diogo Gomes recalled late in life, describing to the globe maker Martin Behaim a voyage he had made to Guinea in about 1460. "I found it better than the chart," he continued. "It is true that the sailing course can be seen on the chart, but once you get wrong you do not recover your true position."

But a quadrant was only useful for navigating in the northern hemisphere. When the Portuguese crossed the equator they noticed that the Pole Star dipped below the horizon—and that it had no analogue in the southern skies, no single star around which all the others turned. They would have to look elsewhere to determine latitude, they realized, and their astronomers suggested the sun.

It wasn't a new idea. Since at least Ptolemy's time, most famously in the Islamic world, astronomers had studied the skies using a complex instrument known as the astrolabe (*Figure 58*). In essence, the astrolabe was a kind of computer: a portable, two-dimensional version of an armillary sphere (*Figure 29*) that allowed astronomers to observe, mimic, and predict the movements of the moon, the sun, the planets, and the stars. Used to make an accurate sighting of a celestial body on a given day in a given place, the astrolabe could then be manipulated to re-create a picture of the skies at any given time and place, which in turn allowed astronomers to tell the time, predict eclipses, estimate relative distances, and more. Many of the astrolabe's functions were well beyond the grasp of most scholars, much less ordinary sailors. But during the second half of the fifteenth century the Portuguese realized that their mariners could make use of the astrolabe for one relatively simple purpose: to determine their latitude in the southern hemisphere. There was no longer any need for a sighting of the Pole Star. Instead, they could measure the noonday sun's distance from the celestial equator (the imaginary circle in space that extended out from the earth's equator) and could use that information—along with special tables compiled over the centuries by ancient and modern astronomers—to determine their latitude. By the time Columbus arrived in Portugal, they had created what's known as the mariner's astrolabe (*Figure 58*) and were using it, with decidedly imperfect results, to help them navigate in African waters below the equator.

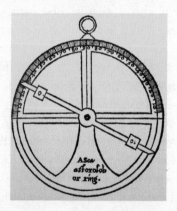

Figure 58. Mariner's astrolabe (1574). A stripped-down version of the astrolabe, the instrument allowed wind to pass through its center and was weighted heavily at the bottom to keep it steady at sea.

With astronomers, quadrants, and astrolabes at their disposal, and now sailing in caravels too, which could not only follow the wind but also head into it, the Portuguese began venturing more confidently into open waters. To ensure his sailors' success, in the early 1480s King João convened a team of cosmographical advisors and gave them the job of refining the art of celestial navigation in African waters. The three men João asked to review Columbus's proposal were a part of this team—as was, some indications suggest, Martin Behaim. It's possible, therefore, that Behaim got the idea for both his globe and his proposed voyage of western discovery directly from Columbus himself.

Working with an unprecedented range of latitudinal observations, which extended from northern Europe all the way to southern Africa, João's advisors soon came up with a newly precise estimate of the size of a degree of latitude. For centuries many astronomers and geographers had assumed a degree corresponded to $56\frac{2}{3}$ miles—and that's the figure Columbus used when making his proposal to João. But by that time the Portuguese, drawing upon their wealth of new astronomical observations, had revised this estimate upward. A degree, their experts said, corresponded to $66\frac{2}{3}$ miles (a figure much closer to the actual length, about 69 miles). João and his advisors had a sound theoretical reason for rejecting Columbus's proposal.

The rejection stung Columbus. Bitter, more convinced than ever in the rightness of his ideas, he decided to leave Portugal for good. He would try his luck in Spain.

* * *

THE SAGA OF Columbus's years in Spain has been told countless times: How in the summer of 1485 he arrived penniless in the southern port town of Palos, not far from Seville. How he spent his first night in the country at a nearby Franciscan monastery called La Rábida, which housed travelers free of charge. How that very evening he met the head of the monastery, Fray Antonio de Marchena, who would become his spiritual mentor for life. How Fray Antonio, a learned astrologer and cosmographer, immediately embraced his ideas. How Columbus stayed at La Rábida for some five months discussing his plans in intimate detail with Fray Antonio, who subsequently used his connections in Seville to get Columbus access to the Spanish court. How almost nobody at the court took Columbus and his ideas seriously, because, according to one contemporary, "he was a foreigner, and was poorly dressed, and with no greater support than a friar." How Columbus nevertheless managed to present his proposal in person to Isabella and Ferdinand, the Spanish Sovereigns, on January 20, 1486. How the Sovereigns, at the time fighting the Moors of Granada, referred him to their own council of advisors. How for the next six years Columbus doggedly followed the court around Spain, suffering bouts of extreme poverty and hardship, as he tried to win support for his plan. And how the Sovereigns repeatedly turned him down, but not without intimating each time that he might consider trying again a bit later.

Six years is a long time to wait, especially for a sailor itching to get back to sea. Few details survive of how Columbus occupied himself during those years, but one thing is certain: he began to read.

When Columbus arrived in Spain, according to his contemporary Andrés Bernáldez, who knew him in Seville, he was "a man of great intelligence, though with little book learning." This lack of a formal education pained Columbus. He knew that part of the reason he had failed to win support for his proposal in Portugal was that he hadn't been able to spar with King João's cosmographers. When called upon to respond to their scholarly objections, he had probably been able to do little more than refer time and again to Toscanelli's letter. Now, in Spain, again considering the prospect of doing battle with learned cosmographers on their own intellectual terrain, he resolved to prepare himself better and embarked on a crash course of self-study.

Fray Antonio probably got him started. The library at La Rábida contained some ten thousand volumes, and Fray Antonio knew its contents well. During the five months that Columbus stayed with him at the monastery, Fray Antonio had ample opportunity to introduce Columbus to the principles of academic cosmography—and to steer him specifically toward texts that might help him support and flesh out his ideas.

Nobody knows exactly what Columbus read at La Rábida. Even the contents of the library are a mystery; no catalog survives for the collection, which was dispersed long ago. Nevertheless, it is possible to make an educated guess about what Columbus would have found in a well-stocked fifteenth-century Spanish Franciscan library. The Bible, of course, and scores of other works that in one way or another discussed the East; the commentaries of the Church Fathers; the histories and literature of ancient Rome; the great classical and medieval encyclopedias; the *Historical Mirror* of Vincent of Beauvais, which contained condensed accounts of early Franciscan and Dominican missions to Asia; the *Book of Knowledge of All Things*, itself possibly the work of a Spanish Franciscan; the widely popular *Voyage of Saint Brendan the Abbot* and *Mandeville's Travels*; the works of

Ptolemy and Strabo, both of which appeared in print repeatedly during the 1470s and 1480s; and perhaps—surely the most thrilling find of all for a man who had now spent years poring over Toscanelli's letter—a copy of Marco Polo's Book. That the Book would have found a place in a sizable Franciscan library is not at all unlikely. The Latin abridgment of the Book by Friar Pipino was still circulating widely in the fifteenth century, a work specifically produced so that "the hearts of some members of the religious orders may be moved to strive for the diffusion of the Christian faith, and by divine aid to carry the name of our Lord Jesus Christ, forgotten among such vast multitudes, to those blinded nations among whom the harvest is indeed so great, and the laborers so few."

After he left La Rábida, Columbus began to build a small library of his own. This was a radical departure from tradition. Books had never been intended for the masses; they were rare and precious objects, laboriously copied by hand onto expensive sheets of vellum and then guarded as treasures by monks, scholars, and aristocrats. But Columbus arrived on the Iberian Peninsula just as the first printed books did, and that changed everything.

* * *

JOHANNES GUTENBERG PRODUCED his first Bible in Mainz, Germany, in 1454 or 1455, and word soon spread beyond Germany about the potential of the printing press. Leon Battista Alberti, for example, wrote admiringly of "the German inventor who has recently made it possible, by making certain imprints of letters, for three men to make more than two hundred copies of a given original text in one hundred days." By the early 1460s printing presses had begun to spread to many of Europe's important cities, although not everybody understood what they were. In 1465 the secretary of the Vatican Library still felt it necessary to describe the advantages of the new invention to Pope Paul II. "Every poor scholar can purchase for himself a library for a small sum," he explained. "Those volumes that heretofore could scarce be bought for a hundred crowns may now be procured for less than twenty, very well-printed and free from those faults with which manuscripts used to abound, for such is the art of our printers and letter makers that no ancient or modern discovery is comparable to it."

Columbus belonged to the first lay generation to benefit from the spread of printing, and he made the most of the opportunity that this offered him. After arriving in Spain he acquired a number of newly printed books, almost all of which concerned geography, and for the rest of his life he kept them at his side as trusted companions. He didn't just read his books; he engaged them in conversation, scribbling notes to himself in the margins, calling out statements he agreed with, testily objecting to others. Several of his books survive, and together they provide invaluable information about how Columbus tried to build his case in Spain—and, later, after he had finally crossed the ocean, how he struggled to make sense of what it was that he had found on the other side.

One of Columbus's favorite books, published in 1477, was the *Historia rerum ubique gestarum*, or *History of Matters Conducted Everywhere*—one of the earliest of all printed guides to geography. Written in the aftermath of the Council of Florence by the Italian humanist Aeneas Sylvius Piccolomini, who would go on to reign as Pope Pius II from 1458 to 1464, the work surveyed traditional medieval ideas about the world, and updated them with references to Ptolemy, Strabo, and even Niccolò Conti. Its quintessentially humanist aim, Piccolomini wrote, was "matching modern with ancient geography." The book consists of two parts, one devoted to Asia, the other to Europe. Columbus, naturally, read the former with great avidity, making a total of 861 different notes in the margins.

Those notes show Columbus drawing on his own experiences as a sailor to question traditional ideas. When Piccolomini observes that many authorities consider parts of the world uninhabitable, Columbus writes, "The opposite has been proved in the south by the Portuguese, and in the north by the English and the Swedes, who navigate in these parts." This he backs up repeatedly in the book with references to his own travels to Ireland, to Iceland, and especially to the fort at Elmina, in Guinea. When Piccolomini notes that Ptolemy declared the Indian Ocean to be closed, Columbus makes another objection. "Julius [Solinus]," he writes, "teaches that the entire sea from India to Spain around the tip of Africa is open." Columbus takes an active interest in Niccolò Conti's travels ("Note the voyage of this Niccolò"), and in general is captivated by Piccolomini's description of the Far East: he calls out or highlights numerous references to Cathay, the

Great Khan, and many of the traditional marvels and monsters of Asia. Cathay exists at the eastern edge of India, he tells himself, and Spain lies to the east of it, across the ocean. This sets up one of his most memorable notes: a description of two Asiatic-looking corpses that had washed ashore during his visit to Ireland—possibly flat-faced Sami from Scandinavia, or even Inuits from Greenland. "Men from Cathay have come eastward," he writes. "We have seen many signs of this, especially in Galway, in Ireland: a man and his wife, of strange and wonderful appearance, taken out of two dugout canoes."

The Book of Marco Polo was another of Columbus's favorites. He owned a copy of Friar Pipino's abridged Latin translation, published either in 1485 or 1486, and his notes in the margins highlight precisely the elements of the Book that Toscanelli had called attention to in his letter. "The Great Khan sent delegates to the Pope," Columbus writes at one point. "Perfumes," he writes at another. "Pearls, precious gems, golden fabric, ivory." Elsewhere he calls out references to "gold mines" and to "pepper, nuts, nutmeg, cloves, and other spices in abundance." The Great Khan, Cathay, Mangi, Cipangu: Columbus reads about them all like a man obsessed. The port city of Kin- sai, in particular, captivates him. "Biggest city in the world," he tells himself. "Quinsay is 25 miles from the sea," he writes in another note—the Sea of Cathay, that is, which Polo, of course, had gone out of his way to connect to the Sea of England and the Aegean.

This sort of talk is music to Columbus's ears. He avidly consumes what Polo had to say about the Sea of Cathay and its thousands of islands. More than once he writes the words "Ocean Sea" in the margins, two simple words that speak volumes about his aspirations. And when Polo describes a giant fleet of ships belonging to the Great Khan that ply the waters of the region, he dreamily glosses the passage with this note to himself: "15 thou- sand ships in the Ocean Sea."

Columbus owned one book that he consulted and treasured even more than Piccolomini's *History* and Polo's Book: *The Image of the World*, by Pierre d'Ailly. Manuscript copies of the text, written in about 1410, had been cir- culating ever since the time of the Council of Constance, but Columbus, as usual, owned a copy of the work's first printed edition, published sometime between 1480 and 1483.

Columbus spent countless hours going through *The Image of the World*. The book provided him with an accessible introduction to all of the standard elements of medieval geographical theory: the spherical cosmos, the five climate zones, the three-part known world, various theories about the ratio of land to water on the earth, the debate about the Antipodes, the sacred geography of the East, and much more. The book was just what Columbus needed: it surveyed, summarized, and quoted the writings of many ancient and medieval authorities, sparing Columbus the need to search out the originals. He could take the book with him and study it and make notes in it (898 in all) wherever he went, and when anybody challenged his ideas he could flip to various passages in the book that seemed to back up his ideas.

Paradoxically, many of the earliest printed books gave medieval texts and ideas a new lease on life just as those ideas were becoming obsolete. Most of the geographical works printed before 1500 perpetuated a theoretical view of the world that both explorers and humanist scholars were rapidly deconstructing. The same held true for cartography: the first printed maps to appear, in the 1470s, were medieval diagrams and charts, simple Christian *mappaemundi*, and Ptolemaic atlases. This is easy to explain, however. Just as they do today, publishers in the late fifteenth century were responding to demand. They rushed into print whatever they had at hand, and whatever they thought would sell.

Nowhere is this paradox more evident than in Columbus's copy of *The Image of the World*. Hot off the presses in the 1480s, it peddled a conception of the world that dated back not just to 1410 but even much earlier. Many parts of the work amounted to little more than a distillation of the works of earlier authorities, notably Roger Bacon. Not once did d'Ailly make any reference to the travels of modern Europeans, and he only slightly acknowledged Ptolemy's *Geography*, while completely ignoring its revolutionary potential. Nevertheless, *The Image of the World* retained its authority during the final decades of the fifteenth century. This serves as a useful reminder that even after the Age of Discovery was well under way, most Europeans continued to think about the world much as they always had. This certainly applied to Columbus. He didn't intend to expand the picture of the world. He just wanted to fill in some of its blank spots.

The notes Columbus wrote in *The Image of the World* reveal him to be an eager but sometimes disputatious student. He highlights the definitions of some introductory concepts: *sphere, cosmography, geography, eclipse.* He repeatedly calls out references to the roundness of the earth. When d'Ailly declares, in his Ptolemy supplement, "The length of the land toward the east is much greater than Ptolemy admits . . . [and] the habitable earth on the side of the Orient is more than half the circuit of the globe," Columbus heavily underlines his remarks and adds, "The sea cannot cover ¾ of the earth." When d'Ailly lists various estimates of the size of a degree and suggests that the figure of 56⅔ miles *may* be the best, Columbus latches on to the remark and invests it with certainty not present in the original. "Note," he writes, "that any degree at the equator corresponds exactly to 56⅔ miles."

Just as he did in *The History of Matters Conducted Everywhere,* in *The Image of the World* Columbus takes issue with traditional zonal theories about the earth. The torrid zone, he writes emphatically, "is not uninhabitable," because the Portuguese go there today. "It is even very populated." He returns to this idea repeatedly, often citing his own firsthand knowledge. In one particularly long note on this subject, he records having been present at the royal court in Lisbon when Bartolomeu Dias delivered his report to King João about the Cape of Good Hope. "He described and depicted the voyage league by league on a marine chart," Columbus writes, "so as to lay it out before his Highness the King. I took part in all of this."

This is an especially revealing note. In it Columbus also writes that Dias used an astrolabe to determine that the cape's latitude was 45 degrees (the same erroneous figure used by Henricus Martellus on his maps). Then, relying on that figure, he goes on to state that Dias sailed some 3,100 leagues south from Portugal—a very long distance indeed, and one that suits Columbus very nicely. D'Ailly and Marco Polo both have already told him that the known world wraps around much of the globe, and now Dias has shown that it also extends much farther south than previously thought. Columbus's notes crackle to life in this part of *The Image of the World.* Most of the world *has* to be land, he concludes—and what little ocean is left must be "completely navigable."

*　*　*

WITH PRINTED BOOKS and learned authorities as his allies, Columbus now felt able to debate Isabella and Ferdinand's advisors on their own terms. And in making his case to them he deployed his new sources so selectively and so imaginatively that he came up with perhaps the smallest estimate of the earth's size ever made. The actual distance straight around the world from the Canaries to Japan is about 10,600 miles; Columbus claimed it was 2,400. The actual distance from the Canaries to China is about 11,700 miles; Columbus claimed it was 3,500. Like their Portuguese counterparts, the Spanish cosmographers reviewing Columbus's proposal had a reasonably good idea of the circumference of the earth, and they knew his numbers couldn't be right.

They had another important reason for disagreeing with Columbus: the scholastic model of the cosmos, in which the earth emerged from only one side of the watery sphere. "The earth lies in the water like a light ball," one Spanish authority wrote in 1484, summing up the theory, "or like an apple in a basin full of water, of which only the summit appears above the water." Sacrobosco had modeled the earth this way, in his *Sphere*, which was still widely taught (*Figure 26*); Pierre d'Ailly had done the same, in the *Image of the World* (*Figure 43*); and plenty of other maps were in circulation that reinforced the idea, including those of a sort that would later belong to Columbus's son Ferdinand (*Plate 8*). To the cosmographers considering Columbus's proposal, this model of the earth taught an obvious lesson. To try sailing halfway around the globe or more on the open sea would be pure folly. "The only way to explore," they told Columbus, according to Las Casas, "was to hug the coast, just as the Portuguese had been doing in Guinea." It was a hard idea to let go of.

Columbus had another obstacle to contend with. In Spain, as in Portugal, his timing was unlucky. During the 1480s hopes had been rising that the Moors might at long last be expelled from Granada, and Isabella and Ferdinand were devoting most of their time, energy, and financial resources to the effort. Focused as they were on dreams of completing the Reconquista, they had little reason to choose to give much attention to a foreign sailor proposing to sail west in search of the Indies. So the Sovereigns turned Columbus down. Columbus went back to them repeatedly to make his case, but each time they turned him down—gently, it seems, signaling that perhaps when the struggle against the Moors had come to an end they might reconsider. But when Columbus came to them after they finally had defeated the Moors, at the Battle of Granada, in January of 1492, the Sovereigns decided to reject him with finality. This time, according to Las Casas, they gave him "no prospect that they might at some future date reconsider their decision."

At this point the story takes on a Hollywood-like quality. Dejected but unbowed, Columbus is said to have left the Spanish court, mounted a mule, and set off toward Córdoba, already mulling over plans to take his proposal to the king of France. But some ten miles from Granada a court messenger who had been sent out after him at a gallop overtook him. He had startling news: Luis de Santángel, the court treasurer and an old friend of Columbus's, had convinced Isabella and Ferdinand to back Columbus's enterprise of the Indies after all.

Three months later Columbus had it in writing. "By our command," the Sovereigns declared on April 30, 1492, Columbus was to set forth on a mission "to discover and acquire certain islands and mainland in the Ocean Sea." The enterprise would be dangerous, the Sovereigns continued, but the rewards would be great. If Columbus found the said islands and mainland he would legally become their governor—and would earn himself the right to be called Spain's Admiral of the Ocean Sea.

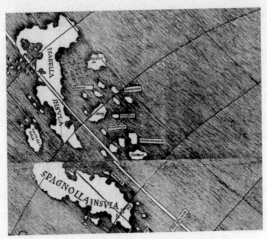

• Cuba, Hispaniola, and the Caribbean •

CHAPTER FOURTEEN

THE ADMIRAL

He saw so many islands that he couldn't count them . . . and he says
that he thinks that these islands are those innumerable ones that are
found on the world maps at the end of the Far East.

—Bartolomé de Las Casas, summarizing the November 14, 1492,
entry in Columbus's now-lost Journal of the First Voyage

$\mathcal{I}$N MAY, COLUMBUS traveled to the port town of Palos, not far from
Fray Antonio's monastery at La Rábida, and there he spent a busy summer
preparing for his voyage: leasing ships, hiring crew, gathering provisions,
studying books and maps. By the end of July he had three fully outfitted
vessels—the *Niña,* the *Pinta,* and the *Santa María*—and a crew of some
ninety sailors under his command. On Friday, August 3, 1492, at about
half an hour before sunrise, he and his crew caught the ebb tide out of Palos
harbor and set out for the Canary Islands.

Columbus sailed first to the Canaries for several good reasons. One was
legal: according to the recently signed Treaty of Alcáçovas, the Canaries

were the only islands in the Western Ocean that Spain had rights to; everything else belonged to Portugal. But Columbus had more practical reasons, too. Many maps of his time—Toscanelli's chart, Martellus's world maps, Behaim's globe—suggested that Antilia, Cipangu, and Kinsai all lay on about the same latitude as the Canaries. The most reliable way of finding those places, Columbus knew, would be to sail to the Canaries and then run down the latitude to the west.

Columbus spent the better part of a month in the Canaries making final preparations for his ocean crossing, and during this period he kept a journal. Although the original is lost, Bartolomé de Las Casas and Ferdinand Columbus both would quote and summarize it in their own biographies of Columbus. What they preserved in doing so has become the source of almost everything now known about Columbus's first voyage across the Atlantic.

Columbus began his journal with a carefully crafted preamble, addressed to Isabella and Ferdinand. Situating himself prominently and self-consciously at the end of a long line of medieval European explorers, missionaries, and crusaders, and borrowing language directly from the Toscanelli letter, Columbus reminded the Sovereigns of the opportunity that had been lost, some two centuries earlier, when the pope had failed to dispatch to the Far East the Christian teachers that the Great Khan had begged them to send. Now, at last, in their wisdom, the Sovereigns were seeing fit to make amends. "Your Highnesses," Columbus wrote, "as Catholic Christians and Princes devoted to the Holy Christian Faith and the propagators thereof, and enemies of the sect of Mahomet and of all idolatries and heresies, resolved to send me, Christopher Columbus, to the said regions of India, to see the said princes and peoples and lands and [to observe] the disposition of them and of all, and the manner in which may be undertaken their conversion to our Holy Faith, and ordained that I should not go by land (the usual way) to the Orient, but by the route of the Occident, by which no one to this day knows for sure that anyone has gone." Columbus went on briefly to describe the journey he had just made from Palos to the Canaries and then, as he contemplated sailing west into the unknown, concluded by making a solemn promise to the Sovereigns—and himself.

I intend to make a new chart of navigation, upon which I shall place the whole sea and lands of the Ocean Sea in their proper positions under their bearings, and set down everything as in a real picture, by latitude north of the equator and longitude west; and above all it is very important that I forget sleep and labor much at navigation, because it is necessary, and the which will be a great task.

* * *

MANY OF THE people who knew Columbus left records of him as an active and capable mapmaker. Las Casas wrote that Columbus supported himself during his time in Spain by "making or drawing nautical charts." Andrés Bernáldez recalled Columbus as "very skilled in the art of cosmography and the mapping of the world." Several of the sailors who joined Columbus on his voyages across the ocean echoed this assessment. "The Admiral," one of them would testify after Columbus's death, "made a sphere and maritime charts . . . in such a way that those who went with him learned many matters."

When Columbus left the Canaries for good, on September 9, he took one of his own maps with him: a map showing the ocean he proposed to cross and the lands he proposed to discover. The first reference to the map appears in his journal entry for September 25, when—according to Las Casas's summary of the entry—Columbus and Martin Alonso Pinzón, one of his captains, had begun to discuss their location by zipping a chart back and forth across a line they had strung temporarily between the *Santa María* and the *Pinta*. It was a chart, Las Casas wrote, "on which it seems the Admiral had depicted certain islands"—almost certainly a reference to Antilia and its sister island, Satanezes, and probably Saint Brendan's Isle as well. Both men felt that they must be in the vicinity of the islands depicted on the map, but that northeasterly ocean currents must have sent their ships wide of them. By October 3 the islands still hadn't appeared. Figuring that he had "left astern the islands that were depicted on his chart," Columbus then resolved to spend no more time looking for these islands—noting, as Las Casas put it, that "his object was to reach the Indies."

By this time Columbus and his crew had been sailing out of sight of land for twenty-four days. On the first day out at sea, aware that his crew had doubts about actually reaching the Indies, Columbus decided that he would keep two logs of the journey, one for his own private use and one in which he would deliberately underestimate the progress the expedition made each day—"so that if the voyage were long," Las Casas wrote, paraphrasing Columbus's journal entry for September 9, "the people would not be frightened and dismayed."

The early days of the voyage passed without incident, and in less than a week the crew began clutching hopefully at any suggestion that they were approaching land. Extracts from Las Casas's abstract of the journal provide a sense of how the days passed for Columbus and his men as they sailed day after day through uncharted waters.

FRIDAY, 14 SEPTEMBER: He sailed that day and night their course to the W, and made 20 leagues. He reckoned somewhat less. Here they of the caravel *Niña* said they had seen a tern and a boatswain bird, and these birds never depart from land more than 25 leagues.

SUNDAY, 16 SEPTEMBER: He sailed that day and night on his course to the W. . . . Here they began to see many bunches of very green weed . . . whereby all judged that they were near some island, but not the mainland, according to the Admiral, who says: "Because I make the mainland to be further on."

FRIDAY, 21 SEPTEMBER: The day was for the most part calm, with some wind later on. . . . At dawn they saw so much weed that the sea seemed to be a solid mass of it; it came from the W. . . . Saw a whale, which is a sign that they were near land, for they always stay near.

SUNDAY, 23 SEPTEMBER: The sea fell flat and calm, [and] the people grumbled, saying that since there was no heavy sea, it never would blow hard enough to return to Spain. But afterwards the sea made up considerably, and without wind, which astonished them, commenting on which the Admiral here says, "Thus very useful to me was the high sea, [a sign] such as had not appeared save in the time of the Jews, when they came up out of Egypt [and grumbled] against Moses, who delivered them out of captivity."

SUNDAY, 7 OCTOBER: He sailed on his course to the W. . . . A great multitude of birds passed over going from the N to the SW. It was thought, therefore, they were either going to sleep ashore, or fleeing the winter which in the lands whence they came might be supposed to be coming on. . . . For these reasons the Admiral decided to abandon his W course, and to turn the prow WSW.

WEDNESDAY, 10 OCTOBER: He sailed to the WSW. . . . Here the people could stand it no longer and complained of the long voyage; but the Admiral cheered them as best he could. . . . He added that it was useless to complain, he had come [to go] to the Indies, and so had to continue it until he found them.

Fortuitously, the following day, just as mutiny was brewing, signs of land began to appear again: shore birds, some fresh clumps of seaweed, a hand-carved stick, a board, a piece of cane, a branch with flowers still on it. Lookouts on all three ships eagerly began to scan the horizon, and at last, at two hours after midnight on the morning of October 12, a crewman on the *Niña* shouted out that he had just caught sight of land.

Not knowing what sorts of coastal hazards to expect, Columbus ordered his ships to heave to offshore until dawn. A more pregnant pause in the history of discovery is hard to imagine. When daybreak finally came, the sailors got their first look at what they had found: a small island, flat and green, that appeared to be some six miles away. Columbus ordered his ships to approach the island, and as they advanced it came into focus, revealing thick forests, beaches of gleaming-white coral sand, tall trees, numerous streams, and flocks of parrots. In search of a beach on which they could make a landing, all three ships were warily skirting a ring of coral reefs when something else came into view. "Presently," Las Casas wrote, "they saw naked people."

And so came the irrevocable moment of first contact: a meeting of cultures and microbes that over the years has been peered at and examined over from every possible angle and in every possible light. Columbus rowed ashore, unfurled a royal standard, and formally took possession of the island for the Spanish Sovereigns in the presence of several of his officers—a bit of

European pomp that must have bewildered the islanders who had gathered on the beach to gawk at the oddly dressed, foul-smelling, scurvy-ridden beings who had just arrived on their shores. Following a custom established decades earlier by the Portuguese in Africa, Columbus and his men then began handing out trinkets—glass beads, tiny round falconry bells, small red caps—that they had brought along with them for just this sort of encounter.

The gifts broke the ice. At least as Columbus described it, the encounter from that point on was a giddy one on both sides—a day of laughter, curiosity, conversations in pantomime, and much gift giving. But even during those first hours Columbus was already having thoughts that didn't bode well for his new hosts. "They ought to be good servants," he wrote in his entry for that day. "I believe that they would easily be made Christians, because it seemed to me that they belonged to no religion. Please our Lord, I will carry off six of them at my departure to your Highnesses, that they may learn to speak."

* * *

THAT FIRST NIGHT, thousands of miles away from home, one question above all surely concerned Columbus. Where exactly *was* he?

He had reached Asian waters, in the beginnings of the East—that much he knew. This new island, which he had named San Salvador even though its inhabitants called it Guanahaní, was situated precisely where mapmakers such as Toscanelli, Martellus, and Behaim had all suggested the great archipelago of Far Eastern islands should begin. The islanders on Guanahaní had confirmed this impression, intimating to Columbus that numerous other inhabited islands lay nearby—something Columbus would confirm for himself in the days that followed. "I . . . saw so many islands," he wrote just two days later, "that I could not decide where to go first; and those men whom I had captured made signs to me that they were so many that they could not be counted."

Everything Columbus had learned about the Far East in Spain and Portugal had led him to expect to sail through a maze of islands before reaching the Asian mainland. So now he drew the only conclusion he could: he had reached the Sea of Cathay and its thousands of islands, and was somewhere in the vicinity of Cipangu.

Gold was what Columbus had traveled so far to find, and on his second day on Guanahaní, wasting no time, he began actively looking for it. Soon enough he noticed that tiny ornamental pieces of gold hung as jewelry from the noses of some of the islanders. Undaunted by the language barrier, he pressed them for the source of this gold, which, he reported, they openly indicated to him "by signs." The source was an island to the south of Guanahaní. "There was a king there," Columbus understood the islanders to be telling him, "who had great vessels of it and possessed a lot."

Cipangu! It had to be. The time for idling on Guanahaní was over. "To lose no time," Columbus wrote at the end of his journal entry that day, "I intend to go and see if I can find the Island of Cipangu."

*　　*　　*

FOR THE NEXT week Columbus wound his way south through what we know today as the Bahamas, stopping briefly on various islands to claim and explore them but always remaining on the lookout for gold. "I do nothing else," he would write on October 19, "but go on to try and run across it."

The initial signs were promising. On October 16 he was told—or so he thought—that he had reached the vicinity of "a mine of gold." The next day, on an island he named Fernandina, in honor of King Ferdinand, he noted the presence of what appeared to be houses built "in the manner of Moorish tents"—a sure sign he had arrived in the Far East. (Martin Behaim's globe shows many such tents on Cipangu and in eastern Asia.) Columbus was in high spirits. On Sunday, October 22, at anchor off an island that he had named Isabela, in honor of the Spanish queen, he summed up his plans for the coming days. First he would see if he could find gold on Isabela. Then, he wrote,

I shall . . . depart for another much larger island, which I believe must be Cipangu, according to the description of these Indians I carry, and which they call Colba [Cuba] . . . and beyond this is another island, which they call Bofío [Haiti], which they also say is very big; and the others which are between we shall see as we pass, and according as I shall find a collection of gold or spicery, I shall decide what I have to do. But in any case

I am determined to go to the mainland and to the city of Kinsai, and to present Your Highness's letters to the Great Khan, and to beg a reply and come home with it.

Two days later, on October 24, Columbus set out to find these two big islands, which are indeed the two largest in the Caribbean, lying just to the south of the Bahamas. "This night at midnight," Columbus wrote, "I weighed anchors from the island Isabela to go to the island of Cuba, which I heard from that people is very great, and with a lot going on, and therein gold, spices, big ships, and merchants. . . . I believe that it is the island of Cipangu, of which are related marvelous things; and on the globes that I saw, and in the delineations on the world map, it is in this region."

When Columbus and his men reached the north shore of Cuba four days later, they found no giant junks laden with gold and spices, no teeming ports full of merchants, no palaces of gold. Instead they found only a sparsely inhabited coastline that for miles rose up sharply up from the sea. Eventually it flattened out and led to the mouth of a wide and navigable river, which they followed inland, finding it to be lined with mangroves, palm trees, tall grasses, and "little birds that sing very sweetly." But the only signs of human habitation they encountered along the river were two empty thatched huts, a few fishing nets and hooks, and, memorably, "a dog that didn't bark."

Columbus remained buoyant, at least in his journal. Describing this new river, he wrote that he had "never beheld so fair a thing." Before long he once again managed to find locals who led him to believe—surely by simply nodding "yes" to whatever leading questions he was putting to them, in a language they couldn't understand—that "mines of gold and pearls" lay off to the west. Off he went again in search of them, following the coast west for mile after mile but never finding what he was looking for. Eventually he realized he had a problem: the coast he was sailing along was far longer than his maps told him Cipangu's should be.

On the big Martellus map (*Figure 56*) and the Behaim globe (*Figure 57*), both of which depicted a vision of the Far East that Columbus clearly had in mind as he explored his Indies, Cipangu is far taller than it is wide. Columbus therefore expected that Cuba's north coast would make a sharp turn to the south relatively quickly after he began sailing along it. But it didn't. Cuba is far wider than it is tall. Its north coast spans a distance of more than seven hundred miles, and for most of that distance, as it moves west, it gradually rises toward the north instead of turning sharply south.

Columbus began to reconsider Cuba. Beyond Cipangu to the west on maps like Martellus's and Behaim's, an eastward-extending portion of Asia reaches out into the Sea of Cathay. Inevitably, Columbus began to wonder if this place he was exploring wasn't an island at all but instead was this eastern portion of the Asian mainland. Martin Pinzón, the captain of the *Pinta*, had begun to have similar thoughts. On October 30, having conducted his own interviews with the locals, he reported to Columbus "that this Cuba was a city, and that that land [the one they were coasting along] was a very great continent that trended far to the N." This trend conformed very nicely to the contours of the Asian mainland as depicted on maps like Martellus's and Behaim's, and two days later Columbus was convinced. "It is certain," he wrote, "that this is the mainland, and that I am before Zaitun and Kinsai."

Cipangu would have to wait, he decided. It was time to find the Great Khan.

Columbus had long dreamed of and planned for this phase of his journey. With him from Spain he had brought several copies of a generic letter of introduction from the Spanish Sovereigns, complete with a blank space in which he could fill in the name of the Great Khan, Prester John, or any other powerful Asian ruler he might encounter. "To the Most Serene Prince _____," the letter began, and after a flurry of salutations it came to the point.

From the statements of certain of Our subjects and of others who have come to Us from Your Kingdoms and Domains, We have learned with joy of Your esteem and high regard for Us and Our nation, and of Your great eagerness to be informed about things with Us. Wherefore we have resolved to send you Our Noble Captain, Christophorus Colon, bearer of these, from whom You may learn of Our good health and Our prosperity, and other matters which We ordered him to tell you on Our part. We therefore pray You to give good faith to his reports as You would to Ourselves, which will be most grateful to Us; and on Our part We declare ourselves ready and eager to please You.

From Our City of Granada, 30 April, 1492.

Columbus had brought an interpreter with him too. He was Luís de Torres, a converted Jew who knew "Hebrew and Aramaic and also some Arabic"—languages that many Europeans, familiar with the accounts of Marco Polo and the early missionaries to Asia, assumed would be readily understood at a cosmopolitan Far Eastern court. Columbus gave Torres one of the letters of introduction, told him what he "had to say on the part of the Sovereigns of Castile," and then sent him ashore in search of the Great Khan.

Torres and a few companions tromped inland for two days. Eventually they reached a village of some fifty houses. The villagers received their odd guests warmly ("feeling them to ascertain if they were of flesh and bones . . . and begging them to stay,"), but Torres and his companions quickly realized that they had not found what they were looking for. As soon as they were able they left the village and retraced their steps back to the coast, where Torres glumly reported to Columbus that during the four days he had been away he had seen "nothing that resembled a city."

Columbus had kept himself busy during Torres's absence. Working with the distance estimates he had recorded each day in his journal during the ocean crossing, he determined that he had sailed 1,142 leagues from the Canaries, or well over three thousand miles—a distance that confirmed to him that Cuba was indeed far west enough to be part of the Asian mainland. He had had more conversations with the locals, too, which led him to believe that Haiti, not Cuba, was where he would find gold—"infinite amounts" of it. They had also told him that beyond Haiti he would find "men with one eye, and others with dogs' heads who ate men." These details, along with a number of others that Columbus would record in his journal, so closely echo passages about the Indies found in *Mandeville's Travels* that it seems safe to conclude that Columbus had the book with him and was routinely consulting it for clues about where he might be. (Columbus's son Fernando and the Spaniard Andrés Bernáldez would both later make a point of noting that Columbus had studied the *Travels*.)

So Columbus stopped traveling west and left the question of Cuba's identity unresolved. Neither he nor anybody else on his expedition had actually seen enough to be able to declare with certainty if Cuba was an island or a part of Asia—but if it *was* a part of Asia, Columbus knew, he had good reason not to follow it to the northwest. Many of his books and maps told him that if he traveled north from the vicinity of Cathay he would find wild beasts, monstrous races, icy wastes, marauding Tartars, and perhaps even the kingdom of Gog and Magog. If he made his way south, on the other hand, gradually moving toward the equator just as Marco Polo had done, he would enter some of the wealthiest, most civilized, and most temperate parts of the whole world.

Weighing his options, Columbus made the obvious choice. He gave up on Cuba, turned his ships around, and began sailing to the southeast, toward Haiti.

* * *

THE ISLAND CAME into view on December 6. Lush and mountainous, it was a feast for the eyes. "Seeing the grandeur and beauty of this island, and its resemblance to the land of Spain, although much superior," Columbus named it La Ysla Española—or Hispaniola, as the name would soon be rendered in Latin. For more than two weeks he and his men made their way east along the island's north coast, handing out more trinkets, interviewing more islanders ("every day we understand these Indians better, and they us"), and confirming that the island did indeed produce gold, which many of the islanders wore as jewelry and traded with Columbus's men for almost nothing. A gift that Columbus received on December 22 from a local chief—"a belt that, in place of a purse, bore a mask that had two large ears of hammered gold, as well as the tongue and the nose"—gave him a strong feeling he was finally nearing the gold's source. Two days later, on Christmas Eve, a few helpful locals boarded the *Santa María* and began naming specific places on Hispaniola where gold could be found, and one of the places they mentioned, Columbus wrote in his journal with evident delight, was "Cipangu, which they called Cybao."

As Columbus understood it, Cybao was an inland region on Hispaniola that lay not far off to the east. To get there as quickly as possible, Columbus ordered his ships to sail through the night on Christmas Eve. The conditions were perfect: calm waters, balmy air, clear skies. At 11 P.M., having set the *Santa María*'s course for the night and feeling "secure from shoals and rocks," Columbus finally turned over command of the ship to his crew and went below for a few hours of much-needed sleep. The evening was a pleasant, languorous one, and it lulled the sailors on deck into a drowsy stupor. Once they knew Columbus was asleep, they handed the tiller over to a ship's boy, told him to steer straight, and stretched out under the stars to steal a few hours of sleep for themselves—and not long afterward, in the early hours of Christmas morning, the *Santa María* scraped onto a coral reef and ran aground. Gentle rolling swells swung the vessel's stern around and began thumping its beam against the reef. Columbus and all hands sprang on deck, desperate to dislodge the ship, but soon its sides split open, and the cause was lost.

For the rest of the day Columbus and his men worked to salvage everything they could from the *Santa María*. They were greatly aided in their efforts by a local king named Guacanagarí, who at first light mobilized his people in large canoes and sent them out to the *Santa María* to ferry the Europeans and their wares to safety—a gesture that made a big impression on Las Casas. "Observe the humanity of the Indians," he wrote in summarizing Columbus's account of the shipwreck, "toward the tyrants who have exterminated them."

During the following few days, Columbus spent much time with Guacanagarí, who consoled him with talk of huge deposits of gold in the region—so much gold that Columbus soon began to consider his shipwreck a divinely ordained stroke of good fortune. Had the grounding of the *Santa María* not brought him in contact with Guacanagarí, he would have sailed right by what he was looking for. But the wreck had forced him to salvage what he could from his ship, and he and his men had now begun to create an impromptu settlement that promised to bring great rewards. "So many good things came to hand," Columbus wrote on December 26,

that in truth it was no disaster but great luck; for it is certain that if I had not run aground, I should have kept to sea without anchoring in this place . . . nor on this voyage should I have left people here, nor had

I desired to leave them could I have given them good equipment, or so many weapons or supplies, or materials for a fortress. . . . Now I have given orders to erect a tower and fortress [using the wood from the *Santa María*] . . . not that I believe it to be necessary for these people, for I take it for granted that with this people that I have I could conquer all this island, which I believe to be bigger than Portugal and double the number of inhabitants, but they are naked and without arms and very cowardly.

Columbus had in mind creating something like what he had seen when he had sailed to Elmina with the Portuguese: a fortified colonial outpost that would serve as the access point to gold-producing territories and a base for future trading and exploration. The fortress at Elmina had even been built with wood salvaged from scrapped Portuguese ships.

A day later, in his journal, Columbus laid out his plan. He would first supervise the building of the fortress, which he decided to call La Navidad, in honor of the Christmas shipwreck. Then he would leave a contingent of sailors there with enough supplies to last a year and would hurry back to Spain with news of his discoveries. This would allow him to outfit a second and much larger expedition that would return to Hispaniola as soon as possible—a true colonial venture that would allow him to realize one of his grandest dreams. Las Casas captured the details in his summary of what Columbus wrote that day.

He says that he hopes to God that on his return, which he intends to make from Castile, there would be found a cask of gold, which those whom he left behind would have obtained by barter, and that they would have found the mine of gold and the spicery, and that in so great quantity that the Sovereigns within three years would undertake and prepare to go and conquer the Holy Sepulchre, "for so," says he, "I declared to Your Highnesses that all the gain of this my Enterprise should be spent in the conquest of Jerusalem."

Columbus acted quickly. On January 2, with construction of his new fort now well under way, and with thirty-nine of his men chosen to stay behind, he ordered the sails raised on the *Niña* and began the long voyage back to Spain.

* * *

ON THE JOURNEY home Columbus composed a letter to the Spanish Sovereigns. In it he announced a "great triumph": he had reached the Indies. "And there," he continued, "I found very many islands filled with people without number, and of them all have I taken possession for Their Highnesses." He then summarized the main events of his voyage: the arrival at Guanahaní, the naming of various islands in honor of the Sovereigns, the unsuccessful attempt to find the Great Khan in Cuba, the Cuban coast's undesirable northward trend, the discovery of Hispaniola. Columbus described Hispaniola in rapturous detail: its great riches, its natural beauty, its submissive population. "In this Española," he wrote, skipping over the shipwreck of the *Santa María* entirely, "in the most convenient place, and in the best district for the gold mines, and for all trade both with this continent [Europe] and with that over there belong to the Great Khan, where there will be great trade and profit, I have taken possession of a large town, to which I gave the name La Villa de Navidad, and in it I have built a fort and defenses." His voyage had necessarily been hasty, he told the Sovereigns, but if they would agree to send him back, this time with more ships and men, he would soon be able to bring them great quantities of gold, cinnamon, mastic, aloe, rhubarb, and "a thousand other things of value"—including as many pagan slaves as they would care to order. He also pledged once again to devote himself and his newfound riches to the recovery of Jerusalem—or at least so it would seem based on a recently discovered version of this letter. "Seven years from today," the letter reads, "I will be able to pay Your Highnesses for five thousand cavalry and fifty thousand foot soldiers for the war and conquest of Jerusalem, for which purpose this enterprise was undertaken."

After he arrived back in Portugal on the storm-battered *Niña*, on March 4, Columbus immediately made arrangements for his letter to be sent to the Sovereigns. He planned to return to Spain as soon as possible, of course, to deliver the news of his discoveries in person. But first he had to explain himself to King João.

The king was not happy to hear that a sailor in the service of Spain had claimed rights to the Indies—*his* Indies. "He believed," the Portuguese court chronicler Rui de Pina would record, "that this discovery was made within the seas and boundaries of his Lordship of Guinea, which was prohibited." Moreover, according to de Pina, João "understood from the treaty that he had with the Sovereigns [the 1479 Treaty of Alcáçovas] that the acquisition belonged to him." To deal with this unexpected problem, the king's advisors devised a drastic plan. Columbus should be discreetly assassinated, they told the king, because "the prosecution of this enterprise by the Sovereigns of Castile would cease with the death of the Discoverer."

João didn't bite. Perhaps this was because he knew Columbus personally. Perhaps it was because he knew that Columbus's letter was already on its way to the Sovereigns. Or perhaps it was because he heard that Martin Pinzon and the *Pinta*—which had become separated from Columbus and the *Niña* during the ocean crossing—might already have reached Spain. This was a problem, he evidently decided, that he would resolve through more conventional, diplomatic channels. And so he received Columbus graciously, listened attentively to all he had to say, helped him make repairs to the *Niña*, and then allowed him free passage to Spain.

Columbus's letter reached the Sovereigns long before Columbus himself did, sometime in mid to late March. They read it with delight and immediately ordered that it be printed and distributed, to confirm their discovery and ownership of the new islands. They also wrote a short letter back to Columbus, who at the time was parading his way triumphantly toward them across Spain, showing off his Indian captives, his parrots, and his gold to the crowds of people who came out to see him as he passed though their towns. "We have seen your letters," the Sovereigns told Columbus, "and we have taken much pleasure in learning whereof you write." Aware of the need to more firmly establish their claim to his discoveries, they urged him to come find them as quickly as possible. "As you can see," they wrote, "the summer has begun, and you must not delay in going back."

Columbus needed no reminder. He reached the court in the third week of April, and by September, with the full backing of the Sovereigns, he had outfitted seventeen ships and had pulled together a crew of more than twelve hundred men—sailors, officers, navigators, mapmakers, cooks, doctors, craftsmen, workmen, weavers, bureaucrats, trusted old friends, and others, including several monks whose job it would be to build the first church on Hispaniola and start converting the locals. This time Columbus was going to bring everything necessary to get a large new settlement established: herds of sheep, cows, goats, pigs, and chickens, and enough wine, olive oil, vinegar, salted meat, flour, seeds, and other supplies to last at least a year.

It was the most ambitious colonial enterprise that medieval Europeans had ever undertaken.

* * *

WORD OF COLUMBUS'S discovery spread quickly that spring and summer. What attracted the most attention were the islands that Columbus had mentioned in his letter, not his tentative mentions of Cathay and the Great Khan. As usual, Italians living on the Iberian Peninsula helped spread the news by sending letters home. On April 9 a merchant who signed himself Hanibal Ianuarius wrote from Barcelona to Milan to report that "one called Columbus" had sailed west for thirty-four days, and that, "the world being round," Colombus had discovered a great many islands of the Indies— including "two of great size, the one larger than England and Scotland, the other larger than all of Spain." The first printed Latin edition of Columbus's letter, published before the end of April, conveyed this same idea in its title, which referred to "the Islands newly discovered in the Indian Ocean."

But not everybody believed that Columbus had reached the Indies. Some observers felt he had found nothing more than new Canary Islands. "The king of Spain discovered many islands again this year," the Siennese chronicler Allegretto Allegretti wrote in his entry for April 25, "that is, in the Canaries." Likewise, on June 15, the poet Giuliano Dati published a verse adaptation of Columbus's letter in Italian—designed to be recited aloud as news in Italy's piazzas—and titled it *The History of the Discovery of the New Islands of the Indian Canaries.* Dati, no doubt skeptical of Columbus's grand claims, left out all talk of Cipangu or the Asian mainland.

The Italian humanist Peter Martyr d'Anghiera had his doubts, too. Martyr worked at the Spanish court and had been present when Columbus first arrived bearing the news of his discovery. Throughout the months and years that followed, Martyr would keep his friends and colleagues informed about the discoveries being made in the Western Ocean. "A certain Colonus," he wrote to a friend in October of 1493, "has sailed to the western Antipodes, even to the Indian coast, as he believes. He has discovered many

Figure 59. First contact: the title page illustration
from an early edition of Columbus's *On the Islands Recently
Discovered in the Indian Ocean* (1494).

islands that are thought to be those of which mention is made by cosmographers, beyond the eastern ocean and adjacent to India. I do not wholly deny this, although the size of the globe seems to suggest otherwise."

Printers in several European cities soon published Columbus's letter, and it became something of an early bestseller, adorned in some editions with woodcuts that crudely illustrated the moment of first contact (*Figure 59*). But despite the considerable renown that Columbus now enjoyed, while he made preparations for his second voyage he was not hailed anywhere as the discoverer of a previously unknown continent, or as a man who prompted a revolution in geographical thinking. He had discovered some new islands in the Western Ocean and was making big claims about them. That was all.

* * *

THE SOVEREIGNS MOVED quickly to have Columbus's islands recognized internationally as theirs. To that end they wrote on April 15 to Pope Alexander VI—a Spaniard by birth, and a man who owed them favors—and requested that he affirm their ownership of the islands. The pope duly complied, in a bull dated May 3. King João and the Portuguese soon challenged the document, and a week of tense negotiations between the Spanish and the Portuguese followed. The result was a second bull, known as *Inter caetera II*, that famously proposed to divide the Atlantic between Spain and Portugal "by drawing and establishing a line from the Arctic pole, namely the north, to the Antarctic pole." All non-Christian lands discovered to the west of the line would belong to Spain; all those to its east would belong to Portugal. The bull even decreed exactly where that line should fall—"one hundred leagues toward the west and south from any of the islands commonly known as the Azores and Cape Verde."

The pope hadn't come up with this line out of the blue. The Spanish Sovereigns had fed it to him. "This sea belongs to us to the west of a line passing through the Azores and the Cape Verde Islands," they had declared on May 28, "and extending from north to south from pole to pole." But the idea didn't originate with the Sovereigns, either. It came from Columbus— or at least so it would seem from a letter the Sovereigns sent to him on September 5, in which they referred to "the line which you said ought to come in the Bull of the Pope."

Columbus believed the hundred-league mark to be a fundamentally important one. It represented the point in his ocean crossing at which he felt he had observed a dramatic shift from European to Asian conditions: the point at which the weather grew milder, the magnetic variation of his compass shifted, the great expanses of seaweed in the Sargasso Sea began to appear, and more. "When I sailed from Spain to the Indies," he would later write to the Sovereigns, "I found straightway on passing 100 leagues to the west of the Azores a very considerable change in the sky and the stars, and in the temperature of the air and in the waters of the sea; and I used much care in verifying this." Las Casas provided his own gloss on this phenomenon, noting that shipborne lice disappeared like magic just after the line had been crossed going west, only to return "in great and disturbing numbers" immediately after the line had been crossed in the other direction— "as if they waited upon us."

King João didn't like this second bull any better than the first. Something, nobody knows quite what, had led him to believe that valuable lands lay undiscovered across the Ocean in the southern hemisphere, to the southeast of Cuba and Hispaniola, and he felt that they were exclusively his to discover. His demand to the Sovereigns, therefore, was that they "grant him 300 leagues more to the west, beside the hundred which they had granted before."

And so on the negotiations went. What was necessary for their resolution, the Sovereigns soon recognized, was a good map—exactly the kind of map that Columbus had promised to make for them at the outset of his first voyage. Columbus had yet to produce any kind of map for them, probably because he had yet to figure out how to make the reality of what he had discovered accord with the geography of the Far East as he imagined it. But the Sovereigns remembered his promise. "If the sea chart you undertook to produce is now finished," they wrote in their September 5 letter, "we should be pleased to receive it forthwith. We should also be pleased if you would now set in motion your departure, so that by the grace of God the chart may be completed with as much dispatch as may be, for, as you are aware, it is vital to the successful outcome of the negotiations."

* * *

COLUMBUS AND HIS grand colonial fleet left Spain for the Canaries in early October, and on October 13 they began the ocean crossing. This time Columbus set a course somewhat to the south of Hispaniola. His captives had described for him a long chain of islands that swung out in a southwesterly arc from Hispaniola, at one point even using dried beans to assemble a rough map of the islands for him, and Columbus hoped to shorten his journey by first reaching them. From these islands (today known as the Antilles, a name related to Antilia), Columbus planned to island-hop his way back north to Hispaniola.

The journal Columbus kept on his second voyage hasn't survived, but details of the voyage have been reconstructed using the accounts of some of his shipmates. On November 3, after an apparently uneventful crossing, the fleet reached the island today known as Dominica—a name given to it by Columbus. For the next three weeks the fleet gradually sailed north and west, stopping in briefly at Martinique, Guadeloupe, Antigua, Nevis, Saint Kitts, Saint Croix, Puerto Rico, and other islands as it headed for Hispaniola. The sight of the fleet must have bewildered the locals: seventeen giant ships flying huge sails and an array of colorful flags, carrying hundreds and hundreds of strange-looking people and animals, and occasionally barking thunderously with their cannons.

On November 27 the fleet finally reached La Navidad, only to discover that the fortress there had been burned to the ground. No giant cask of gold awaited Columbus, and not even one of the men he had left on Hispaniola appeared to greet him. Eventually he was able to piece together what had happened: left to their own devices, his men had launched a series of increasingly violent raids on nearby villages, carrying off by force whatever food, women, and gold they could find. The villagers had decided to fight back, and soon enough Columbus's men, greatly outnumbered and already in poor health, had all been killed.

Columbus decided to rebuild his settlement elsewhere on the island. On January 2 he chose a site to the east of La Navidad, closer to the presumed gold mines of Cibao. There he ordered his men to begin the construction of a small town, to be named Isabela, and sent out a party in search of Cibao and its gold. And with construction of Isabela and the search for gold under way, he at last sat down to draw a map for the Sovereigns. The map that he drew has disappeared, but a letter that he sent to the Sovereigns in January of 1494 described it in detail.

> Your Highnesses will see the land of Spain and Africa and, in front of them, all of the islands found and discovered on this voyage and the other [first] one; the lines that run lengthwise show the distance from east to west, the others that run lengthwise show the distance from north to south. The spaces between each line represent one degree, which is comprised of fifty-six and two-thirds miles that correspond to fourteen and one-sixth of our sea leagues; and thus you can count from west to east, like from north to south, the said number of leagues. . . . And so that you can see the distance of the route from Spain to the beginning, or end, of the Indies, and see the distance between particular lands, you will find on said chart a red line that runs from north to south and passes over the

island of Isabela . . . beyond which line are the lands discovered on the other [first] voyage; and on this side of the line lie the other ones from [this voyage]; and I have hope in Our Lord that each year we will have to enlarge the map, because we will continuously discover [new lands].

Notably absent from this description, of course, is any mention of Cipangu and the lands of the Great Khan, but that's not surprising: Columbus still hadn't worked out exactly where they were.

That spring Columbus's search parties reached Cibao, an inland region in the eastern part of Haiti, and there they did indeed find small alluvial deposits of gold. But still nothing that they found corresponded to what they had read and heard about Cipangu. Columbus began to feel that Haiti must be something else. But what? The only way to answer this question, Columbus realized, was to begin to piece together a bigger picture of the region. And so on April 24 he set out again for Cuba—"to discover the mainland of the Indies."

This time Columbus chose to follow Cuba's south coast. He and his men sailed west for hundreds of miles, once again without finding an end to the land, and the journey confirmed for Columbus what he had already come to believe: that Cuba was indeed a part of the Asian mainland. He sailed so far along the coast, in fact, that according to Andrés Bernáldez, with whom he stayed after returning from the second voyage, he decided that he "had reached a point very near the Aurea Chersonese."

Anybody who had studied geography in the fifteenth century would have understood what Columbus was talking about. A name meaning "Golden Peninsula," the Aurea Chersonese was the name that had been use by Pliny, Ptolemy, and countless medieval authorities to describe the gold-producing regions of the Malay Peninsula. It lay on the eastern side of the Indian Ocean, and Ptolemy had clearly identified it on his world map. Columbus owned a copy of the edition of the *Geography* that had been printed in Rome, and as he looked it he must have had a thrilling thought. If he indeed could reach the Aurea Chersonese, at the end of Ptolemy's east, then, as Andrés Bernáldez wrote, "he would be able to return to Spain by the East, coming to the Ganges and thence to the Arabian [Persian] Gulf." He could sail, in other words, all the way around the world.

Columbus explored Cuba and nearby islands for weeks, discovering Jamaica in the process, but eventually he realized he would have to return to Hispaniola. His supplies were short, his crew's morale was low, and he had a new colony to govern. So in June, at a point probably only some fifty miles from the western end of Cuba, he made his crew all swear solemnly in written depositions that Cuba was not an island, and turned his ship around.

Columbus would insist until his dying day that Cuba was not an island. But as he sailed back to Hispaniola that summer, he had to admit to himself that he had yet to find any proof that he had reached the lands described by Marco Polo. Perhaps he needed to think differently about the geography of the Far East, he decided. Perhaps he should take another look at the *mappaemundi*—and focus on the sacred geography of the Bible.

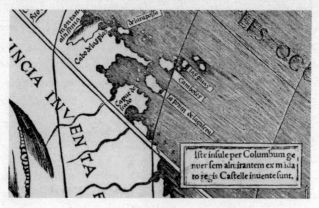

CHAPTER FIFTEEN

CHRIST-BEARER

It pleased our Redeemer to send his obedient apostles through various parts of the earth to teach the truth of our holy faith. . . . The Divine Providence sent the great Thomas from Occident to Orient to preach in India our holy Catholic faith, and you, my lord, were sent in the opposite direction. . . . Thus will be accomplished the supreme prediction that the whole world shall be under one shepherd and one law. . . . The great mission you have undertaken ranks you as an apostle and ambassador of God.

—Jaime Ferrer de Blanes to Christopher Columbus
(August 5, 1495)

IN THE TENTH century B.C., King David captured Jerusalem, unified the Jews, and made the city their capital. He also dreamed of building a great temple in the city to house the Ark of the Covenant, but he died before work got under way. His son Solomon went on to build the temple, however, and when the time finally came for him to give it its finishing touches, he relied on gold—lots of gold. "Solomon overlaid the house within with pure gold," the Old Testament records, "and he made a partition by the chains of gold before the oracle; and he overlaid it with gold. And the whole house he over-laid with gold, until he had finished all the house: also the whole altar that was by the oracle he overlaid with gold."

Some of the gold came from the Queen of Sheba, who had traveled from "the uttermost parts of the earth" to pay tribute to Solomon. But most of it came from a distant eastern land known in the Bible as Ophir. It was a region that many subsequent writers, in antiquity and in the Middle Ages, would come to associate with the Aurea Chersonese, or the Malay Peninsula.

The Old Testament records that Phoenician mariners regularly sailed from the Holy Land to Ophir in the service of David and Solomon, and that they were often joined by a great fleet of ships from Tarshish, an east-ern kingdom that the medieval Christian tradition would identify as the home of one of the three Magi. The voyage from the Holy Land to Ophir and back took the mariners three years, but it was time well spent: when they returned they brought back great bounties of "gold, and silver, ivory, and apes, and peacocks."

Columbus knew this story well. He liked it so much that he copied a lengthy account of it into the back of his edition of *History of Matters Conducted Everywhere*, near his copy of Toscanelli's letter. In the version Columbus copied down for himself, Solomon summons "many navigators and other specialists in maritime sciences" and orders them to do something that Columbus himself felt he had been summoned to do by the Spanish Sovereigns: that is, "to navigate . . . to the place then called Ophir, which is now called the Gold Country [the Aurea Chersonese], which is in India, to gather gold there." Columbus even factored this story into his calculations about the size of the Western Ocean. "The kingdom of Tarshish," he wrote in his copy of *The Image of the World*, "is at the limit of the East, at the limit of Cathay. It was in this country, at the place called Ophir, that Solomon and Jehosaphat [a later king] sent ships which brought back gold, silver, and ivory. . . . Note that the king of Tarshish came to the Lord at Jerusalem, and spent a year and thirteen days en route, as the blessed Jerome has it." If it took more than a year to sail from the Red Sea east across the Indian Ocean to Ophir, Columbus reasoned, then the distance involved had to be great. And this led him to a logical conclusion: sailing *west* to Ophir from Spain, which itself was already more than two thousand miles west of the Holy

Land, could not possibly take long. The distance involved had to be small, just as his other calculations had already told him.

The Ophir story intoxicated Columbus. By sending out brave mariners to bring back gold from the Far East, Solomon had made Jerusalem into the holiest and most glorious of all the cities in the Bible; now Columbus was embarking on a similar mission and intended to do the same. When exactly Columbus began to think along these lines is impossible to determine, but his ideas didn't emerge in a vacuum. By the end of the fifteenth century an apocalyptic tradition had arisen in Spain that identified King Ferdinand as a new King David who would unite the world's true believers, win back Jerusalem, and at last rebuild Solomon's temple, this time for Christ. Ferdinand even went along with the idea; after he and Isabella had defeated the Moors, he took on the empty title King of Jerusalem.

Two distinct but overlapping kinds of geography guided Columbus, then, when he began to sail across the Atlantic. One was a literal, empirically based conception of the world, as laid out on his marine charts and Ptolemaic-style maps. The other was a figurative, spiritually based conception, as laid out in the Bible: the idea, as Roger Bacon had put it, that "corporeal roads signify spiritual roads." As a student of Pierre d'Ailly, who himself had borrowed so extensively from Bacon, Columbus believed that his voyages to the East would take him back into the geography of the Bible—which is why, it would seem, that by the time of his second voyage he had begun to entertain the idea that Hispaniola and Cipangu and Ophir might all be one and the same island.

An early sign of this shift in Columbus's thinking appears in his copy of Pliny's *Natural History*. Columbus consulted the work regularly as he tried to identify the plants, animals, and minerals that he was finding in the Indies. "Amber," he wrote at one point alongside Pliny's treatment of the subject, "is certainly found in India under the ground, and I have had it dug out of several hills in Feyti [Haiti], Ophir, or Cipangu, which I afterwards named Hispaniola." Columbus also flirted with the idea that Hispaniola might be the biblical Sheba, known in Latin as Saba. One of his companions on the second voyage, a Genoese nobleman named Michel de Cuneo, recalled in a letter to a friend that when the fleet was approaching Hispaniola, Columbus explicitly made this connection in a speech to his men.

Before we made the big island he spoke these words: "Gentlemen, I wish to bring us to a place whence departed one of the three Magi who came to adore Christ, the which place is called Saba." When once we made that place and asked the name of the place it was, he answered us that it was called Sobo. Then the Lord Admiral said that it was the same word, but that they could not pronounce it correctly.

Another companion of Columbus's on the same voyage also recalled that Columbus had associated Hispaniola with Saba. "The Admiral," he wrote to a friend in Italy, "sent [two noblemen] with a company of light-armed troops to the interior of the land of the Sabaeans, to make their way to King Saba. It is believed that these are the Sabaeans from whom frankincense is obtained and who are mentioned by our histories and foreign chronicles."

Ophir, Saba, Tarshish, Cipangu: Columbus believed that Hispaniola might be any one of these places, or perhaps all of them at once, a single land that over the ages had been known by a succession of different names. Even several years later, after having devoted much study to this question, Columbus still hadn't worked out an answer. Instead, he had identified even more biblical possibilities. "This island is Tarshish," he wrote in a letter to Pope Alexander VI in 1502, "it is Chittim, it is Ophir and Ophaz, and Cipangu, and we have named it Hispaniola."

* * *

As had been the case during his first voyage, Columbus ended up having little time to ponder abstract geographical questions during his second voyage. When he returned to Hispaniola in the fall of 1494, after having given up on the exploration of Cuba's south coast, he found his new colony of Isabela in a state of turmoil and near collapse. His men were sick and dying, the locals were openly rebelling again, and famine had set in. In Isabela itself, rival factions of colonists had formed in Columbus's absence, adding to the level of unrest, unhappiness, and uncertainty. Almost everybody at Isabela did agree on one thing, however: Columbus was to blame.

Back in Spain the news was better. The Sovereigns and King João had finally signed an agreement that summer, known as the Treaty of Tordesillas, which decreed the drawing of a line of demarcation that would indeed divide the Western Ocean between them, by means of "a boundary or straight line" that was to be "determined and drawn north and south, from pole to pole, on the said Ocean Sea, from the Arctic to the Antarctic pole . . . at a distance of three hundred and seventy leagues west of the Cape Verde Islands." All new lands to the east of the line would belong to Portugal; all new lands to the west would belong to Spain; and finally, after months of diplomatic wrangling, Spain's rights to Hispaniola were uncontested.

But just when everything seemed to going the Sovereigns' way, word began to reach Spain that all was not well on the island. The Sovereigns quickly dispatched a royal commission to Hispaniola to investigate. By the end of the year they had a report in hand that confirmed their worst fears.

The report, backed up by the stories of disillusioned colonists who had begun to return from Hispaniola to Spain, threatened to ruin Columbus's reputation and to end the Sovereigns' interest in funding the colony on Hispaniola. By early 1496, Columbus had recognized the gravity of the situation, and by June he was back in Spain, hoping to plead his case.

But now the aura of triumph was gone. There was no flaunting of bounty and captives, no joyful procession through the Spanish countryside. Instead Columbus hastened solemnly to Ferdinand and Isabella, wearing only the robe and sandals of a Franciscan, and beseeched them to give him the time and the money to bring the situation under control—and to find the gold that he *knew* was out there. The tactic worked. The Sovereigns heard Columbus out, gave him the benefit of the doubt, and eventually agreed to send him on a third official voyage—but this time they expressed no particular sense of urgency. When it came to organizing his third voyage, he would be largely on his own.

The job ended up taking Columbus two years. When he was finally ready to set sail again, in May of 1498, he had only six ships at his command. "Nobody wants to live in these countries," the Genoese nobleman Michel de Cuneo had written about Hispaniola and other islands, and the news had clearly gotten around. The crew Columbus had managed to pull together consisted mostly of conscripted prisoners.

* * *

WHILE COLUMBUS LANGUISHED in Spain, the business of discovery continued apace. By the early 1490s, it seems, the men of Bristol had finally relocated their Island of Brasil, in the form of a long, sparsely populated and heavily forested expanse of coastline, probably some part of modern-day Newfoundland or Labrador. But they devised no elaborate geographical theories about their discovery. All that mattered to them was that they had discovered a rich new source of fish and timber. They did talk about their discovery, however—and in Spain, where they regularly sailed on trading missions, this talk soon caught the attention of a Venetian merchant named Giovanni Caboto, more commonly known today as John Cabot.

Cabot first appears in the historical record in about 1490, in the archives of Valencia, where he is identified as *Johan caboto monecalunya venesia*. This John Cabot, the archives reveal, met twice with King Ferdinand in the first half of 1492, to present a proposal for the modification of Valencia's harbor. This puts Cabot in the Sovereigns' orbit in early 1492—which makes it possible, although not provable, he encountered Columbus there. The idea isn't implausible: both were Italian sailors, after all, and both were negotiating with the Sovereigns at about the same time about maritime ventures.

Cabot seems to have still been in Spain in the spring of 1493, during the period when Columbus, having returned from his spectacularly successful first voyage, was parading slowly through Spain with his gold, his natives, and his parrots. Cabot may well have seen Columbus pass by, or heard word of his feat. But Cabot was an educated and well-traveled man—had been to Mecca on a spice-trading mission and he knew his Marco Polo well— and he seems to have decided fairly quickly that Columbus was all talk. Columbus hadn't come home with any valuable spices and hadn't found any of the places or peoples described by Polo; all he had done was discover a few islands on the way to Asia. But the men of Bristol, on the other hand— they had actually made an important discovery, he seems to have decided, although almost nobody was talking about it. If the long coastline that they had discovered at much higher latitudes was, a mainland, as it appeared to be, then it had to be some desolate northern portion of the Asian coastline. This gave Cabot an idea. While Columbus and his colonists were frittering away the resources of Spain on Hispaniola, Cabot could sail west from England to this newly discovered mainland, follow it south, and eventually reach the lands of the Great Khan.

The next reference to Cabot in the historical record appears in early 1496, in a letter sent by the Spanish Sovereigns to their ambassador in

England. The ambassador had apparently reported to the Sovereigns that Cabot had arrived in England, and in their letter the Sovereigns make it clear that they know him—a good sign that he is indeed the same Cabot who had met with Ferdinand earlier in the decade. "In regard to what you say of the arrival of one like Columbus," they wrote, "for the purpose of inducing the King of England to enter upon another undertaking like that of the Indies, without prejudice to Spain or to Portugal, if he [the English king] aids him . . . the Indies will be well rid of the man." Cabot, it would seem, had been pestering the Sovereigns with requests that they back his venture west to the Indies, but the Sovereigns—already burned by one Italian who had made big promises about sailing off to the west in search of the Great Khan—wanted nothing to do with him.

Cabot had better luck with King Henry VII of England. On March 5, 1496, Henry granted Cabot the official right to make discoveries for England in "the eastern, western, and northern sea"—and within months Cabot was sailing west. Foul weather and a lack of food forced him to turn back, but in May of 1497 he set out again, in a ship called the *Matthew*—and this time, somewhere along the coast of Newfoundland or Labrador, he arrived at and explored a stretch of heavily forested coastline, one that corresponded to what the men of Bristol had described. By August Cabot was back in England making grand pronouncements, just as Columbus had done before him, about having reached the Asian mainland. Just days after Cabot's return, a Venetian living in England recorded the news in a letter home. "That Venetian of ours," he wrote,

> who went with a small ship from Bristol to find new islands has come back and says he has discovered mainland 700 leagues away, which is the country of the Great Khan, and that he coasted it for 300 leagues and landed and did not see any person. . . . His name is Zuan Talbot [John Cabot], and he is called the Great Admiral, and vast honor is paid to him, and he goes dressed in silk, and these English run after him like mad.

Almost nothing is known about Cabot's voyage. None of his letters or journals survives (if he even wrote any). Unlike Columbus, he had no admirers who thought to record the details of his life and achievement for posterity. But a map does survive that seems to illustrate what Cabot thought he had found. Drawn according to Ptolemaic principles in the early 1500s by a Dutch explorer named Johannes Ruysch, who some scholars believe accompanied Cabot on his first voyage, the map shows an eastward-jutting mainland directly to the west of Europe, which would appear to be the region visited by Cabot: contiguous with that mainland, but farther to the southeast, it shows the lands of the Great Khan. Hispaniola, Cuba, and other islands discovered by Columbus also appear at the bottom left, with Cuba's status (island or mainland?) conveniently obscured by legend-bearing scroll (*Figure 60*).

Figure 60. The northern discoveries of John Cabot. Detail from the world map of Johannes Ruysch (1508). The British Isles are on the right. Jutting toward them from the left is a peninsula consisting of Greenland (top) and, below it, *terra nova*. At the top left, connected to the peninsula, is the realm of Gog and Magog, and, near the top left corner, the Chinese city of Quinsai. Columbus's islands are at the bottom left.

King Henry seems to have been very pleased; like the Sovereigns in 1493, he rushed to send his "one like Columbus" back to Asia on a second voyage, to solidify the English claim on the lands he had discovered. Another Italian living in England at the time recorded the king's plans, in a letter sent to the duke of Milan on December 18, 1497.

His Majesty here has gained a part of Asia, without a stroke of the sword. There is in this kingdom a man of the people, Messer Zoane Cabot by name, of kindly wit and a most expert mariner. Having observed that the Sovereigns first of Portugal and then of Spain had occupied unknown lands, he decided to make a similar acquisition for his Majesty. . . . This Messer Zoane has the description of the world in a map, and also in a solid sphere, which he has made, and shows where he has been. . . . Before very long they say that his Majesty will equip some ships . . . and they will go to that country and form a colony. By means of this they hope to make London a more important mart for spices than Alexandria.

Cabot set out on his second voyage in May of 1498, as soon as the weather was once again suitable for sailing in northern waters. Full of high hopes, he sailed west from England with five ships at his command—and was never heard from again. Some historians believe he perished during the crossing, but others argue that he reached North America and made his way south, perhaps as far south as the Caribbean. The evidence for this is slender but suggestive: a letter of authorization given by the Spanish Sovereigns in 1501 to Alonso de Hojeda, a captain who had already sailed with Columbus. Hojeda, they wrote, was authorized to make a voyage of his own—"toward the region where it has been learned that the English were making discoveries . . . so that you may stop the exploration of the English in that direction." The Sovereign later even granted Hojeda six leagues of land on Hispaniola "for the stopping of the English"—a remark that has led some students of the period to suggest that Cabot and his men were captured or killed by the Spanish not far from Hispaniola.

If indeed this ever happened, no such news ever reached England. All anybody knew was that Cabot had disappeared. In 1512 an English chronicler named Polydore Vergil summed up what he assumed to have been Cabot's achievement—and his fate. "He is believed," Vergil wrote, "to have found the new lands nowhere but on the very bottom of the ocean."

* * *

COLUMBUS HIMSELF GOT word of Cabot's first voyage sometime in the winter of 1497–98. It came in a letter from John Day, an English merchant working in Spain. The letter in fact is addressed only to the "Lord Grand Admiral," but because only two men in Spain at the time could have claimed that title, and because only one of them, Columbus, had a special interest in Atlantic discoveries and the geography of the East, Day's correspondent is generally considered to have been Columbus.

Day opened his letter by acknowledging receipt of a letter from Columbus. The men had clearly already been discussing geographical matters. Day wrote that he had been unable to find for the admiral a copy of the *Inventio fortunata*, a now-lost work thought to have described the voyage of an English Franciscan into Arctic waters, but he did have something else the admiral had asked for: a new complete edition of Marco Polo's Book. (The Latin edition that Columbus already owned was a condensed version.) Not only that, Day wrote that he had managed, as requested, to make "a copy of the land that has been found"—that is, a map of Cabot's discoveries. "From the said copy," Day wrote, referring to a copy of the map he had included with his letter, "your Lordship will learn what you wish to know, for in it are named the capes of the mainland and the islands, and thus you will see where land was first sighted." Day provided an account of the voyage and then brought his letter to a close with one of the few known references to the discovery that the men of Bristol had made before Cabot made his, in 1497. "It is considered certain," he wrote, "that the cape of the said land was found and discovered in the past by the men from Bristol who found 'Brasil,' as your Lordship well knows."

Columbus didn't believe Cabot's claim to have found the lands of the Great Khan. "Cathay is not so far to the north as the map shows," he noted at one point in his copy of *The Image of the World*—a remark that may represent his indignant reaction to the news of Cabot's discovery, or to what he saw on John Day's map. But Cabot had found a mainland—and a big one at that. This only reinforced Columbus's conviction that he, too, had reached mainland Asia. And because Marco Polo and other early European travelers to Asia had reported finding gold and spices in equatorial latitudes, not in the north, Columbus chose to ignore Cabot's discovery. Instead, he decided that on his third voyage he would sail even farther south than he had before.

A new sense of political urgency goaded Columbus on. The previous summer the Portuguese had launched a new effort to reach the Indies by sailing around Africa—their first attempt since Bartolomeu Dias had rounded the Cape of Good Hope. On July 8, 1497, four ships under the command of Vasco da Gama had set out for Calicut, the Indian port successfully reached by Pêro da Covilhã almost a decade before. Da Gama carried letters of introduction addressed to the ruler of Calicut and to Prester John, and his mission was simple: he was "to make discoveries and go in search of spices."

The fate of da Gama's expedition was still unknown in Europe when Columbus began his third voyage, on May 30, 1498, but Columbus knew that a race to the Indies was now on. As he had done twice before, he sailed first to the Canaries, but then he sailed even farther to the south, to the Cape Verde Islands, from where, on July 4, he began his third Atlantic crossing. The journal he kept of the voyage is again lost, but an abstract of it survives in the work of Bartolomé de Las Casas.

Columbus ordered three of his ships to sail directly to Hispaniola, but he and his remaining three ships set a course for the southwest, toward the equatorial latitudes where medieval alchemists believed gold, gems, spices, and medicinal drugs all "grew" in great quantities. It wasn't just the prospect of finding riches, however, that drew him in this direction. According to Las Casas, Columbus pursued the course he did on his third voyage because he intended "to test the theory of King João of Portugal"—about which nothing else is known—"that there was a great landmass to the south."

* * *

The voyage took twenty-seven days, and introduced Columbus to the mid-Atlantic Doldrums. "On Friday, 13 July, the wind failed," Las Casas wrote, summarizing Columbus's journal, "and he came into such great, vehement, burning heat that he feared lest the ships catch fire and the people perish. So suddenly did the wind cease and the excessive and unusual heat come on that there was no one who would dare to go below to look after the casks of wine and water, which burst, snapping the hoops of the pipes; the wheat burned like fire: the bacon and salt meat roasted and putrefied." The ships were becalmed for a week, during which time

Columbus "suffered an attack of gout and insomnia," but eventually they picked up a fair breeze and continued on their way—and on July 31, at about 10 degrees north of the equator, a lookout shouted out from the crow's nest on Columbus's ship that he had caught sight of land. It turned out to be another large inhabited island. Columbus promptly named it Trinidad, in honor of the Holy Trinity.

Columbus and his men explored the south coast of Trinidad in early August, and as they did they caught sight of what appeared to be another island to the west. When they sailed over to explore it, what they found pleased Columbus greatly. "No lands in the world can be greener and lovelier and more populous," he wrote. "The climate is also excellent, for since I have been on this island it has been cold enough every morning, I say, for a lined coat, although it is so close to the equator." The inhabitants called this place Paria; Columbus named it Gracia.

The inhabitants of Paria also pleased Columbus. They had fine manners, fair skin, handsome bodies, and long, smooth hair that was, he observed, "cut in the manner of Castile." They wove brilliantly colored textiles and wore articles of clothing, and some even bound their heads with what the ever-hopeful Columbus assumed was "a Moorish shawl." They cultivated the land, rich in aromatic trees and exotic fruits, and kept monkeys and parrots as pets. Best of all, they wore jewelry made of gold and pearls.

In the days that followed, Columbus and his men explored the coast of Paria, in search of the source of this gold and these pearls, and as they progressed they gradually realized that they were sailing in a vast gulf that seemed to be ringed by several large islands. What was more, the sea enclosed by this gulf was "all sweet," almost like a lake. The mouth of a great river had to be nearby—and one of Columbus's ships soon found it. Returning from a bit of independent reconnaissance, the ship's sailors reported that they had found "a vast gulf and four large openings therein, which appeared to be small gulfs, at the end of each of them a river."

Four great rivers on one single island, all filling a huge gulf with sweet water? It didn't make sense. "Neither the Ganges, nor the Euphrates, nor the Nile carried so much fresh water," Columbus noted in his journal. "Greatly did the Admiral long to find the key to the mystery," Las Casas added.

By August 13 he had come up with an answer. The several "islands" ringing the gulf were actually all parts of a single coastline. Based on the amount of fresh water pouring into the gulf, the land behind that coastline had to be very big—so big, in fact, that it couldn't be an island at all. "I have come to believe," Columbus wrote, "that this is a mighty continent that was hitherto unknown."

It was a prophetic guess. Columbus and his men had reached the coast of South America: specifically, the delta of the great Orinoco River, which spills out from modern-day Venezuela into a gulf still known today as the Gulf of Paria. But just four days later, on August 17, Columbus changed his mind. He hadn't found a *new* continent, he decided. He had found the oldest one of them all.

He had reached the outskirts of the Earthly Paradise.

* * *

COLUMBUS HAD READ extensively about the Earthly Paradise by the time he made his third voyage. Along with Ophir and Tarshish, it formed a central part of his spiritual map of the East. Much of what he knew about the place came from Pierre d'Ailly. "The Earthly Paradise is a pleasant spot," d'Ailly had written in *The Image of the World*, in a passage that Columbus marked up enthusiastically,

> located in certain regions of the East that are separated by land and sea from our inhabited world. It is so high up that it abuts on the sphere of the moon, and the . . . waters that descend from this very high mountain form a vast lake; it is said that these falling waters are so noisy that they make the inhabitants of the region deaf. . . . From this lake . . . flow the four rivers of Paradise: the Pishon, that is, the Ganges; the Gihon, which is no other than the Nile; and the Tigris and the Euphrates.

All of the great Christian authorities, among them Saint Augustine, the Venerable Bede, Isidore of Seville, and Thomas Aquinas, had left behind similar descriptions, and in the fourteenth century the English Benedictine Ranulf Higden had distilled their conclusions in his *Polychronicon*, or *Universal Chronicle*—a popular work first printed in 1482, just when Columbus was beginning to immerse himself in his studies. But Higden, writing after Europeans had begun to explore Asia, also described the Earthly

Paradise with a new kind of geographical specificity. "The learned conclude," he wrote, "that the Earthly Paradise is located in the farthest East, and makes up a sizable part of the earth's mass, being no smaller than India or Egypt, for the place had been intended for the whole of the human race, if man had not sinned."

By Columbus's time this was the reigning view. The Earthly Paradise was a real place in the world, and therefore could be located with precision by travelers and geographers. There was the memorable account of Sir John Mandeville, of course, which Columbus knew well. But by the fifteenth century even mathematical geographers were getting into the act. One authoritative table of geographical coordinates, for example, put the Earthly Paradise at the top of its list of the world's places and assigned it a latitude of 0 degrees (the center of the earth) and a longitude of 180 (as far away as geometrically possible from where the west "began").

Columbus is often ridiculed for having decided in the Gulf of Paria that he had reached the outskirts of the Earthly Paradise. But given the geographical information he had to work with, his conclusion had a certain logic to it. He had reached a giant continent at the very end of the East, near the equator, and there he had found an agreeable climate, gentle breezes, trees that produced wonderful fruits and fragrances, a giant land, and a vast freshwater lake fed by four great rivers. He had discovered something else too: a mysterious shift in his compass readings at the hundred-league mark. This was the normal shift in magnetic variation that takes place as one crosses the Atlantic, caused by the fact that the world's physical and magnetic north poles don't coincide. (The geometrical point around which the earth turns is not exactly the same as where a compass points.) But Columbus didn't know that. In his mind this shift had a powerful significance, which he explained in a rambling letter that he wrote to the Sovereigns on October 18, 1498.

I have always read that the world, both land and water, was spherical, as the authority and researches of Ptolemy and all the others who have written on this subject demonstrate and prove, as do the eclipses of the moon and other experiments that are made from east to west, and the elevation of the North Star from north to south. But I have seen this discrepancy [magnetic variation], as I have said. I am compelled, therefore, to come to this view of the world: I have found that it does not have the kind of sphericity described by the authorities, but that it has the shape of a pear, which is all very round, except at the stem, which is rather prominent, or, that is, as if one had a very round ball on one part of which something like a woman's teat were placed, this part with the stem nearest to the sky, lying below the equinoctial line in this Ocean Sea, at the end of the East. I mean by the end of the East the point where its land and islands terminate. To confirm all this I cite all the arguments written above about the line that passes from north to south 100 leagues west of the Azores. For in crossing this to the westward the vessels keep rising gradually toward the sky and then enjoy milder weather.

This passage has opened Columbus up to even more ridicule than his claim to have reached the vicinity of the Earthly Paradise. But once again, in making this conclusion he wasn't actually straying far from cosmographical ideas that were already in the air. In *The Divine Comedy*, for example, Dante put the Earthly Paradise on top of what he called Mount Purgatory—which he placed on the equator at the beginning of the East, exactly halfway around the globe from Jerusalem. Dante's world, in other words, was pear-shaped, just like Columbus's. Some copies of *The Divine Comedy* that circulated during Columbus's lifetime even contained illustrations that gave this idea visual form (*Figure 61*).

Did Columbus have some such illustration in mind when he proposed his pear-shaped world? It's at least possible, given that Columbus read cosmographical works voraciously, sold printed books imported from Italy, and had access to Spain's great Franciscan libraries. What's very clear, at least, is that by the time of his third voyage Columbus understood the Earthly Paradise as a real place of great symbolic power, located at the zero point of space and time. The Earthly Paradise was where history and geography had begun, and it was where they both would come to an end—as the influential theologian and *mappamundi*-maker Hugh of Saint Victor had written at the end of the twelfth century.

The order of space and the order of time seem to be in almost complete correspondence. Therefore, divine providence's arrangement seems to

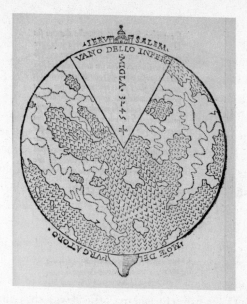

Figure 61. The pear-shaped world, as Columbus might have seen it illustrated.
From Dante's *Divine Comedy* (1506). Jerusalem and the Inferno are at the top;
Europe and Africa are on the left; Asia is on the right; and at the bottom is
Mount Purgatory, on top of which Dante located the Earthly Paradise.
Note the New World emerging to the left and right of Mount Purgatory.

have been that what was brought about at the beginning of time would
also have been brought about in the East—at the beginning, so to speak,
of the world as space—and then, as time proceeded towards its end, the
center of events would have shifted to the West, so that we may recognize
out of this that the world nears its end in time as the course of events has
already reached the extremity of the world in space.

By crossing the ocean, connecting East and West, and locating the outskirts
of the Earthly Paradise, Columbus was helping to bring about the end of
the world.

* * *

BY MID-AUGUST, JUST as he was beginning to suspect that he had discovered something big in the Gulf of Paria, Columbus realized that he would

have to stop his explorations. He and his men had been at sea for more than ten weeks, their provisions were running out, and their ships were too big to explore coastal waters effectively. Columbus himself was very sick, and he knew he had urgent matters to attend to on Hispaniola. So on August 17 he ordered his ships to leave the Gulf of Paria and began sailing north.

He reached Hispaniola at the end of the month, only to discover that things had gone from bad to worse. The colonists were sick and mutinous, the locals were increasingly hostile, and gold remained elusive. The whole colonial venture on the island was fast becoming a fiasco. Columbus would stay on the island for the next two years, attempting to reassert his authority and to restore order by fighting and rounding up runaways, executing mutineers, and enslaving more islanders. But the problems were now simply too large for him to control. By 1500 the Sovereigns once again had lost faith in him and had even begun to wonder if he had reached the Indies at all. One event in particular had pushed their thoughts in this direction: in 1499, Vasco da Gama had returned to Portugal with the remarkable news that he had sailed around Africa, crossed the Indian Ocean, and reached Calicut.

Word of da Gama's return had traveled fast, and there was little doubt that da Gama's achievement far surpassed Columbus's. "An enterprise has been carried out," one Italian chronicler wrote at the time, "that arouses the admiration of the whole world. The spices that should, or used to, go to Cairo by way of the Red Sea are now carried to Lisbon, and with this the Sultan [of the Turks] has lost some five or six thousand ducats a year, and the Venetians the same."

Even as a mariner, da Gama had outdone Columbus. He had traversed some six thousand miles of open water, at one point sailing for three months out of sight of land—a wide southwesterly swing out into the Western Ocean designed to avoid the African coast and to find the westerly winds that had blown Dias around the Cape of Good Hope. The whole journey had been far longer and more difficult than Columbus's, and at the end of it da Gama had found precisely the riches he had hoped to find. He had returned to Portugal with gold, gems, and spices, whereas Columbus had returned to Spain with nothing but some trinkets, a few naked captives, and a head full of dreams.

Increasingly unsure of Columbus, in the summer of 1500 the Sovereigns asked one of their advisors, Francisco de Bobadilla, to sail to Hispaniola to investigate the situation. When Bobadilla arrived, in late August, he was greeted with the macabre spectacle of seven Spanish colonists hanging lifeless on the gallows, executed not long before on Columbus's orders. The sight didn't inspire confidence, and in the weeks that followed Bobadilla witnessed firsthand the violence, insurrection, and mismanagement that were rife on Hispaniola. By October he had seen enough. Although the Sovereigns had named Columbus governor of the Indies for life, Bobadilla took over his quarters, had him arrested, and shipped him home in chains.

*   *   *

BACK IN SPAIN, humiliated, aggrieved, and brimming over with not unjustifiable self-pity, Columbus launched a campaign to win back the rights and privileges that the Sovereigns had granted him as governor of the Indies. He proceeded methodically, collecting in a single volume all of the agreements he had made with the Sovereigns since 1492. This volume, known today as the *Book of Privileges*, became the centerpiece of a legal case he made for his rehabilitation, but the case went nowhere during his lifetime. Resolution only came in 1563, almost sixty years after his death, when his heirs finally reached a settlement with the Spanish government.

At the same time that Columbus was compiling a legal case for his rehabilitation, he was also compiling a theological one. He had played a starring role, he felt, in the cosmic drama that Europe's chroniclers and mapmakers had been describing for centuries and he wanted the Sovereigns to recognize that fact. By leading the grand procession of human history right across the ocean and back to the place where time and space began, he had made it possible for both Rome and Christ to extend their dominion to the very ends of the earth, and this made him nothing less than an apostle of God. "The truth is," Las Casas would later write, echoing Columbus's own sense of himself, "that he was the first to open the gates of the Ocean Sea, in order to bear our Savior Jesus Christ over the waves to those remote realms and lands until now unknown." It was for this reason, Las Casas continued, that Columbus bore the name Christopher—*Christo-ferens*, that is, "Christ-Bearer."

After his third voyage, Columbus made the name *Christo-ferens* an official part of his signature. He also began compiling what's known today as *The Book of Prophecies*—a haphazard and disorganized collection of quotations, drawn from the Bible and the works of ancient and medieval authorities, all designed to bolster the theological case for himself as a figure whose deeds had been foretold, whose voyage across the ocean had brought humanity closer to salvation, and whose discovery of the farthest East had brought history itself closer to its end. The opening lines of the work succinctly describe its contents. "Here begins the book, or handbook," they read, "of sources, statements, opinions, and prophecies on the subject of the recovery of God's Holy City and Mount Zion, and on the discovery and evangelization of the islands of the Indies, and of all other peoples and nations. To Ferdinand and Isabella, our Hispanic rulers."

References to Solomon, Ophir, and Tarshish abound in *The Book of Prophecies*—more evidence that Columbus saw himself as reenacting that ancient story. "Surely the isles shall wait for me," read one typical passage, borrowed from the Book of Isaiah, "the ships of Tarshish first, to bring thy sons from far, their silver and their gold with them, unto the name of the Lord thy God, and to the Holy One of Israel." Columbus also cited plenty of nonbiblical sources to make his case. At one point, notably, he quoted a letter that the Sovereigns had received in 1492 from a delegation of Genoese ambassadors. The Sovereigns had just defeated the Moors, and the ambassadors were now urging them to continue on to Jerusalem, insisting that it was the Sovereigns' sacred destiny to launch a new Crusade. "Joachim the Abbot of southern Italy," Columbus wrote, quoting the letter, "has foretold that he who is to recover again the fortunes of Zion is to come from Spain."

Columbus would repeat this prophecy elsewhere in his writings and clearly took it to heart. He did the same with the famous passage in the *Medea* in which Seneca (who also happened to be from Spain) prophesied that "Tethys will disclose new worlds, and Thule no more be the ultimate." Columbus cited this passage several times in his writings, but he always quoted a slightly different version of the text. A corrupt edition was circulating widely in Europe during the fifteenth century, and in this edition, thanks to a fortuitous scribal error, Tethys, a Greek god of the sea, became Typhis, the mariner who in Greek mythology had guided Jason and the Argonauts in search of the Golden Fleece. The transformation suited Columbus perfectly. Seneca now appeared to be predicting that a figure almost exactly like Columbus—an expert mariner who has embarked on a difficult ocean voyage in search of gold for a great king—would be the one to loose the chains of things and disclose new worlds. Whenever he cited this passage, Columbus also replaced Seneca's plural "worlds" with the singular "world"— a sleight of hand that suggests that Columbus in his messianic phases wasn't beyond pandering to the apocalyptic dreams of the Sovereigns.

The biblical prophets, the Greeks, the Romans, the learned Christian authorities of the Middle Ages—everybody, it seemed, had foretold the special role that Columbus was now playing in history. And history itself was now swiftly coming to an end. In a draft letter to the Sovereigns that he included in the *Book of Prophecies*, Columbus described exactly when the end would come.

> From the creation of the world, or from Adam, until the advent of our Lord Jesus Christ there were five thousand, three hundred and forty-three years, and three hundred and eighteen days, according to the calculation by King Alfonso, which is considered to be the most exact. Following Pierre d'Ailly, in the tenth heading of his *Explanation of the Agreement of Astronomy with Biblical and Historical Records*, if we add to these years an additional one thousand, five hundred and one years of waiting, this makes a total of six thousand, eight hundred, forty-five years of waiting for the completion of the age.
>
> According to this calculation, only one hundred and fifty years are lacking for the completion of the seven thousand years that would be the end of the world.

In other words, my most noble Sovereigns, it's high time you sent your Christ-bearer back to the Indies.

* * *

COLUMBUS WOULD EVENTUALLY get his way—but not before news of other discoveries had reached Europe. The word was that the English were continuing to explore the northern landmass that John Cabot had reached in 1497, and that they might even be encroaching on Spanish territories in the south. Even more troubling, the Portuguese, now under the rule of King Manuel I, were making important new discoveries of their own.

Manuel had wasted no time in following up on the success of Vasco da Gama's voyage to India. By early 1500 he had a fleet of thirteen ships ready for a new voyage and had appointed Pedro Álvares Cabral as their commander. Cabral set sail in early March. Following the route pioneered by da Gama, he swung far west out into the Atlantic on his way south, and in late April, well below the equator, he bumped into something entirely unexpected: another long expanse of inhabited coastline that ran from north to south as far as the eye could see. Cabral had little time to stay and figure out what it was that he had found. But he did recognize that his discovery was important. King João had been right, it seemed; a great land did lie across the Atlantic to the southwest. Not only that, Cabral and his men were able to determine that this land extended so far out east into the Atlantic, toward Africa, that by rights, according to the Treaty of Tordesillas, it belonged to them. Cabral therefore claimed possession of it for Portugal, named it Land of the True Cross, and sent one of his ships home to report the news.

Cabral's name didn't last long. Some Europeans began using the name Land of the Parrots, for obvious reasons, and others—after learning that large quantities of brazilwood grew on its shores—began using the name that the land still bears today: Brazil.

The Portuguese were also venturing into northern waters at about the same time. The most famous of these expeditions was that of Miguel and Gaspar Corte-Real, two brothers who owned land in the Azores and had connections with the men of Bristol. In 1501 the brothers sailed north from Lisbon, only to find their passage eventually blocked, in the vicinity of Greenland, by "huge moving masses of solid snow"—icebergs. They then turned to the west and not long after found themselves along a long stretch of coastline, once again probably some part of Labrador, Newfoundland, or even Nova Scotia. When they returned to Lisbon that fall and told stories of what they had found, rumors began to fly. "The men of the caravel believe that the above-mentioned land is a continent," one Italian living in Lisbon wrote on October 18, just nine days after the brothers had returned. He continued, "They also believe that it joins with the Antilles, which were discovered by the kings of Spain, and with the Land of Parrots, recently found by the ships of this king that were going to Calicut."

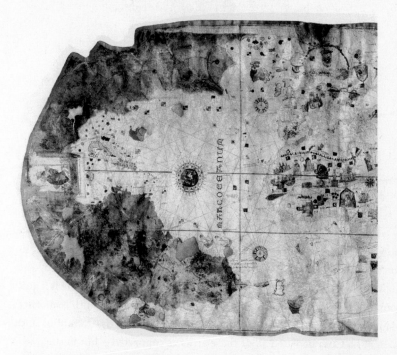

Figure 62. Detail from the earliest surviving chart of the New World: the La Cosa chart (circa 1500). Europe and Africa are at the far right; the islands discovered by Columbus are at far left, in the middle; and the dark landmasses around those islands are North and South America. The map shows the New World falling entirely on the Spanish side of the line of demarcation between Spain and Portugal.

This idea of connectedness also began to appear on maps at the turn of the century. The earliest surviving marine chart of the New World—drawn in about 1500 by Juan de la Cosa, a Spanish pilot who had accompanied Columbus on his second and third voyages—illustrates the idea very clearly (*Figure 62*). Despite its relatively poor state of preservation, the chart provides a revealing look at how the earliest visitors to North and South America, Columbus notably among them, were beginning to piece together a picture of what lay on the far side of the Atlantic.

Figure 63. The Christ-bearer: Saint Christopher carrying the infant Christ
across the water, symbolizing the achievement of Columbus in crossing the Atlantic.
Reconstructed detail from the westernmost part of La Cosa chart (*Figure 62*).

The chart's main point of interest (aside from its newly accurate depiction of Africa, based on da Gama's recent voyage) appears in the west. In the north is a mainland representing the various lands discovered by the men from Bristol, John Cabot, and the Corte-Reals; in the middle is a gulf containing the islands discovered by Columbus; and in the south is another mainland, this time representing the lands discovered by Columbus and Cabral. La Cosa depicted all of these areas as connected—but in the middle, directly to the west of Cuba and Haiti, he covered up the coast with an image of Saint Christopher bearing the infant Christ across the water (*Figure 63*). He did this for two reasons. The first was to associate Columbus and his Atlantic crossing with an ancient Roman Catholic legend in which a pagan named Reprobus had carried the infant Christ across a dangerous river, and had been rewarded for his deed by being baptized as Christopher, or the Christ-bearer. The second reason was more practical: it allowed La Cosa to leave open the possibility that a narrow strait passed through the area covered by Saint Christopher—one that opened up a western passage to the Indian Ocean, the Aurea Chersonese, and beyond. Columbus believed firmly in the existence of this strait, and he promised the Sovereigns that if he could find it he could open up a route to India and Calicut that was far shorter than the eastern route around Africa now being used by the Portuguese. It was this promise, not his messianic ramblings, that seems finally to have convinced the Sovereigns to send Columbus back across the Atlantic on a fourth voyage.

Columbus set sail in May 1502, bearing a letter that he was to present to the Portuguese if he managed to reach India and find them there. In terms of pure adventure, this voyage would turn out to be the most remarkable of all: Columbus survived a violent hurricane at sea; he discovered and explored a large stretch of the Central American coastline, from modern-day Honduras down to Panama; he found gold and signs of advanced civilization in the region, and decided he had reached the Aurea Chersonese; he endured a long and demoralizing year marooned on Jamaica; and he survived a harrowing fifty-six-day return journey to Spain, in November of 1504, during which his mast broke into four pieces.

But he never did find his strait, and once back in Spain, he had to confront the fact that his star had faded for good. His health was poor, Queen Isabella had died, King Ferdinand cared little for him, and his legal case was going nowhere. Already he was remembered in Europe only dimly, if at all, as a Genoese sailor who, like Lanzarotto Malocello before him, had discovered a few new Atlantic islands. Less than two years later he was dead—and Europe was abuzz with the news that another Italian had made a dramatic discovery across the ocean.

His name was Amerigo Vespucci.

• Amerigo Vespucci •

CHAPTER SIXTEEN

AMERIGO

I trust Your Magnificence will have heard the news brought by the fleet
[of Vasco da Gama]: I do not call such a voyage discovery but merely
a going-to-discovered-lands, since, as you will see by the map, they . . .
sail along the entire southern part of Africa, which is to proceed along a
route discussed by all the authorities in cosmography.

—Amerigo Vespucci to Lorenzo di Pierfrancesco de' Medici
(July 1500)

*I*N EARLY 1503, printers in Venice, Paris, and Antwerp almost simultane-
ously began to publish copies of a small Latin pamphlet bearing the title
Mundus novus, or *New World*. The pamphlet described a voyage of discovery made not long before by the Florentine merchant Amerigo Vespucci,
and its reception was electric. Within two years new editions of the work
had been printed in Augsburg, Basel, Cologne, Munich, Nuremberg, Rome,
Strasbourg, and other European cities. Twelve different editions appeared
in Germany alone, a stark contrast to the reception of the 1493 Columbus
letter, which had been printed in the country only once.

Mundus novus took the form of a letter from Vespucci to his patron in Florence, and its opening paragraph made a sensational claim.

> Amerigo Vespucci to Lorenzo di Pierfrancesco de' Medici, with many salutations.
>
> In the past I have written to you in rather ample detail about my return from those new regions which we searched for and discovered with the fleet, at the expense and orders of His Most Serene Highness the King of Portugal, and which can be called a new world, since our ancestors had no knowledge of them, and they are an entirely new matter to those who hear about them. Indeed, it surpasses the opinion of our ancient authorities, since most of them assert that there is no continent south of the equator, but merely that sea which they called the Atlantic; furthermore, if any of them did affirm that a continent was there, they gave many arguments to deny that it was habitable land. But this last voyage of mine has demonstrated that this opinion of theirs is false and contradicts all truth, since I have discovered a continent in those southern regions that is inhabited by more numerous peoples and animals than in our Europe, or Asia or Africa, and in addition I found a more temperate and pleasant climate than in any other region known to us, as you will learn from what follows, where we shall briefly write only of the main points of the matter, and of those things more worthy of note and record, which I either saw or heard in this new world, as will be evident below.

The letter's tone and substance in fact were sensational from beginning to end. The initial ocean crossing was fraught with peril. "We had forty-four continuous days of rain, thunder, and lightning," Vespucci wrote, "so dark that we never saw sunlight in the day, nor clear sky at night. Fear so overwhelmed us that we had almost abandoned hope of all survival." Only thanks to the heroic efforts of Vespucci himself had the expedition managed to reach its destination.

> If my companions had not relied on me and my knowledge of cosmography, there would have been no pilot or captain on the voyage to know within five hundred leagues where we were. Indeed, we were wandering

with uncertainty, with only the instruments to show us accurate altitudes of the heavenly bodies, those instruments being the quadrant and the astrolabe, as everyone knows. After this, everyone held me in great honor. For I truly showed them that, without any knowledge of sea charts, I was still more expert in the science of navigation than all the pilots in the world.

The expedition hadn't just briefly visited this new world. It had also followed its coast south for thousands of miles: "Part of that new continent," the letter continued, "lies in the torrid zone beyond the equator. . . . We sailed along the shore until we passed the Tropic of Capricorn and found the Antarctic Pole, 50 degrees above their horizon, and we were 17½ degrees from the Antarctic Circle itself." This—the news that a vast unknown land lay not just to the west but also so very far to the *south*—was the letter's most astonishing revelation.

Then there were the people. "Everyone of both sexes goes about naked," Vespucci wrote, "covering no part of the body, and just as they issued from their mothers' wombs so they go about until their dying day." This nakedness led to appalling sexual mores: "Their women, being very lustful, make their husbands' members swell to such thickness that they look ugly and misshapen; this they accomplish with a certain device they have, and by bites from certain poisonous animals." He had other, even more shocking details to report. "Human flesh is common fare among them," Vespucci noted, adding, "I myself met and spoke with a man who was said to have eaten more than three hundred human bodies; and I also . . . saw human flesh hanging from house-beams, much as we hang up bacon and pork." Such horrors notwithstanding, the inhabitants of this new world seemed to live in a beguiling state of prelapsarian innocence. "They have no . . . private property," Vespucci wrote, "but own everything in common: they live together without a king and without authorities, each man his own master. They take as many wives as they wish, and son may couple with mother, brother with sister, and in general men with women as they chance to meet." Disease was rare and easily cured with medicinal roots and herbs that grew wild; people lived to be 150 years old. Even the land itself had distinctly idyllic qualities: its climate was temperate, its waters abundant, its trees fragrant, its fruits delicious, and its animals colorful and wild. "If anywhere in the world there exists an Earthly Paradise," Vespucci concluded, speaking literarily rather than literally, "I think it is not far from those regions."

Vespucci acknowledged at the end of his letter that there was much more to say. He had written a fuller treatment of his voyage already, he claimed, but at present it was in the hands of the king of Portugal. He had also made two earlier visits to this new world in the service of Spain, he reminded Lorenzo, and on all three of his voyages he had kept a diary—"so that," he wrote, "if ever I am granted the leisure, I may gather together all these marvels one by one and write a book, either of geography or of cosmography, so that my memory will live on for posterity." He planned soon to embark on a fourth voyage, he added, "to search for new regions to the south."

Perilous ocean wanderings, tempests at sea, an enlightened and valiant hero, revolutionary discoveries, nakedness, bestiality, sex, incest, cannibalism, a paradisiacal new world—the Vespucci letter had it all. Readers clamored for copies, and publishers kept churning them out. By 1506 twenty-three different editions were in print. The letter was a bestseller. But it was also almost certainly a fake.

*   *   *

HERE BEGINS WHAT historians despairingly describe as "the Vespucci problem": the enormously vexing possibility that Vespucci didn't write the letters published under his name.

Vespucci did make voyages of discovery, and he did send letters home describing them. That much is clear. Three complete letters undeniably written by Vespucci himself survive today, dating from 1500, 1501, and 1502. Known to scholars as the "familiar letters," they tell stories that overlap in many ways with those told in the printed letters, but they also differ from them in highly problematic ways. The familiar letters describe not four but only two voyages: the first for Spain and the second for Portugal. They deliver their news in a much less sensational, self-aggrandizing, and prurient style. And they do not announce the discovery of a new world but only a previously unknown part of Asia.

Ever since the European exploration of the Atlantic had gotten under way in earnest, in the middle of the fourteenth century, Italian merchants living on the Iberian Peninsula had been sending letters home describing new discoveries, and their letters had been copied, adapted, and passed around in both mercantile and literary circles. This is what happened when Florentine merchants in Seville sent home their letter describing the 1341 expedition to the Canary Islands. Their account soon fell into the hands of Boccaccio, who recorded his own humanist interpretation of it in his journal, and his version of it quickly took on a life of its own as *Of Canaria and Other Islands*—a short work that helped Europeans think in dramatically new ways about geography. The same thing seems to have happened with Vespucci's letters. Vespucci sent the letters home knowing that they would be passed around, read, and discussed by many of Florence's merchants and humanists—and indeed they were. The reason that his three familiar letters survive, along with a fragmentary fourth letter of uncertain date, is that several Florentines who came across them copied them into their notebooks. Inevitably, printers soon got wind of the letters and sensed an opportunity: they could add some spice to Vespucci's letters, publish them quickly and cheaply, and make a tidy profit. Even five hundred years ago, publishers recognized that sex and horror sell.

Because Vespucci's published letters are based on what Vespucci actually wrote, they're not outright forgeries. They contain valuable details about his voyages of discovery, and in some cases they furnish important and apparently reliable firsthand information that can be found nowhere else. But because they have so clearly been doctored they have to be treated with great caution. The first avenue of approach for anybody trying to make sense of Vespucci and his voyages of discovery, therefore, has to be his familiar, not his published, letters.

* * *

VESPUCCI WAS BORN in Florence in the early 1450s, and from an early age he inhabited two different but overlapping worlds.

One was the prosaic and specialized world of commerce. In his youth Vespucci was steered toward a career in business by his father, a notary who belonged to an undistinguished branch of a well-connected family. Most young merchants-in-training in Florence had little time or need for the liberal arts. They went to school above all to study the one discipline that would allow them to make nimble calculations of profit and loss: mathematics. Teachers taught them practical solutions to the problems they were likely to encounter in their working lives—problems like the following, drawn from a secondary-school textbook of the time. At issue is the less-than-scintillating question of how to determine the volume of individual barrels, which in the fifteenth century didn't come in standard sizes.

There is a barrel, each of its ends being 2 bracci in diameter; the diameter at its bung is $2\frac{1}{4}$ bracci, and halfway between bung and end it is $2\frac{2}{9}$ bracci. The barrel is 2 bracci long. What is its cubic measure?

This is like a pair of truncated cones. Square the diameter at the ends: $2 \times 2 = 4$. Then square the median diameter: $2\frac{2}{9} \times 2\frac{2}{9} = 4\frac{76}{81}$ [sic]. Add them together: $8\frac{76}{81}$. Multiply $2 \times 2\frac{2}{9} = 4\frac{4}{9}$. Add this to $8\frac{76}{81} = 13\,2\frac{31}{81}$. Divide by $3 = 4\frac{112}{243}$.... Now square $2\frac{1}{4} = 2\frac{1}{4} \times 2\frac{1}{4} = 5\frac{1}{16}$. Add it to the square of the median diameter: $5\frac{5}{16} + 4\frac{76}{81} = 10\frac{1}{129}$. Multiply $2\frac{2}{9} \times 2\frac{1}{4} = 5$. Add this to the previous sum: $15\frac{1}{129}$. Divide by $3: 5\frac{1}{3888}$. Add it to the first result: $4\frac{112}{243} + 5\frac{1}{3888} = 9\frac{1792}{3888}$. Multiply this by 11 and then divide by 14: the final result is $7\frac{23600}{54432}$. This is the cubic measure of the barrel.

This was one of Vespucci's worlds. But he also inhabited the more rarefied world of Florence's intellectual and artistic elite.

Florence was in the full bloom of its cultural and intellectual flowering when Vespucci was born. Cosimo de' Medici informally ran the city; Poggio Bracciolini was its chancellor; Leon Battista Alberti was designing some of its most notable public monuments; Paolo Toscanelli was in his prime. But a new generation of notable Florentines, all rough contemporaries of Vespucci's, was emerging. Among them were Lorenzo the Magnificent, a grandson of Cosimo who himself would soon run the city; Niccolò Machiavelli, who would become the chancellor of Florence in 1498 and would employ Vespucci's oldest brother; Sandro Botticelli, who lived next door to Vespucci and would paint members of his family; and Leonardo da Vinci, who knew Vespucci's family and is reported to have sketched a now-lost portrait of Vespucci himself.

Through his father's brother, Giorgio Antonio, one of the city's most respected scholars, Vespucci had privileged access to this world of thinkers, artists, and ideas. Giorgio Antonio was a member of the Dominican convent of San Marco—a center of humanist studies in Florence that received much of its funding from Cosimo de' Medici—and he gave lessons there to the children of elite Florentine families. When Amerigo was old enough to begin serious studies, Giorgio Antonio took him in. He taught him Latin, made him read the works of ancient philosophers and modern poets, and, needless to say, introduced him to the humanist study of geography.

Geography was a Florentine obsession during the years that Amerigo studied with his uncle. The Council of Florence was still a recent memory, the cult of Ptolemy was alive and well, editions of Strabo and Marco Polo were proliferating, and Poggio and Toscanelli were tracking the progress of the Portuguese in Africa. The humanists were working harder than ever to reconcile ancient and modern descriptions of the world, and they had more material than ever to work with. The idea of the globe as a fully navigable, inhabitable, and knowable place was coming into being.

Despite the efforts of his uncle, Amerigo seems to have been a mediocre student at best. Both he and Giorgio Antonio knew that he was destined for a career in business and would never become a scholar. Given the other demands on his time and what he knew about his nephew's abilities, Giorgio Antonio probably didn't even devote a whole lot of time or energy to teaching Amerigo. But he does seem to have made a special effort, as a good humanist, to convince Amerigo that there was more to being a merchant than just buying and selling. This seems to have been his aim when he made Amerigo translate into Latin the following passage, which survives in the young Amerigo's student notebook.

Going back and forth to many distant lands, where by talking and trading one can learn many things, not a few merchants have become wise and learned, something that cannot be explained in a few words. Moving about and making inquiries concerning the world, whose limits we have not yet completely ascertained, they can furnish valuable advice by word and association to those who come to them in search of counsel, or clarification of some doubt concerning matters of business and custom.

* * *

VESPUCCI BEGAN HIS working life in earnest in the 1480s, when he entered
the service of Lorenzo di Pierfrancesco de' Medici. This Lorenzo is not to
be confused with Lorenzo the Magnificent. Lorenzo di Pierfrancesco—
Amerigo's Lorenzo—was one of Lorenzo the Magnificent's younger cous-
ins, and the two belonged to rival branches of the de' Medici family.

Vespucci would maintain a working relationship with Lorenzo for
almost two decades. He began by handling minor domestic matters—stock-
ing household supplies, ordering casks of wine, buying the occasional precious stone—but gradually he began to work as a freelance business agent
of sorts. He made trips for Lorenzo to other towns in Italy, looking to turn a
profit on anything he could: fish, pigeons, cherries, salt, mustard, mulberry
seed, bed curtains, carpets, silverware. Evidently he acquitted himself well,
because in 1489 Lorenzo sent him to Spain.

The trip would be a turning point in Vespucci's life. Lorenzo asked
Amerigo to look into the possibility of a business partnership with a certain
Giannotto Berardi, a Florentine merchant and slave trader who had long
lived in Spain. "Inform yourself there of his character," Lorenzo wrote to
Vespucci on the eve of his trip, "whether he is a good man in whose hands
our firm would be secure." No other record of this trip exists, but the two
men must have hit it off; by the end of 1492, Vespucci had left Florence for
good, settled in Seville, and joined Berardi's firm as a junior partner.

His timing couldn't have been better. The Spanish had just defeated
the Moors, and the Sovereigns were now busy ruthlessly expelling the
Jews from Spain—a move that had quite suddenly opened up new business opportunities for Italian merchants there. And Berardi himself had
just made a risky investment in a potentially lucrative new venture: the first
voyage of Columbus.

At first the investment seemed to have paid off. After Columbus had
returned in triumph in 1493, Vespucci and Berardi helped him outfit his
second voyage, arranging for the purchase of bulk stores of sea biscuit,
wine, cheese, oil, and vinegar—business that no doubt required the measurement of many barrels. But Columbus, of course, never brought home
the riches he kept promising, and backing him ruined Berardi, who died in
1495. In his will Berardi made a point of listing Columbus's debts to him,

and gave the job of recovering them to his executors, one of whom he listed as "Amerigo Vespucci, my agent."

This put Vespucci and Columbus in a close orbit. Perhaps they already knew each other, or perhaps Berardi's will brought them together for the first time. One way or another, despite the awkward circumstances, they seem to have become friends. Instead of forcing Columbus to repay his debts, Vespucci fell under his spell and began to dream of sailing west himself.

He finally got his chance in 1499. By this time the Sovereigns were seriously disillusioned with Columbus and had decided to end his monopoly on western voyages of discovery. Official licenses for new voyages, they decreed, would now be available to other sailors. One of the first men to apply was Alonso de Hojeda—the Spanish captain who had sailed with Columbus on his second voyage, and whom the Sovereigns would later commend for stopping the discoveries of the English in the Indies.

Hojeda presented the Sovereigns with an appealing plan. Columbus had very recently discovered the Gulf of Paria, he knew, and had reported finding both gold and pearls there. Not only that, Columbus had sent the Sovereigns a chart of his discoveries. Hojeda proposed to outfit a small fleet of exploring vessels; to sail back to the Gulf of Paria, using Columbus's map; and to bring back a bounty of gold and pearls—all while Columbus himself continued to languish in administrative hell on Hispaniola.

The Sovereigns liked the idea. Hojeda received the very first license that the Sovereigns would grant after ending Columbus's monopoly on the exploration of the Indies, and in May of 1499 he set sail with four vessels under his command. Accompanying him on the journey, he would later testify, were Juan de La Cosa, the pilot and chart maker who had also sailed with Columbus, and a certain "Morigo Vespucci."

* * *

A LITTLE MORE than a year later Vespucci was back in Seville. On July 18 he sent a letter to Lorenzo di Pierfrancesco de' Medici in Florence describing his travels—the first of his familiar letters. "The present letter," he began, "is to give you the news that I returned about one month ago from the Indian regions. . . . I believe Your Magnificence will be pleased to hear of all that took place on the voyage, and of the most marvelous things I encountered upon it."

Vespucci started by describing the outgoing part of his voyage. "I departed with two caravels on 18 May 1499," he wrote, "to go off and make discoveries in the western regions by way of the Ocean Sea; and I set my course along the African coast, navigating by way of the Fortunate Islands, now called the Canary Islands. And then, when I had secured all necessary provisions . . . we set sail from an island known as Gomera, and turning our prows to the southwest, sailed for twenty-four days."

I departed, I set my course, I secured provisions: Vespucci went out of his way at the outset of his letter to imply that he had played an important role on the voyage. But it's not at all clear that he did. His name has yet to be found on any official document concerning the expedition, and many scholars argue that he tagged along as nothing more than an adventure tourist or commercial scout, eager to look into the truth of all those stories of gold and pearls. Vespucci never mentions Hojeda or any of his crewmates by name in this letter, which makes the skeptics even more skeptical; not only did he inflate his own sense of importance, they feel, he also shamelessly wrote his fellow explorers right out of the story.

Maybe. But there are other ways of explaining Vespucci's choices. He may have chosen not to mention his fellow sailors simply because he knew their names would have been of no interest to Lorenzo and other Florentines. Or perhaps he really did reinvent himself as something of a cosmographical expert after Berardi's death. This latter possibility is certainly what Vespucci himself wanted his readers to believe.

Vespucci and his fellow sailors made landfall on the twenty-fourth day of their voyage, he told Lorenzo, at a point several degrees below the equator. In front of them was what appeared to be a mainland that ran northwest to southeast. An important decision now confronted them. Should they turn right and head for the Gulf of Paria, or turn left and see what new discoveries they might make?

The answer they came up with, it seems, was both. Hojeda, under contract to find gold and pearls as quickly as possible, took two ships and headed north, while Vespucci—motivated, if he can be believed, by loftier concerns—took command of the other two ships and began sailing south.

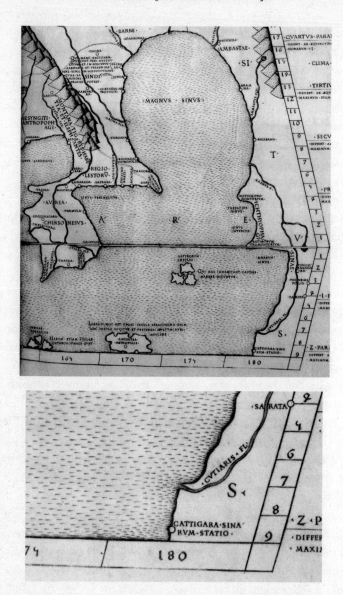

Figures 64 and 65. Top: southeast Asia as Vespucci imagined finding it. From the 1478 Rome edition of Ptolemy's *Geography* (For whole map, see Figure 54.). Cattigara is at the bottom right; the Sinus Magnus is the large gulf at the top center; and Aurea Chersonese is the peninsula on the left. *Bottom:* Cattigara, at the southeastern limits of Ptolemy's Asia.

"It was my intention," he told Lorenzo, "to see whether I could round a cape of land that Ptolemy calls the Cape of Cattigara, which is near the Sinus Magnus. For in my opinion it was not far from there, to judge by the degrees of latitude and longitude."

Anybody who knew the *Geography* would have known what Vespucci was talking about (*Figures 64 and 65*). Cattigara was a legendary eastern trading center dating back to antiquity. Ptolemy had placed it almost at the very end of his East—at 8 ½ degrees south and 177 degrees west, not far from the Aurea Chersonese. But Vespucci couldn't have had the traditional Ptolemaic picture of Asia in mind when he made this remark, because Ptolemy had described the Indian Ocean as closed. This made Cattigara and the Sinus Magnus inaccessible to sailors from Europe. Since Vespucci thought he *could* reach them, he clearly had something else in mind—something a lot like the world as Henricus Martellus and others had begun to map it in the late 1480s. Martellus located Cattigara, the Sinus Magnus, and the Aurea Chersonese exactly where Ptolemy had put them, but his Indian Ocean was open, bounded to the west by a circumnavigable Africa—and the imaginary Dragon's Tail (*Plate 9*).

Vespucci's remark should now make more sense. Basing his judgment on an estimate of how far he had sailed in twenty-four days, and on a vision of the Far East as Martellus had mapped it, he had determined that he had sailed far enough west to have reached the Dragon's Tail, and far enough south to have almost reached its tip. Cattigara and the Sinus Magnus had to be nearby.

Vespucci set out to find them. For days he followed the coast steadily to the southeast, taking in a succession of remarkable sights: a riot of unfamiliar vegetation; sprawling coastal wetlands and the mouths of impossibly large rivers; dense, fragrant forests teeming with colorful birds and frightening reptiles; a night sky unlike anything visible from Europe; and obvious signs of human habitation. After an unspecified amount of time, he and his fellow sailors explored the mouth of an especially large river (a foray that made them the first Europeans to visit the Amazon), and not long afterward they encountered a fierce offshore current that prevented them from continuing any farther south. "Given the little progress we were making and the great danger of our situation," Vespucci wrote, "we decided to turn our prows northwest, and sail northward."

A powerful ocean current like the one that Vespucci described does indeed exist off the coast of modern-day Brazil: the Guiana stream, which flows not far offshore from a promontory known as Cape São Roque. The cape is located at about 5 or 6 degrees south of the equator, and it may well be the southernmost point reached by Vespucci on this voyage. But it's also possible that something more than just a powerful current made Vespucci turn around. Cape São Roque lies at the easternmost limit of South America, on a stretch of coast even closer to Africa than the one that Cabral would soon discover farther to the south. Even some pretty rough dead reckoning would have told Vespucci and his Spanish colleagues that in following the coast to the southwest they had crossed the line of demarcation in the vicinity of the cape—and had entered a region that belonged to Portugal.

*   *   *

VESPUCCI AND HIS men headed back north. They sailed across the equator and stopped briefly at Trinidad, where they went ashore and discovered a race of cannibals—although Vespucci's familiar account of them was far less sensational than the one that appeared in the *Mundus novus* letter. "Almost the majority of this race, if not all, live off human flesh," he wrote, "and of this fact Your Magnificence can be certain. They do not eat one another, but navigate in certain vessels of theirs, called *canoes*, and they go to neighboring islands or lands in search of prey from the races that are either their enemies or different from them. . . . These are a people of quite courteous disposition and fine stature; they go about completely naked."

Vespucci sailed on. First he explored the Gulf of Paria, where he and his men received a warm welcome from the locals, who fed them, plied them with exotic drinks, and gave them gifts of gold and pearls. They then made their way back out into open water, only to discover that just above the Gulf of Paria, the coast took a turn to the west. They had reached the top of South America.

This was thickly settled territory. "Sailing along the coast," Vespucci wrote, "each day we discovered an infinite number of people." On one island he noted the presence of "houses built with great skill upon the sea, as in

Venice"—the source of the name Venezuela, which derives from the Italian for "Little Venice." What particularly impressed him was that none of the peoples he encountered seemed to wear any clothes. "They all go about naked as they were born, without the least shame," he told Lorenzo incredulously, but he then once again exercised a kind of restraint not on display in the *Mundus novus* letter. "Yet to relate in full what little shame they have," he wrote, "would mean broaching improper matters; better to be silent about it." Many of the people the Europeans encountered made welcoming overtures, but others, Vespucci noted, "did not want our friendship." This led to a series of lopsided skirmishes in which, as Vespucci put it dryly, "We would put them to rout, and kill many, and pillage their houses."

By now Vespucci and his fellow sailors had been at sea for months, and time was taking its toll: supplies were running low; the ships were damaged and leaky; and the crew, living on an allowance of only six ounces of sea biscuit a day, were beleaguered and restive. Vespucci decided it was time to abandon his explorations, and to head north across open water toward the Spanish colony on Hispaniola.

But not before trying to determine exactly how far to the west he had sailed. Dead reckoning gave him a rough idea; otherwise he and his men wouldn't have known that Hispaniola lay to their north. But Vespucci wanted something more precise. He wanted to know his exact longitude, and if his letter to Lorenzo can be believed, he devised a way to figure it out.

His idea was this: If he could manage to catch sight of an unusual celestial event—a planetary or lunar conjunction, say—he could record the exact time it took place. By then consulting the reasonably reliable astronomical tables he had with him, he could determine how many hours earlier the event was supposed to have taken place in Europe. That difference, in turn, would be easy to convert into a distance measured in degrees, since the heavens revolved 15 degrees around the earth every hour.

Vespucci got a chance to put his theory to the test on August 23, 1499. That night, he told Lorenzo,

there was a conjunction of the moon with Mars, which, according to [one of his tables], was to occur at midnight, or half an hour before. I found that when the moon rose on our horizon, an hour and a half after the sun had set, the planet had passed into the east, which is to say that the moon was about a degree and some minutes to the east of Mars, and at midnight was five and a half degrees to the east, more or less. Thus, setting up the proportion "If twenty-four hours equal 360 degrees, what do 5 ½ hours equal?," I find the answer to be 82 ½ degrees; and such was my longitude from the meridian of Cadiz [in Spain]. For, giving 16 ⅔ leagues to each degree, I reckoned that we were 1,366 ⅔ leagues, or 5,466 ⅔ miles, west of Cadiz.

To hear Vespucci tell it, gauging the size of the earth was little different from gauging the size of a barrel. But scholars are deeply divided about this passage. Some argue that it reveals Vespucci as the first mariner to employ what is known as the lunar method of determining longitude; others argue that the whole passage is nothing but puffery designed to make Vespucci seem knowledgeable and important. Whatever the truth, the estimate itself was far from perfect. Vespucci had actually traveled only about 60 degrees west of Spain.

Vespucci offered only a few details about the rest of his voyage. He and his men reached Hispaniola in seven days. They stayed for two months, repairing and resupplying their ships. They made a brief foray north into the archipelago of islands discovered by Columbus on his first two voyages, where they captured 232 slaves. And then they returned to Spain.

At the end of his letter Vespucci did what presumably he had been doing in his business correspondence with Lorenzo for years: he gave his patron an accounting. "We were thirteen months on this voyage," he wrote, "encountering great dangers and discovering endless Asian land." He had sailed 6 ½ degrees south of the equator, and 84 degrees west of Cadiz; he had seen a dizzying number of strange new plants, animals, people, and places; and he had brought home a small sampling of gold, pearls, and gems, as well as hundreds of slaves. Most significantly, he claimed to have reached the Dragon's Tail—"continental land," he called it, "which I esteem to be bounded by the eastern part of [Ptolemy's] Asia." And already he was making plans for a return voyage, during which he intended to find a route around this land, and sail past Cattigara and the Aurea Chersonese all the way to Taprobane. Like Columbus, Vespucci dreamed of making the full circuit of the earth, and at the very end of his letter he promised to provide Lorenzo with the cartographical evidence of how it could be done.

> I have resolved, Magnificent Lorenzo, that, just as I have given you an account by letter of what happened to me, I shall send you two depictions of the world, made and ordered by my own hand and knowledge: one chart will be a flat rendering and the other a map of the world in spherical form. . . . I believe they will be to your liking, especially the globe; for, not long ago, I made one for their Highnesses the Sovereigns, and they prize it highly. . . . There is no lack in your city of those who know how the world is depicted, and some may wish to emend something in what I have done; nonetheless let him who would emend me wait for my coming, as it may be that I shall defend myself.

*   *   *

VESPUCCI DID INDEED soon make a second voyage. But this time he sailed with the Portuguese.

Vespucci's second familiar letter, dated June 4, 1501, informed Lorenzo of the switch. At the time he wrote, Vespucci was in Cape Verde, on Africa's west coast, preparing to depart for points west. "You will have heard, Lorenzo," he wrote, "either through my letter, or through that of our fellow Florentines in Lisbon, that while staying in Seville I was summoned by the King of Portugal, and that he asked me to prepare to enter his service on this voyage."

Vespucci didn't explain why he had decided to sail for Portugal. But it's not hard to imagine why the Portuguese would have sought out Vespucci. He had returned to Spain the summer before, after all, claiming to have discovered a new Asian land that lay at least partly in Portuguese waters. This was news that would quickly have reached his Florentine friends and colleagues in Lisbon, some of whom were intimately involved in the financing and outfitting of Portuguese voyages of discovery. Similar news had reached Lisbon through different channels: at about the same time that Vespucci had returned from his voyage, the ship sent home by Cabral had returned to report his discovery of the Land of the True Cross. King Manuel I had been greatly pleased to learn of this discovery—and so, too, had one of the richest and most powerful figures in Lisbon, the Florentine merchant and slave trader Bartolomeo Marchioni, one of the backers of Cabral's voyage. Marchioni wrote home in 1501 and put a special spin on the discovery for his Florentine audience. "The Portuguese King," he declared, "has discovered a new world."

Eager to learn more about what exactly Cabral had discovered, King Manuel decided to dispatch a reconnaissance mission across the Atlantic, and it was while the planning for this mission was under way that Vespucci got involved. How exactly this happened isn't clear. Perhaps Manuel heard about Vespucci and decided to reach out to him in Spain. But things may also have worked the other way around: Vespucci may have heard about the expedition from his Florentine friends in Lisbon, recognized that it offered him a way to follow the Dragon's Tail into Portuguese waters, and therefore convinced Manuel to invite him along.

However it all transpired, in June of 1501 Vespucci was in Cape Verde. As luck would have it, while he was there he came across two ships from Cabral's fleet, on their way home after having successfully reached India, and this chance encounter prompted Vespucci to write his second familiar letter to Lorenzo. Cabral's ships, he wrote, had reached Calicut, had explored other parts of India, had learned a great deal about the Far East, and had returned with a great stores of spices and gems—"infinite cinnamon," Vespucci reported to Lorenzo, "fresh and dried ginger, much pepper and cloves, nutmeg, mace, musk, civet, storax, benzoin, purslane, mastic, incense, myrrh, red and white sandalwood, lignum aloe, camphor, ambergris, much gum-lac, mummy, indigo, cadmium, opium, hepatic aloe, cassia . . . diamonds, rubies, and pearls."

Such riches! On the eve of his departure from Cape Verde, Vespucci began to dream about bringing home a similar bounty of his own, and about finding a western route to India that was shorter than the long and arduous eastern one then being used by the Portuguese. "I hope in this voyage of mine," he wrote, after describing the many places Cabral's men had seen, "to revisit and traverse much of the aforementioned area, and to discover much more."

But he was getting ahead of himself. He still had the Atlantic to cross and the Dragon's Tail to round. Success was by no means guaranteed. "This voyage I embark upon now is perilous to the security of our mortal existence," he told Lorenzo before signing off. "Nonetheless I make it with a spirit resolved to serve God and the world."

*   *   *

VESPUCCI WOULD BE gone for about a year. He wrote to Lorenzo again not long after he had returned to Lisbon, early in the summer of 1502, and picked up right where he had left off. "When last I wrote to Your Magnificence," he began, "it was from the coast of Guinea, from a place called Cape Verde, by which you were informed of the start of my voyage; and by the present letter you will be told in brief of the middle and end of it." He then launched into a description of the voyage that the *Mundus novus* letter would also describe.

In this letter—the third familiar letter—Vespucci made no claims to have been in charge. He reported straightforwardly to Lorenzo that he had sailed with the Portuguese for sixty-four days to the southwest until he had reached what he called "a new land." Not a new world, as the *Mundus novus* letter would have it, but just a new part of the endless Asian land he had already visited. He did have some stunning news to report, however. "We passed by that land for about eight hundred leagues," he told Lorenzo, "always on a southwest, ¼ course, and we found it full of inhabitants. . . . We traveled so far upon these seas that we entered the torrid zone and passed south of the equator and the Tropic of Capricorn, so that the South Pole stood fifty degrees above my horizon. . . . In conclusion, I was in the region of the Antipodes, on a voyage which covered a quarter of the world."

A giant new inhabited land that extended at least 50 degrees south of the equator on the opposite side of the world? The discoveries of Columbus and da Gama paled by comparison. They had described reaching parts of the world that Europeans had known about for centuries, whereas Vespucci was now describing a part of the world that appeared on *nobody's* map. Not even the Dragon's Tail was supposed to extend anywhere near as far south as Vespucci was claiming to have sailed. Skeptics had every reason to doubt the claim, but Vespucci wasn't the only member of his voyage making it. "The caravels that were sent out last year to make discoveries in the Land of the Parrots and Santa Cruz returned on July 22," the Venetian ambassador to Portugal wrote in a letter he sent home that summer. "The captain states that he has discovered more than 2,500 miles of new coast without having come to the end of it."

Vespucci ended his letter with an apology. Because he had sailed "in the name of discovery . . . and not to seek after profit," he told Lorenzo, he could not yet provide a proper reckoning of this new land's commercial potential. But the initial signs were promising. He had observed brazilwood growing in great quantities and he had seen lots of stones, spices, and medicinal herbs that he assumed were of value. But unlike Columbus, he made no promises that easy riches were just around the corner. "The men," he wrote, alluding to his fellow sailors, "tell of many miracles concerning the gold and other metals and drugs, but I am one of those like Saint [Doubting] Thomas: time will tell."

The letter was short. Vespucci explained to Lorenzo that he had written it only to provide a quick summary of what he had discovered on his voyage. But he also made a tantalizing remark about having already written a much fuller account—the source, presumably, of a similar remark that would appear in the *Mundus novus* letter. "The most remarkable of all the things I beheld on this voyage," he wrote, "I have gathered into a small work, so that . . . I may gain some fame after my death." He had been hoping to send Lorenzo a copy of the work, which he referred to as his *Voyage*, but the king of Portugal had taken it from him. "When he returns it to me," he promised, "I shall send it."

*   *   *

THAT FALL, EVIDENTLY feeling mistreated by King Manuel, Vespucci returned to Seville. "Amerigo Vespucci will be arriving here in a few days," a Florentine merchant living in Seville wrote on October 3, 1502. "He has undergone great hardships and has derived a small profit. He deserves a better fate. The lands he has discovered have been leased by the King of Portugal to a group of new Christians, who have pledged themselves to send out six ships every year and to explore three hundred leagues farther each year." (The role of these "new Christians"—that is, recently converted Jews who had fled from Spain to Portugal—is an aspect of the early colonization of the Americas that deserves more attention.)

Vespucci then disappears from the documentary record for more than two years. When he finally reemerges he is once again in Seville—in the company of Columbus. The evidence for this comes in the form of a letter that Columbus sent to his son Diego on February 5, 1505. Just back from his fourth voyage, Columbus had once again thrown himself back into his campaign to win back the rights he felt the Sovereigns had illegally stripped him of. He had just spent time with Vespucci, who was on his way to the Spanish court and had promised to intercede there in Columbus's behalf. "My Dear Son," Columbus began,

> Diego Mendez left here on Monday, the 3rd of this month. After his departure I spoke with Amerigo Vespucci, the bearer of this letter, who is going to the Court on matters relating to navigation. He always showed a desire to please me and is a very respectable man. Fortune has been adverse to him, as to many others. His labors have not been so profitable to him as he might have expected. He leaves me with the desire to do me service, if it should be in his power. I am unable here to point out in what way he could be useful to me, because I do not know what may be required at Court; but he goes with the determination of doing all he can for me.

Columbus died a little more than a year later. Vespucci seems not to have managed to do anything for him on his visit to the court—if indeed he ever carried through on his promise to help. But Vespucci himself seems to have benefited from his visit to the court; not long afterward he became a naturalized citizen of Castile and began working for the new Casa de Contratación, or House of Trade—the government agency established just two years earlier by Queen Isabella to oversee trade with Spain's new overseas possessions.

One of the House of Trade's most important functions was to consider geographical and navigational questions. What were the safest and quickest routes across the Western Ocean? What newly discovered territories held the most commercial promise? How did the bits and pieces of geographical information being brought back to Spain by different expeditions fit together? Who had already been where, and when? What was the full extent of the territories discovered to date, not just by the Spanish but also the Portuguese and the English? Where exactly should the line of demarcation be drawn? In what direction should new voyages of discovery be sent? It was surely these matters that drew Vespucci to the Spanish court.

Juan de La Cosa, the pilot and mapmaker who had sailed with both Vespucci and Columbus, had actually already signed on to work for the House of Trade—as a spy. According to an entry in the house's accounting ledger, La Cosa had traveled to Portugal in the summer of 1503 "to find out or learn secretly about voyages the Portuguese made to the Indies." It was a perilous mission. The penalty for anybody divulging Portuguese geographical secrets was severe, as one Venetian observed in 1501 when writing home about Cabral's voyage to the Land of the True Cross and India. "It is impossible to procure the map of that voyage," he complained, "because the king has placed a death penalty on anyone who gives it out."

Despite the dangers, La Cosa seems to have succeeded. Later that summer, the House of Trade paid him handsomely for what he had managed to bring back to Spain—"two mariner's charts," according the house's accounting ledger, "which he gave to Our Majesty the Queen."

The Spanish presumably thought that Vespucci could provide them with even more information. He was an established merchant who had lived in Seville for years and who knew the business of western discovery well. In the service of Spain he had worked for Columbus, he had sailed with Hojeda, and he had visited most of Spain's new territories himself. He had also sailed with the Portuguese, and, indeed, was one of the very explorers who had brought home the new information that the Portuguese were now trying so hard to keep secret. He had seen and mapped the new regions that La Cosa had risked his life to find out about, including that new part of Asia that extended thousands of miles into the southern hemisphere. Not only that, he had described this coast as always heading southwest, and this raised an intriguing possibility—perhaps while sailing for the Portuguese he had crossed *back* across the line of demarcation into regions once again belonging to Spain. Vespucci also clearly knew something of cosmography: he had studied with his uncle, one of the most respected humanists in Florence; he knew how Ptolemy had mapped the world and how modern geographers had updated his map; he knew how to make maps and globes himself, and had actually sent some of his own design to the Sovereigns already; he had experience using the astrolabe and quadrant; and he even claimed to know the longitude of the lands he had explored.

Vespucci, then, had much to offer the Spanish in 1505, whether or not he was everything he made himself out to be in his letters. Few people had as good an idea about the totality of what was emerging on the other side of the Western Ocean. A strange new part of Asia had been discovered, it seemed, and now the challenge—not just for the Spanish and the Portuguese but for anybody interested in geography—was to figure out how to put it on the map.

Plate 1 ◦ A medieval Christian worldview: the Psalter Map (1265). East is at the top.
Christ, holding a T-O globe, surveys all of space and time. Just below him, underneath
the rising sun, are Adam and Eve in the Earthly Paradise, located in the most distant
part of the East, where both geography and history begin. At the center of the map
is Jerusalem; on the far right, under Africa, are the monstrous races; and at
the bottom is the western edge of the world, where the sun sets
and where geography and history come to an end.

Plate 2 ◦ Christ crucified on a T-O map, symbolically bringing together a world divided into three parts: Asia, Europe, and Africa (thirteenth century).

Plate 3 ◦ Saint Brendan and companions (1460). A real-life sixth-century Irish monk, Brendan reportedly sailed west from Ireland with his companions in a tiny boat and discovered the Promised Land of the Saints, a paradisiacal island of apparently limitless extent.

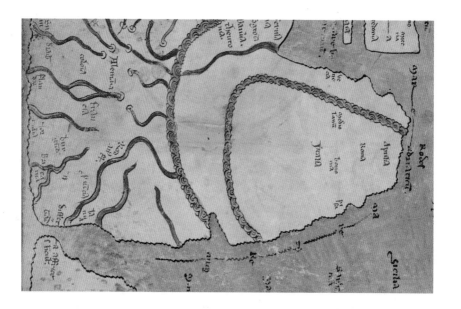

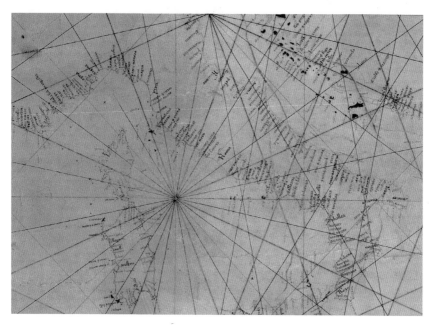

Plates 4 and 5 ◦ Italy before and after the advent of the marine chart.
Top: Italy on a world map by Matthew Paris (circa 1255)—rotated 90 degrees
clockwise to locate north at the top. *Bottom*: Italy on an anonymous
fourteenth-century marine chart.

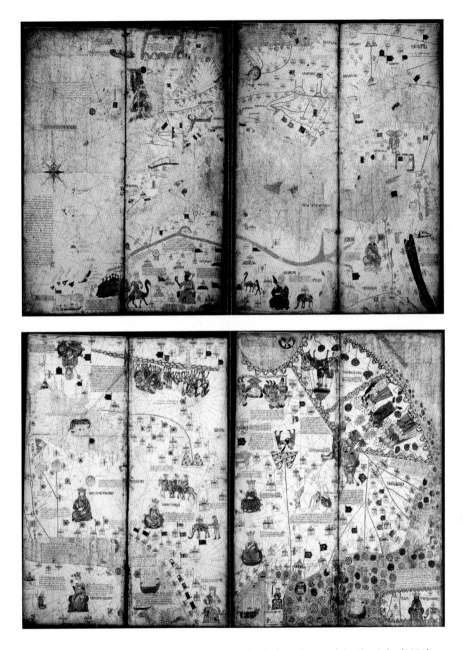

Plate 6 ◦ The most famous chart of the Catalan school: the eight-panel Catalan Atlas (1375). *Top*: the chart's western panels, showing Europe, the Mediterranean, and northern Africa, home to great kings and mines of gold. *Bottom*: the eastern panels, showing the Far East, a region of marvels first described for Europeans by Marco Polo less than a century before. The spice-, gold-, and gem-rich islands at the far right—the Indies—would become a European obsession in the fifteenth century.

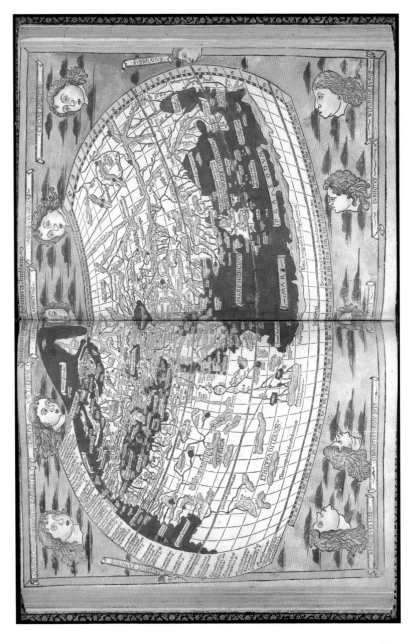

Plate 7 ◦ The second-century world of Claudius Ptolemy, from the Ulm edition of Ptolemy's *Geography* (1482). Lost for centuries but rediscovered in Europe in the early 1400s, the *Geography* explained how to draw maps using latitude and longitude, and contained the names and coordinates of some eight thousand places. The work greatly advanced classical scholars' understanding of ancient geography and helped inspire the European Age of Discovery.

Plate 8 ◦ The three-part world, by Simon Marmion (1460). The known world emerges from one side of the sphere of water, making the existence of land on the back side of that sphere theoretically impossible. Asia is at the top; Europe (*left*) and Africa (*right*) are underneath.

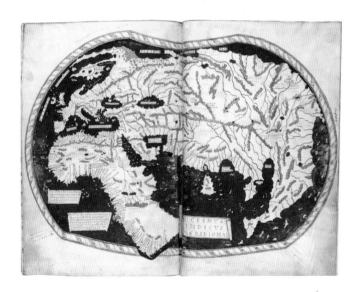

Plate 9 ◦ The world of Henricus Martellus (circa 1489–90). Based on Ptolemy's world map but expanded to include recent Portuguese discoveries in Africa and Marco Polo's Far East, the map shows the known world on the eve of Columbus's first voyage.

Plate 10 ∘ A detail from one of the earliest surviving maps to show any part of the New World: the Cantino chart (1502). The wooded island at the top is probably Labrador or Newfoundland; the coast at the top left margin is North America; the large islands in the middle are Cuba and Hispaniola; and below them, looming large, is South America, known at the time as the Land of the True Cross or Land of the Parrots. The coastlines of Europe and West Africa are at the far right.

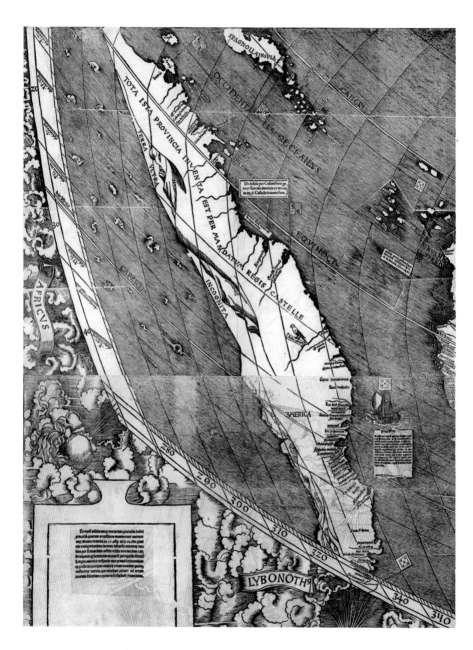

Plate 11 ∘ The fourth part of the world: South America on the Waldseemüller map of 1507, the first map to identify any part of the New World as America—a name coined by the map's makers. The sight of this giant new southern continent on the map, surrounded by water and so far-removed from Europe, Africa, and Asia, may have inspired Nicholas Copernicus to rethink the nature of the cosmos.

PART THREE
THE WHOLE WORLD

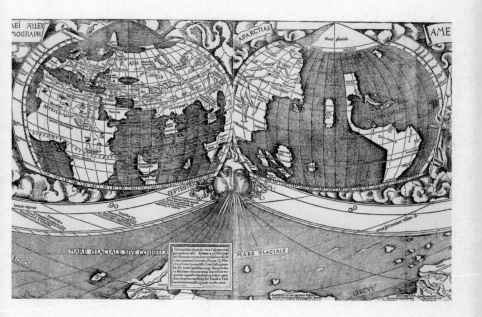

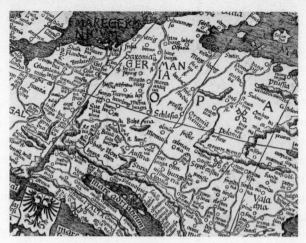

• Germany and the Holy Roman Empire •

CHAPTER SEVENTEEN

GYMNASIUM

*[Concerning] the unknown world recently discovered by the King of
Portugal . . . one will be able to see a more detailed and exact representation of this coast in the Ptolemy (which by the aid of God we shall
soon publish at our expense), revised and greatly augmented by us
and by Martin [Waldseemüller], a most skillful man in such matters. . . . Some verses on this same subject by our Vosgian [Matthias]
Ringmann, printed in the book of Vespucius, are circulating around the
country in the libraries.*

—Walter Lud, *Description of the Mirror of the World* (1507)

GERMANY'S LONGEST RIVER, the Rhine, arises high up in the Swiss
Alps, fed by melting snows and glacial runoff. After leaving the mountains
behind, the river first flows due north, defining the border between Switzerland and Austria; then turns west, at Lake Constance, defining the border between Switzerland and Germany; and then turns north again, at the
Swiss town of Basel, and enters the southern end of the great Rhine Valley.

There it begins to divide France from Germany and travels some seventy-five miles to Strasbourg, the capital of Alsace.

Alsace today is a part of France, but at the beginning of the sixteenth century it was German territory. Strasbourg at the time wasn't even Strasbourg. It was Strassburg—a bustling town of some twenty thousand, located at a geographical crossroads. Up the Rhine to the north were other important German provinces and the Low Countries; to the east were the important German cities of Augsburg, Ulm, and Nuremberg, and beyond them the territories of Austria, Bohemia, and Poland; to the south were Switzerland and Italy; and to the west were France, Spain, and Portugal. The setting literally defined Strassburg: the name derives from the German words for *street* and *town*, and originally connoted something like "town of (intersecting) streets."

Things came together in Strassburg. It was a regional transportation hub, a manufacturing center, and an agricultural entrepôt; its docks and warehouses bustled with the activities of merchants whose business it was to buy and transport goods throughout central Europe. The city was prosperous, but for much of the Middle Ages it was something of a backwater: steady in its traditional ways, focused on basic crafts and trades, conservative in its religion, and little interested in the wider world. But in the final decades of the fifteenth century the town began to come to life in an entirely new way. The book trade arrived, and with it came some of Germany's first humanists.

* * *

HUMANISM EVOLVED IN a very specific political climate in Germany, just as it had done a century before in Italy. When the sixteenth century began, many of Europe's princes considered themselves a part of the Holy Roman Empire: a loose alliance of semi-independent central European fiefdoms, originally brought together by Charlemagne, who in 800 had been granted the title Emperor Augustus by Pope Leo III. The move was political. The pope had hoped that Charlemagne, who had already united many of central Europe's Frankish tribes, might help him in the bigger project of reviving and uniting the defunct western half of the Roman Empire. But medieval Europe proved far too fractious and feudal ever to be united as Leo had hoped. Nevertheless, the move did give official sanction to the idea that central Europe should have an overall ruler whose authority derived from the Church. As a result, by the early 1500s the Holy Roman Empire held at least nominal sway over much of the region and had taken on a noticeably Germanic character—a shift officially acknowledged in 1512, when its name was formally changed to the Holy Roman Empire of the German Nation.

German humanism arose during this growing mood of nationalism and imperial ambition. The movement's leaders argued that in ancient times, the tribes of Germany had come together to form a single nation so powerful and great that it had managed to resist the advances of the Roman Empire. But as the Middle Ages had progressed this great German nation—Germania, the Romans had called it—had broken down into warring factions and had eaten away at itself from within. By the end of the fifteenth century only scattered vestiges and dim memories of a unified Germania remained.

The German humanists considered this a disgrace. They wanted to reunite the divided, downtrodden Germanic peoples of Europe and create a modern successor to ancient Greece and Rome: a new Germania that could become heir to the empire that the Italians, they felt, were proving themselves unfit to rule. To succeed in this effort Germans would have to reacquaint themselves with their shared political and cultural history—and the only way to do that, the humanists argued, was to absorb everything that the ancients had recorded about the history and geography of Germania.

One of the first humanists to popularize this message in Germany was a scholar, professor, and manuscript hunter named Conrad Celtis. In the late 1480s Celtis traveled to Italy, visited a number of important cities (among them Florence, where he moved in the same circles as Giorgio Antonio Vespucci), and returned home appalled. A vast cultural gulf separated the Italians and the Germans, Celtis realized—and true to their imperial Roman heritage, most Italians considered the Germans too barbaric ever to bridge that gulf.

Celtis decided to prove them wrong. In the years that followed he embarked on an almost evangelical mission to plant the seeds of humanism throughout the German-speaking lands. He preached this gospel everywhere he went—most famously, in a rousing lecture he delivered to students and faculty at the University of Ingolstadt in 1492. "Emulate, noble men, the ancient nobility of Rome," he told his audience,

> which, after taking over the empire of the Greeks, assimilated all their wisdom and eloquence. . . . In the same way, you who have taken over the empire of the Italians should cast off repulsive barbarism and seek to acquire Roman culture. Do away with that old disrepute of the Germans in Greek, Latin, and Hebrew writers, who ascribe to us drunkenness, cruelty, savagery, and every other vice bordering on bestiality and excess. Consider it a great disgrace to be ignorant of the histories of the Greeks and Latins, and the height of shame to know nothing about the topography, the climate, the rivers, the mountains, the antiquities, and peoples of our region and our own country: in short, all those facts which foreigners have so cleverly collected concerning us. . . . Assume, O men of Germany, that ancient spirit of yours, with which you so often confounded and terrified the Romans, and turn your eyes to the frontiers of Germany, and collect together her torn and broken territories.

In the years that followed, many Germans would begin studying the history and geography of Germania. Among them was a dreamy young Alsatian poet named Matthias Ringmann, who early in 1505 settled in Strassburg, founded a little school, took a job at a printing press—and threw himself into the study of Ptolemy's *Geography*.

* * *

RINGMANN WAS NO stranger to Strassburg. Born in 1482, he had grown up only twenty miles to the southwest, in Eichhoffen, a tranquil farming village in the foothills of the Vosges. He came from peasant stock, but something drew him into the world of letters, and by 1498 he had left home and enrolled at the University of Heidelberg, a leading center of German learning and national feeling, where Conrad Celtis himself had taught just the year before. Ringmann went on to study at the University of Paris. During his years in both Heidelberg and Paris he pursued a wide-ranging course of studies: Latin, Greek, classical literature, history, poetics, rhetoric, grammar, mathematics, and cosmography.

Ringmann made a host of new friends while he was at university, and together they created a parallel little humanist universe for themselves. They gave themselves classical nicknames; they conversed and corresponded in Latin; they made pilgrimages to Italy; they sought out refugees from Byzantium and studied Greek; they wrote playful poems stuffed full of classical allusions and nationalist sentiment; and they took jobs at printing presses, where they helped publish works designed to spread the humanist gospel far and wide.

When Ringmann returned home from his studies, in 1503, he was transformed. No longer little Matthias, he now called himself Philesius Vogesigena, or "Lover of the Vosges"—a name that also called to mind the Greek sun god, Apollo, whom the Greeks had sometimes referred to as Philesius. Ringmann returned singing the praises of Greece and Rome, babbling in classical languages, composing playful verses of his own, and, above all, talking and talking—about how the Vosges Mountains, now in French territory, were rightfully a part of Germania; about how the Franks had been led to culture by the Druids, who could trace their language, learning, and culture back to ancient Greece; about how Germania had been a European power long before Rome; about how the Roman and Byzantine Empires had only been temporary bridges for the transfer of empire and learning from ancient Greece to Germany; and more. To those who had known him before he had gone away Ringmann must have appeared almost unrecognizable. "Our little Thomas talks so profoundly almost no one can understand him," the relatives of another young German humanist would complain a few years later, after he had returned home from his studies full of similar ideas.

Ringmann began studying the *Geography* not long after settling in Strassburg. Like the early Italian humanists before him, Ringmann first turned to the work not because he sought to map the modern world but because he wanted to reconstruct the ancient one. In 1506 he took on another project in Strassburg that would only strengthen his interest in Ptolemy: the first German translation of the *Commentaries* of Julius Caesar, a text that discussed the geography and the peoples of northern Europe at some length.

Working on the *Commentaries*, Ringmann would discover what Petrarch, Boccaccio, and Salutati in Italy had all discovered before him:

that writing about ancient history for a modern audience posed a major challenge, because so many of the place-names had changed. When Ringmann finally published the work in early 1507, he would alert readers to this problem in a foreword—and would use language that sounded a lot like the language Petrarch had used a century and a half earlier in introducing *his* work on Caesar. "A number of provinces that in earlier times were large and wide," Ringmann wrote, "have now become either nonexistent or small in extent. On the other hand, we find a number of lands or dominions now famous and very powerful that used to be either nothing at all or very small. On these grounds, those who read the following books should excuse me if they find names missing." Ringmann didn't just complain about the problem, however. He turned to Ptolemy to help alleviate the confusion, and at the end of his foreword he supplied readers with a list of the Latin place-names used by Caesar, along with their German equivalents.

Gathering a list of the Germanic towns and regions mentioned by Caesar, cross-referencing them with the names used in the *Geography*, determining their latitude and longitude, and correlating them all with the towns and regions of modern Germany must have been painstaking, dreary work. But Ringmann's work on Ptolemy wasn't all drudgery. Sometime in the spring of 1505 he came across an early copy of the *Mundus novus* letter and, after consulting the *Geography*, decided that it announced the discovery of a land not mentioned by Ptolemy.

* * *

RINGMANN READ THE LETTER with astonishment and pride. Amerigo Vespucci—a fellow humanist, born and raised among the learned men of Florence, apparently related to the great scholar Giorgio Antonio Vespucci—had just made the greatest geographical discovery of all time: an entirely new world. What impressed Ringmann most was the *southness* of this discovery. He knew that a Genoese admiral by the name of Columbus had sailed west to the Indies and found islands there, and that in Columbus's wake the Spanish and the Portuguese had sent other voyages of discovery across the Atlantic. But what he had read about those voyages—that Europeans had finally managed to sail to Asia—didn't challenge anything he had learned in his geographical studies. What Vespucci announced in his letter, on the other hand, was something entirely new.

Or was it?

As a well-read classicist, Ringmann knew that many of the ancients had described the existence of a giant unknown continent on the other side of the world. He knew that Virgil, for example, had predicted that Rome's Caesar Augustus would eventually extend his rule beyond both Africa and India to "a land beyond the stars, beyond the paths of the year and the sun, where Atlas the heaven bearer turns on his shoulder the firmament studded with blazing stars." Today that language requires some unpacking, but Ringmann had no trouble understanding it. A land "beyond the stars," he knew, meant a land in the southern hemisphere, where the northern skies were no longer visible. A land "beyond the paths of the year and the sun" meant something even more specific: a land beyond the Tropic of Capricorn, which, at 23 ½ degrees south, was the most southerly point touched by the ecliptic. The reference to Atlas also reinforced this idea of southness; Atlas held up the globe from underneath, with the South Pole touching his shoulder.

Reading Vespucci and thinking about Virgil, Ringmann put two and two together. Virgil had described a great new southern land that lay far across the ocean from the known world—and this was exactly what Vespucci had found. The new era that Virgil had predicted almost fifteen hundred years before had dawned: Rome was rising again, and a new Caesar, the Holy Roman Emperor, was poised to extend his reach over a truly global Christian empire.

In a state of high excitement, Ringmann decided to publish a new edition of the *Mundus novus* letter, under the title *Concerning the Southern Shore Recently Discovered by the King of Portugal*. He also decided to introduce the work with a short letter of his own, addressed to one of his humanist friends, in which he explicitly connected Vespucci's discovery to Virgil's prophecy.

Master Ringmann Philesius to Jacob Braun, his faithful friend.
Virgil, our poet, has sung in his *Aeneid* of a land that lies beyond the stars, beyond the paths of the year and the sun, where Atlas the heaven-

bearer turns on his shoulder the firmament studded with blazing stars. If one should wonder at a thing like this, he will not restrain his surprise when he reads attentively what a great man of brave courage yet small experience, Americus Vespucius, has first related, without exaggeration, of a people living toward the south, almost under the Antarctic pole. There are people in that place, he says (as you shall presently read yourself) who go about entirely naked, and who not only (as do certain people in India) offer to their king the heads of enemies they have killed, but who themselves feed eagerly on the flesh of their conquered foes. The book itself of Americus Vespucius has by chance fallen into our hands, and we have read it hastily and have compared almost the whole of it with Ptolemy (whose maps you know we are engaged in examining with great care), and we have thus been induced to compose, upon the subject of this newly discovered region of the world, a little poem not only poetic but also geographical in character. We send to you, my friend Jacob, this work with another book, so that you may know that you are not forgotten.

Farewell. In haste, from our schools. Strassburg, July 31, 1505.

Ringmann's "little poem" followed. It celebrated the idea—so popular in Florence at the end of the fifteenth century—that Ptolemy's *Geography* enabled readers to rise above the clouds and gaze down at the entire known world. Ringmann began the poem by referring to himself as Philesius—perhaps a coded signal to his humanist friends that his perspective would be as lofty as Apollo's. From that godly vantage point he surveyed the world as Ptolemy had mapped it, zeroing in especially on its most distant regions. *There*, he observed, looking down at Africa, is the fertile Nile Delta. *There* are the Mountains of the Moon. *There* are the hot desert wastes and the lands of the Ethiopians. And *there*, he continued, shifting his gaze to the east, are the Indian regions, the Indian Ocean, and Taprobane. Looking farther east still, he directed his attention toward a part of the earth that even Ptolemy hadn't mapped: the new world discovered by Vespucci. *There*, he wrote, addressing Ptolemy, "is a world not known in your pictures . . . a land far under the Antarctic Pole."

The sensational aspects of Vespucci's letter—the drama on the high seas, the naked savages, the sex—drew many readers to *Concerning the*

Southern Shore. But Ringmann's humanist friends recognized the work for what it actually was: a call to action. *We new Romans have discovered a great new southern land,* Ringmann was saying, *and we must now add it to Ptolemy's world.*

It was time for a new edition of the *Geography.*

* * *

UNTIL THE SECOND half of the fifteenth century, speaking of an "edition" of the *Geography* made no sense. Anybody wishing to consult the work had to seek out manuscripts that had been laboriously copied by hand—line after line, column after column, page after page, map after map. As Cardinal Fillastre had learned at the time of the Council of Constance, illustrated editions of the work were particularly hard to find; most scribes just didn't *do* complicated illustrations, much less cartography. Early copies of the *Geography* generally just reproduced the original Latin translation of the text made by Jacopo Angeli, supplemented with copies of the maps that had appeared in the copies of the text brought to Italy from Byzantium. The work was not constantly being updated and improved, as Ptolemy had insisted it should be.

Change came in the second half of the century. The rise of the study of Greek in Europe was a contributing factor; scholars newly proficient in the language began to recognize the many flaws and inadequacies of Angeli's translation. Some cartographers, too, began to understand that the *Geography* offered them a way not only of reconstructing the ancient world but also of depicting the modern one. In general, thanks to the expanding imperial ambitions of the Church, the Turkish capture of Constantinople, and the search for new ways to reach the East, Europeans were forced to think about geography in starkly new ways.

And then the printing press arrived.

Countless studies have been devoted to how the advent of the printing press transformed the nature of the book in Europe. As Christopher Columbus's story reveals, the transformation gradually democratized reading and learning. But the advent of printing didn't just enable the spread of the printed word. It also enabled the spread of the printed image—more specifically, what one modern scholar has called "the exactly repeatable pictorial statement." Nowhere was this to have more of an impact than in the case of maps.

Reproducing complex images before the advent of printing was an expensive, difficult, and time-consuming job. The challenge had vexed even ancient authorities. While discussing the various ways in which the Greeks had tried to transmit botanical learning from generation to generation, Pliny the Elder made an aside that could just as easily have concerned maps.

There are some Greek writers who have . . . adopted a very attractive method of description, though one that has done little more than prove the remarkable difficulties that attended it. It was their plan to delineate the various plants in colors and then to add in writing a description of the properties that they possessed. Pictures, however, are very apt to mislead, and more particularly where such a number of tints is required for the imitation of nature with any success; in addition to which, the diversity of copyists from the original paintings, and their comparative degrees of skill, add very considerably to the chances of losing the necessary resemblance to the originals. . . . Hence it is that other writers have confined themselves to a verbal description of the plants.

The arrival of the printing press solved this problem. At last, visual information—so critical to both the preservation and advancement of scientific knowledge—could be efficiently, inexpensively, and widely reproduced. Cartographers soon recognized that they could print scores of identically illustrated copies of the *Geography*. The idea of an edition of Ptolemy was born.

The first printed edition of the *Geography* with maps appeared in Bologna in 1477, and a number of others soon followed: Rome in 1478 (the edition owned by Columbus); Florence in 1482; Ulm, Germany, in 1482 and 1486; Nuremberg in about 1490; Rome again in 1490. The *Geography* became available to a much larger and more varied audience than ever before, and the effects were profound. The sudden availability of Ptolemy's maps, as outdated as they were, helped Europeans from all walks of life visualize the world in a newly standardized way. The rapid spread of the *Geography* during the 1470s and 1480s also helped determine the ongoing course of geographical discovery. It put Africa and the Far East into a new perspective and gave European explorers, merchants, and navigators a set of well-known landmarks to hunt for—the Prassum Promontorium, Cattigara, the Sinus Magnus, the Aurea Chersonese—as they sailed off in search of a sea route to the Indies. For the first time, scholars and explorers now found themselves literally on the same page.

Printers published many other kinds of world maps during the late 1400s: simple charts and diagrams, elaborate *mappaemundi*, hybrid world maps that borrowed from marine charts, and more. In the short term this flood of printed maps gave new life to geographical visions that the Age of Discovery was rapidly making obsolete, but in the long term it had the opposite effect. Newly able to compare different kinds of maps side by side, many Europeans were forced for the first time to come to grips, as Cardinal Fillastre had done at the Council of Constance, with the troubling idea that many of Europe's age-old traditions of world mapping were simply impossible to reconcile. Only one kind of map seemed to offer a systematic way out of this maze: Ptolemy's. And so the *Geography*'s way of mapping the world gradually became Europe's way, and many of the cartographical conventions that we take for granted today began to settle into place. A world map, more and more people agreed, should have no overt narrative or symbolic function. It should be a mathematical projection of the spherical earth. It should locate the world's places within a coordinate grid and should allow easy calculations of distance and adjustments of scale. It should put north at the top; it should make the world appear longer than it is wide; it should display the whole picture within a rectangular frame; and it should be corrected and revised constantly to reflect modern reality.

The first printers to update the *Geography* began tentatively. The Ulm edition of 1482, for example, placed the regions of northern Europe that had been unknown to Ptolemy in a little bubble that rose up through the map's northern border (*Plate 7*). This device—breaking the frame of the map to signal a break with Ptolemy—was one that Henricus Martellus would employ less than a decade later, of course, when he extended Ptolemy's Africa to the south to depict the discoveries made by Bartolomeu Dias. Martellus, Behaim, and others also expanded Asia far beyond the limit the *Geography* had assigned to it in the east. By the early 1490s Ptolemy's grid had been stretched in every direction to include new information, and the whole world seemed to be coming into focus.

Then Columbus returned from his first voyage. Ptolemaic mapmakers went into a state of shock, or at least a kind of defensive crouch. Unsure of how exactly they should update their maps, they simply stopped printing them; after the Rome edition of 1490, no new edition of the *Geography* would appear in print for a remarkable seventeen years. During that period anybody interested in maps of the western discoveries would have to look elsewhere for information—to marine charts drawn by the New World explorers themselves.

* * *

ALL OF THE earliest surviving marine charts of the New World, with the exception of Juan de La Cosa's chart of 1500, are Portuguese in origin. They date from between 1502 and 1506—and they are mesmerizing. On the earliest charts, unconnected stretches of new coastline rise up out of the ocean and then, as more information becomes available, those coastlines gradually merge on later charts to form a single continent. The process is not unlike the pencil rubbing of a coin: various disconnected features appear and then gradually resolve themselves into a familiar image.

The earliest of the surviving Portuguese marine charts is actually an Italian copy, smuggled out of Portugal in 1502 by a secret agent named Alberto Cantino. Officially in Portugal as a horse trader, Cantino had actually been sent by the duke of Ferrara to gather information about the ongoing western discoveries. By 1502, Cantino had already sent the duke a number of letters describing Portuguese voyages, but after the rapid and stunning succession of discoveries made by Cabral, Vespucci, the Corte-Reals, and others between 1500 and 1502, the duke evidently decided he needed something more: a map that pulled everything together.

Cantino found him just that. How exactly he managed it isn't known. The risks involved were considerable, given King Manuel's decree that anybody caught leaking cartographical secrets be put to death. Nevertheless, Cantino obtained an elegantly and colorfully drawn copy of an official Portuguese chart, and in the fall of 1502 he delivered it to an associate in Genoa. In a letter sent to the duke from Rome on November 19 of that year, Cantino told the duke that he could expect to receive the chart soon—and that he would like what he saw.

The duke did. The chart's size alone, almost three and a half feet high by more than seven feet wide, was designed to impress. Even more remarkable was what it actually contained: a radically expanded picture of the world that brought together the results of many recent Portuguese voyages of discovery (*Figure 66*).

The chart's picture of the Old World alone made it valuable. Exposing for the first time the newly opened sea route between Europe and India, the chart depicted Africa's east and west coasts with new accuracy, and revealed something that Europeans hadn't previously recognized: that the Indian subcontinent was a peninsula extending south into the Indian Ocean. But this was just incidental to the chart's real purpose, which, as an inscription on its reverse explained, was to introduce its viewers to "the islands recently discovered in the regions of the Indies."

These islands dominated the western portion of the chart. In the north, to the west of both England and Greenland, bearing the title "Land of the King of Portugal," was what looked like a large wooded island: almost certainly the portion of Newfoundland or Labrador visited by the Corte-Reals in 1501. To the west of that island, across an expanse of water, at the very edge of the chart, was a long, thin stretch of north-south coastline described as "part of Asia": the mainland explored by John Cabot and the English in the late 1490s, labeled in the south with the names of Asian places that Columbus assumed he had discovered on the north coast of Cuba. Just below this coast was a cluster of islands, both big and small, identified as "the Antilles of the King of Castile, discovered by Columbus." Here Hispaniola and Cuba are both clearly visible—although Cuba is mistakenly called Isabella, a reference to Isabela, the name of Columbus's settlement on Hispaniola.

None of this would have struck the duke as revolutionary. It all conformed to what Columbus and others had described finding: a number of important new islands to the west of Europe, not far from Asia's east coast. But what the duke saw farther to the south corresponded to nothing he had ever seen on a map before. There, adorned by parrots and trees, and divided in two by the official Spanish-Portuguese line of demarcation, was a hulking continental monster (*Plate 10*).

What went through the duke's mind when he first saw this vast new

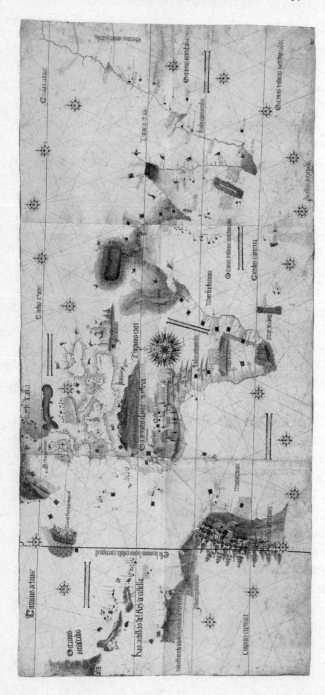

Figure 66. The Cantino chart (1502). On the right, in the east, India appears in its true peninsular form for the first time, reflecting new knowledge recently brought home by the Portuguese. On the left, in the west, are the New World coasts discovered by the Spanish, the Portuguese, and the English. Based on a Portuguese original, the chart places much of South America on the Portuguese side of the line of demarcation.

southern land isn't known. But his reaction was probably similar to the one recorded some years later by the Italian humanist Peter Martyr. Writing to Pope Leo X from Spain, in a work titled *On the New World*, Martyr recalled looking at an early Portuguese marine chart and realizing that the newly discovered southern land literally dwarfed Europe.

> This continent extends into the sea exactly like Italy but is dissimilar in that it is not the shape of a human leg. Moreover, why shall we compare a pigmy with a giant? That part of the continent beginning at this eastern point lying towards Atlas [that is, the part of its east coast that extended into the southern hemisphere] . . . is at least eight times larger than Italy; and its western coast has not yet been discovered.

To boost this map's credibility in the estimation of the pope, Martyr then added an important detail. "It is claimed," he wrote, "that Amerigo Vespucci of Florence assisted in its composition."

Martyr's remark has led some writers to speculate that some of the early Portuguese marine charts of the New World derive from originals drawn by Vespucci himself. The connection has been impossible to prove, but it is at least plausible. The Spanish, for example, had such respect for Vespucci as a navigator and chart maker that in 1508, only two years after they had first hired him to work for the House of Trade, they made him their very first pilot major: the head of mission control, one might say, for the Spanish exploration of the New World. As pilot major, Vespucci was to train all of Spain's pilots and navigators in the art of celestial navigation, and to produce and continually update Spain's master chart of the new discoveries.

Ever the businessman, Vespucci realized that there was money to be made on the side in his new post. He began leaking copies of his official chart—and in 1510 he got caught. Word reached King Ferdinand, who wrote in a fit of pique to his agents in Seville on June 15. They must find Vespucci, he wrote, and make him swear that "henceforth he will not again commit or consent to a proceeding so irresponsible and so promiscuous, but will issue maps only to such persons as the monarch or the House of Trade may order."

If Vespucci leaked official maps in Seville, it's fair to assume that he also leaked them earlier in the decade, after returning to Lisbon in 1501 from his Portuguese voyage to the New World. Demand was high at the time; this was the very period when Juan de La Cosa, Alberto Cantino, and others were in the city avidly trying to acquire charts showing what the Portuguese had just discovered. As a free agent with long experience in the business of western discovery, Vespucci would have recognized the great value of the charts he had brought back with him from the New World, and he would have known that there were foreigners in the city willing to pay handsomely for them.

*   *   *

THE PUBLICATION OF *Concerning the Southern Shore*, in the summer of 1505, seems to have had the effect that Matthias Ringmann hoped it would. Within a year he and a group of area humanists had joined forces, established a small printing press, and begun working on a new edition of the *Geography*. Full of hopes for their modest new press, they gave it a Latin name: the Gymnasium Vosagense.

The Gymnasium came together in Saint-Dié, a quiet church town in the Vosges Mountains, some sixty miles southwest of Strassburg. At the time, Saint-Dié was the religious and administrative capital of the duchy of Lorraine—the easternmost edge of Europe's French-speaking lands. Today the town is known as Saint-Dié-des-Vosges, and its setting is idyllic: a natural mountain basin ringed by pine-cloaked hills and cut through by the swift and muddy Meurthe River. Outside Saint-Dié itself the basin contains a patchwork of forest and pastureland that, viewed from any one of the nearby peaks, reveals itself as a luscious study in different shades of green. Rain and fog settle in often, and church bells ring on the hour, as they have for centuries. The place is peaceful, inviting contemplation.

The animating spirit of the Gymnasium seems to have been Walter Lud, a canon at the church in Saint-Dié. In 1505 Lud was in effect the town's mayor: he administered the affairs of the church and oversaw its building projects; he worked as a personal secretary, chaplain, and advisor to Duke René; and he ran the duke's nearby silver mines. Despite the many demands on his time, Lud also devoted himself to the study of geography and astronomy, observing the skies, studying Ptolemy, making maps, and following the news of the New World discoveries. As a result, sometime in late 1505 or early 1506 he found himself reading *Concerning the Southern Shore*—and agreeing that it was indeed time for a new edition of the *Geography*.

To this end, Lud decided to create the Gymnasium Vosagense, or so the surviving scraps of information about the group's origins suggest. First Lud secured financial backing from Duke René, a boar-hunting military adventurist with an interest in astronomy and the arts, and then he began to assemble his team. He relied on a network of family and friends to the extent possible, but to find scholars who had the time and the expertise to take on the job of revising the *Geography*, he had to look beyond Saint-Dié itself. Having just read *Concerning the Southern Shore*, he knew at least one scholar to turn to: the young Matthias Ringmann.

It was a perfect fit. A native son of the Vosges, Ringmann was a well-connected humanist who had already published a geographical work. He was a writer; he read Latin and Greek; he had studied mathematics and cosmography; and he had experience working with some of Strassburg's most important printers. Best of all, he was already immersed in the study of the *Geography*. Despite all of his qualifications, however, there was one thing that Ringmann couldn't do: make proper Ptolemaic maps. And so Lud turned to another German—a humanist and cleric named Martin Waldseemüller.

* * *

ONLY ONE FAINT TRACE of Martin Waldseemüller has been discovered before his emergence on the scene in Saint-Dié: a handwritten note in the student register of the University of Freiburg that records his matriculation there on December 7, 1490. If he began his studies in his early to mid-teens, as was the norm at the time, Waldseemüller must have been born in the middle of the 1470s, which would suggest he was about thirty years old when Lud brought him to Saint-Dié. By the time he arrived he had already been ordained as a priest in the diocese of Constance and had good contacts among the humanist printers of Basel, which suggests he may have spent time in his twenties working in the town. Waldseemüller was educated among the early German nationalists, and after university seems to have injected himself into the same humanist world as Matthias Ringmann. Like Ringmann, he gave himself a classical-sounding nickname: Ilacomilus, sometimes rendered as Hylacomylus, a coinage involving elements of Latin and Greek that was roughly equivalent to the meaning of *Waldseemüller* in German ("miller of the forest lake").

What brought Waldseemüller to Lud's attention is a mystery. He isn't known to have produced a single map before 1507. And yet, updating Duke René on the progress of the Ptolemy project that year, Lud described Waldseemüller as a master cartographer. Lud had contacts in the printing community in Strassburg, so perhaps he found Waldseemüller there. It's even possible that Ringmann himself brought Waldseemüller to Lud's attention: both he and Waldseemüller were young German humanists, after all; both had been educated in Freiburg and shared some of the same mentors; both had connections to area printers and scholars; and both had a special interest in Ptolemy. In his preface to *Concerning the Southern Shore*, moreover, Ringmann had used the plural when describing his study of Ptolemy ("whose maps *we* are examining with great care"), which makes it conceivable that he and Waldseemüller had already established some kind of working partnership. However they actually met, Ringmann and Waldseemüller would go on to become fast friends and productive collaborators. Ringmann, the literary scholar, would take care of the words; Waldseemüller, the cartographical technician, would take care of the maps.

Lud, Waldseemüller, and Ringmann soon developed an ambitious plan for their *Geography*. As the work's introduction would ultimately explain, a "monstrous chaos" prevailed in existing Latin editions of Ptolemy, and two factors were to blame: the haphazard and uncoordinated attempts of early printers to update his maps for the modern age, and the many corruptions introduced into the work over the centuries by copyists and printers. The situation made careful scholars of classical geography want to pull their hair out. "In many places," the introduction continued, "even the most devoted reader does not know what derives from the moderns and what from Ptolemy himself."

Cleaning up this mess became one of the Gymnasium's main goals. Ringmann and Waldseemüller both agreed that their new *Geography* had to do more than just correct and update what appeared in other Latin editions of the work; it would also hew as closely as possible to Ptolemy's original intentions. To this end they decided to seek out and study whatever old Greek manuscripts of the work they could find. That job alone, they knew, was not for the faint of heart. But Ringmann and Waldseemüller also decided to take on an even more ambitious task. They would attempt to produce something entirely unprecedented: an edition of the *Geography* that contained not one full set of maps but two. The first set would offer classical scholars the best available depiction of the world as Ptolemy and the ancients had known it—the standard world and regional maps made possible by the eight thousand sets of coordinates collected in the *Geography*. In this respect the work had a conventional aim; it was intended for an audience of German humanists whose focus was primarily the reconstruction and revival of the great imperial legacies of the past. But the second set of maps would offer something unprecedented: a full edition of Ptolemaic maps that provided a comprehensive picture of the modern world.

Ringmann and Waldseemüller harbored no illusions about the challenges they faced. Months and even years of trying labor lay ahead. Probably they began their work slowly, with a mixture of excitement and trepidation: receiving instructions from Lud as the project began to take shape; walking up into the Vosges as they discussed their plans; gradually ridding themselves of existing commitments elsewhere; traveling and reaching out to friends and colleagues in search of texts and maps. But just as they were settling in for the long haul, some electrifying news reached Saint-Dié. Like the duke of Ferrara before him, Duke René had managed to obtain a copy of a marine chart from Portugal.

And he had received a personal letter from Amerigo Vespucci.

· The whole world, old and new ·

CHAPTER EIGHTEEN

WORLD WITHOUT END

*It has seemed best, in describing the general appearance of the world,
to put down the discoveries of the ancients and to add what has since
been discovered by the moderns . . . carefully and clearly brought
together, so as to be seen at a glance.*

—The Waldseemüller map of 1507

THE LETTER WAS quite something. Addressed to Duke René personally,
written in French, and dated September 4, 1504, it contained a long and
lively description of four voyages that Vespucci claimed recently to have
made to the New World. It reprised much of what had already appeared in
the *Mundus novus* letter but also provided a trove of new information.

Vespucci began by explaining himself. "The principal cause that moved
me to write you," he told the duke,

> was the request of the present letter-bearer, one Benvenuto Benvenuti, a
> fellow Florentine, very much at Your Magnificence's service, as he shows
> himself to be, and very much my friend. He, on finding himself here in

this city of Lisbon, entreated me to impart to Your Magnificence those
things which I have seen in various regions of the world by virtue of four
voyages which I made to discover new lands: and two were by command
of the exalted King of Castile, Don Fernando VI, to go west over the
depths of the Ocean Sea, and the other two were by command of the
mighty King Don Manuel of Portugal, to go south; Benvenuti telling
me that Your Magnificence would take pleasure in it, and that in this he
hoped to serve you. For which reason I have set about the task, for I am
sure that Your Magnificence counts me among the number of his servants, since I remember how, in the time of our youth, I was your friend,
just as I am now your servant, and how we would both go to hear the
principles of grammar exampled in the good life and doctrine of the venerable religious friar of San Marco, Fra' Giorgio Antonio Vespucci, my
uncle; and would to God I had followed his counsels and doctrine, for, as
Petrarch says, I would be "another man from that which I am."

This was music to Ringmann and Waldseemüller's ears. The Petrarch quote,
the mention of Giorgio Antonio Vespucci, the allusions to the Florentine
community: they all confirmed that this Amerigo Vespucci was indeed
more than just a merchant and a navigator. He was a humanist, a learned
explorer and scholar who, providentially, appeared to have an intimate connection to their own Duke René. And what Vespucci laid out in his letter, Ringmann and Waldseemüller realized, was the most comprehensive
account yet written of the discovery of the New World—"a description,"
as Vespucci put it in the letter, "of the chief lands and of various islands, of
which ancient authors make no mention, but which recently, in the 1497th
year from the incarnation of Our Lord, were discovered in the course of
four ocean voyages." The date 1497 was of particular importance, because
it made clear that Vespucci had first reached the New World a year before
Columbus. Everybody in the Gymnasium agreed: it was a remarkable letter. What they didn't realize, however, was that this letter, like the *Mundus
novus* letter before it, was a fake.

Today this second public Vespucci letter is often referred to as the Letter to Soderini. That's because three separate but virtually identical versions
of the letter have surfaced in which Vespucci addresses not Duke René but

Piero Soderini, who in 1502 had been elected the head of the Florentine Republic. That Vespucci would have written to Soderini rather than René makes perfect sense; René isn't known to have had any Florentines in his service, nor to have studied in Florence in his youth, whereas on both counts Soderini is. Vespucci had another good reason to be writing to Soderini: Lorenzo di Pierfranceso de' Medici had died in 1503. It would only have been natural for Vespucci after his patron's death to try to ingratiate himself with a newly powerful old friend, who, it was well-known, had an interest in foreign lands.

So why was the Gymnasium's copy of the letter addressed to Duke René? The most likely explanation goes something like this. Sometime in 1505, René obtained a copy of the letter, perhaps from friends in Portugal, or perhaps from friends in France. With the letter in hand, he or somebody in his service decided to have a new copy made—a copy in which he, not Soderini, would appear to be the intended recipient. It was this doctored version of the letter that made it to Saint-Dié. But the story gets even more involved, because most scholars now agree that the letter was a fake on another level. It may have been part of what one expert on Vespucci has called "a cultural and political operation, exquisitely Florentine in character, which glorifies Vespucci by indirectly attacking Columbus."

The letter certainly does seem constructed to glorify Vespucci at Columbus's expense. It describes four voyages: a number never once mentioned in any of Vespucci's surviving private letters, but one that matches the number of voyages made by Columbus. Improbably, the letter makes three of Vespucci's four voyages begin on the very same day: May 10. The first voyage it describes, supposedly made with the Spanish in 1497–98, amounts to little more than a collection of details drawn from Vespucci's "second" Spanish voyage (the 1499–1500 voyage to Venezuela and Haiti), details that have been repurposed, embellished, and backdated to make Vespucci appear to have reached Paria before Columbus. The first voyage is described at far greater length than the other three, too, which naturally makes it come across as the most important. Conversely, the final voyage, supposedly made with the Portuguese in 1503–1504, receives only the most cursory treatment. It comes across as an afterthought, tacked on only so that Vespucci will appear to have made as many voyages as Columbus. Additionally, some of the details that the letter *does* provide about this supposed fourth voyage derive very obviously from the experiences and writings of Columbus himself. At one point there's even a shipwreck followed in short order by the building of a fort, the decision to leave a small group of sailors behind to man a small colony, and a final treacherous ocean voyage back to Portugal.

All of which is to say that Vespucci himself almost certainly didn't write the "Letter to Soderini." Unfortunately for those who like clarity, this doesn't necessarily mean that Vespucci didn't write to Soderini at some point, or that Vespucci wasn't actually the source of much of what appears in the "Letter to Soderini." The problem—the Vespucci problem—is that in this letter, as in all of Vespucci's letters, it's impossible to figure out quite what is what. But the members of the Gymnasium never felt there was a problem. They assumed the letter was real, and they read it with fascination.

*   *   *

ALTHOUGH VESPUCCI'S NEW letter was stuffed full of captivating detail, it contained virtually no useful geographical information. This must have greatly vexed Martin Waldseemüller as he scoured the letter for clues about how to map the New World. The one notable exception came near the end of the letter's account of Vespucci's third voyage—the same voyage described in the *Mundus novus* letter. To begin with, the two letters tell much the same story about this voyage. Vespucci and his Portuguese colleagues make their ocean crossing, decide to turn south, and follow the coast for hundreds of miles. But when, by his own reckoning, Vespucci reaches 32 degrees south, his account suddenly—and literally—veers off in a new direction.

Since we had already been ten months on the voyage, and since we had not found any sort of mineral in this land, we decided to leave it, and to go and try some other region in the sea. And holding counsel, we decided that we should follow that course which seemed best to me, and full command of the fleet was entrusted to me. . . . Once we had obtained our provisions, we set out from this land and started our course to the southeast. . . . And we sailed by this wind until we found ourselves at such high latitude that the South Pole stood full 52 degrees above our horizon. . . . A tempest began to rage so violently that we had to furl the sails completely and course on in the strong wind with bare masts, for these were southwesterly winds, with very high seas and the air very turbulent; and this storm was so fierce that the entire fleet was in great fear. The nights were very long, for we had a night on 7 April of fifteen hours. . . . And sailing in this storm, we sighted new land on 7 April, along which we ran some twenty leagues; and we found it to be all exposed coast, and we saw no harbor or people.

This passage is one of the most famous and fought-over in the entire literature of New World discovery. Some readers simply take the whole thing at face value. They assume that Vespucci did indeed take command of the expedition at about 32 degrees south and that he then ordered his ships out to sea on a southeasterly course, toward points unknown. In this scenario the southern land that he and his men eventually reached can only have been one place: the island of South Georgia, which lies all on its own in the southern Atlantic at about 54 degrees south, more than a thousand miles to the east of the southern tip of South America. But this interpretation has its problems. In all of his other surviving letters, familiar and public, Vespucci goes out of his way to say that his expedition hugged the coast and followed it to the southwest—a version of events, backed up by the few independent accounts of the voyage, that makes much more sense. The whole point of the expedition, after all, was to learn as much as possible about this vast new southern land, and Vespucci in his other letters repeatedly noted his interest in sailing west around it in order to pioneer a new route to Cattigara, the Spice Islands, and India. Why abandon that officially sanctioned quest, backed by investors who expected results, in favor of a rogue venture that would carry the expedition with no known objective southeast into the frigid and stormy open ocean at latitudes below even the southern tip of Africa?

Perhaps, some argue, this whole passage is a deliberate deception: an attempt to hide where Vespucci really went. Perhaps Vespucci and his fellow sailors began following the coast to the southwest fully confident that they were exploring a coastline on the Portuguese side of the line of demarcation. But sometime not long before they reached 32 degrees south they realized that the coast had pulled them far enough to the southwest that they had crossed into Spanish territory. (The coast of Portuguese-speaking Brazil in fact today meets the coast of Spanish-speaking Uruguay at about 34 degrees south.) They realized, in other words, that they had no right to travel any farther. Unwilling to give up their search for gold, gems, spices, and a sea route around the continent, however, they decided to flout the Treaty of Tordesillas, to continue following the coast—and, later, to hide what they had done.

This latter possibility, although nothing more than speculation, has much to recommend it. It helps explain the sudden and strange decision to hand command over to Vespucci, for example. By putting an Italian in charge of the expedition and by having him make the decision to sail south, the Portuguese officers on the voyage would have absolved themselves of official accountability for having strayed into Spanish waters. Even the storm performs an important function in this scenario. Whatever and wherever the coast was that Vespucci reported finding at 52 degrees south, he and his men could report that they arrived there by chance, not by design.

The members of the Gymnasium had none of this on their minds when they began scanning Vespucci's new letter. Comparing it with the *Mundus novus* account, they focused on one basic question.

What exactly had Vespucci found?

* * *

DUKE RENÉ'S MARINE charts were the obvious place to seek answers. There was good reason to assume that Vespucci himself had had a hand in preparing them; the *Mundus novus* letter intimated as much. "If my companions had not relied upon me and my knowledge of cosmography," Vespucci had written, "there would have been no pilot or captain on the voyage to know within five hundred leagues where we were."

How did the duke manage to obtain his marine charts? Nobody knows. He had loose family connections to the Portuguese court and had several mariners in his service, so perhaps he worked one of those channels. He also supported a number of scholars, clerics, and noblemen, many of whom had ties to Portugal and Italy, and any one of whom might have acted as a sort of buyer's agent for him, just as Alberto Cantino had done for the duke of Ferrara.

Matthias Ringmann himself represents an intriguing possibility. He is known to have made a brief journey across the Alps to Carpi, in northern Italy, in the fall of 1505, where he met with his friend Gianfrancesco Pico della Mirandola—a minor nobleman who maintained strong ties with the humanist communities in both Italy and Germany. Mirandola was a friend of Conrad Celtis, among other prominent German humanists, and his uncle was the famous Giovanni Pico della Mirandola, who in 1486 had written one of the most important manifestos of the Italian Renaissance, the *Oration on the Dignity of Man.* Having just published *Concerning the Southern Shore*, brimming over with curiosity about the New World, Ringmann would have had plenty of motive and opportunity during his stay in Carpi to put questions about Vespucci to Mirandola, and to seek out marine charts showing the extent of the recent Portuguese discoveries. The Cantino chart itself was only some forty miles away at the time, in Ferrara, where the Mirandola family had regular interactions with the duke, and where Ringmann himself would travel on a second visit to Italy, in 1508.

None of the actual marine charts that reached Saint-Dié in 1505 has survived. However, Waldseemüller is known to have relied on one particular chart in studying the geography of the New World, and the evidence suggests overwhelmingly that it was identical to a chart that does survive: a copy of a Portuguese original, drawn by the Genoese mapmaker Nicoly de Caverio in 1504 or 1505 (*Figure 67*).

The Caverio chart is huge, measuring almost four feet by six and a half feet. Like the Cantino chart, it showcases the many recent Portuguese geographical discoveries—the ocean route around southern Africa, the true peninsular shape of the Indian subcontinent, the emerging contours of the New World. The map's western side bears a strong likeness to what appears on the Cantino chart, but it provides more detail, depicting more coastal features, recording more place-names, and—for the first known time on a marine chart—providing a scale of latitudes. This last feature would have been of special interest to Waldseemüller, because it would have allowed him to pinpoint the southern limit of Portuguese New World exploration at about 35 degrees south, approximately where Vespucci said he had turned away from the coast and headed out to sea.

When Ringmann and Waldseemüller first unrolled their version of the

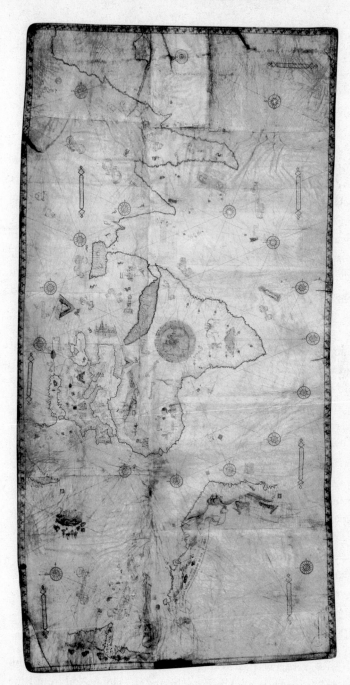

Figure 67. The Caverio chart (circa 1504–1505). This chart, or one almost identical to it, provided Martin Waldseemüller with the information he would use in mapping the New World.

chart, they must have been transfixed. Reading about the New World was one thing, but seeing it—especially in conjunction with a scale of latitudes that presumably derived from Vespucci's own observations—was something else entirely. Stitching together a picture of the entire globe suddenly seemed possible. The chart tugged at their imaginations in even grander ways, too. What the dreamy Ringmann saw when he looked at the chart was not just a new place but a new time: the future. Virgil had prophesied the discovery of a great new southern land on the other side of the world, and now here it was. The world was whole, and it belonged to the new Romans.

Surveying the world as a whole from this perspective—the world of space and time that Christ so often surveyed at the top of the *mappae-mundi*—catapulted Ringmann into an imperial frame of mind. It allowed him to imagine uniting the world not just geographically but politically and religiously. He put his hopes into writing in 1511, after having spent time pondering a majestic map of Europe that Waldseemüller had just finished. "Looking at this same map of Europe," he wrote to Waldseemüller,

> and considering how powerful is Spain, how rich and warlike is France, how great is Germany and how robust are her men, how strong is Great Britain, how brave is Poland, how valiant is Hungary, and (in leaving unmentioned many states not to be despised) how rich, courageous, and experienced in the military art is Italy, I could not but regret most grievously the cruel, harmful, and dreadful wars that our princes wage, their perpetual dissensions and personal hates, disputing forever among themselves questions of territory, sovereignty, supremacy, and self-aggrandizement, while they leave the Turk and the enemies of our faith to spill Christian blood, destroy cities, devastate countries, burn churches, carry off our daughters, violate our wives, and commit the greatest crimes. On the other hand, if they would but give up these serious and perilous quarrels, these enmities and hates, if adopting peace and uniting their forces they would take up arms against the common enemy, they would easily subjugate the entire world and cause the blessed Savior . . . to be the object of worship by all nations.

The Caverio chart was a stunning testament to Europe's newfound ability to expand its reach in all directions around the globe. And from Ringmann's perspective as a German humanist, it was obvious who could—or at least who *should*—rise to the challenge of uniting Europe and presiding over the global Christian empire: Maximilian I, the leader of the Holy Roman Empire. Ringmann and Waldseemüller must have had such thoughts in mind when they first came together in Saint-Dié several years earlier. What better way to celebrate the potential greatness of this new Caesar, and perhaps to goad him and the humanists of Europe into action, than by designing a new map that would show the full reach of his coming global empire?

With their knowledge of Ptolemy, their privileged access to the Vespucci letter and the Caverio chart, and their new printing press, the members of the Gymnasium realized that they were uniquely qualified to make this prospect a reality. Deciding to act quickly, while the Vespucci letter and the Caverio chart were still fresh news, they put their time-consuming *Geography* project on hold. Instead they decided to produce a geographical package consisting of three parts: a huge new map of the whole world, dedicated to Maximilian I, that would sum up ancient and modern geographical learning; a tiny version of that map, printed as a series of globe gores that could be pasted onto a small ball, creating the world's first mass-produced globe; and a sort of users' guide to those two maps, titled *Introduction to Cosmography*, that would also include a full Latin translation of the duke's new Vespucci letter.

It was a profound moment in the history of cartography—and in the larger history of ideas. Ptolemy had shown his readers how to rise up above the clouds and look down at the known half of the world; modern explorers had finally begun to explore and describe its other half; and now, with their book, their map, and their gores, the members of the little Gymnasium Vosagense were about to make it possible for everybody, not just a divine figure hovering above the earth, to understand the makeup of the cosmos, to see the entire world at a glance, and to hold the globe in their hands.

*   *   *

AND SO THE work began. Walter Lud and his nephew Nicholas oversaw the operation, and Walter's friend Jean Basin de Sandaucourt began translating the Vespucci letter from French into Latin. But the bulk of the work—the design of the map and the globe, and the writing of the *Introduction to Cosmography*—fell to Waldseemüller and Ringmann.

Ringmann took the lead in writing the book. Libraries today credit Waldseemüller as the author, but the book actually names no author, and Ringmann's fingerprints appear all over it. The author, for instance, quotes the *Commentaries* of Julius Caesar, the very first German translation of which Ringmann would publish in early 1507. He demonstrates a familiarity with ancient Greek, a language that Ringmann, alone of the members of the Gymnasium, knew well. He embellishes his writing with snatches of verse by Virgil, Ovid, and other classical writers, a literary tic that characterizes all of Ringmann's writing. And the one contemporary writer mentioned in the book, whom the author refers to as "our poet Gallinarius," turns out to have been a young German humanist named Johannes Hänlein, a known friend of Ringmann's.

Ringmann the writer, Waldseemüller the mapmaker: it was a natural division of duties, and the two men are known to have teamed up in precisely this way in 1511, when Waldseemüller printed his map of Europe. Accompanying the map at the time of its publication was a little booklet titled *Description of Europe*, and in this case there's no doubt about who wrote the book. Dedicating his map to Duke René's successor, Duke Antoine of Lorraine, Waldseemüller wrote, "I humbly beg of you to accept with benevolence my work: with an explanatory summary prepared by Ringmann."

Why dwell on this question of authorship? Because whoever wrote the *Introduction to Cosmography* almost certainly coined the name America (which would have been pronounced "Amer-eeka"). Here, too, the balance tilts in Ringmann's favor. Consider the famous passage in which the author steps forward to explain and justify the use of the name.

> These parts have in fact now been more widely explored, and a fourth part has been discovered by Amerigo Vespucci (as will be heard in what follows). Since both Asia and Africa received their names from women, I do not see why anyone should rightly prevent this [new part] from being called Amerigen—the land of Amerigo, as it were—or America, after its discoverer, Americus, a man of perceptive character.

This sounds a lot like Ringmann, who is known to have spent time mulling over the reasons that concepts and places so often had the names of women. "Why are all the virtues, the intellectual qualities, and the sciences always symbolized as if they belonged to the feminine sex?" he would write in a 1511 essay on the Muses. "Where does this custom spring from: a usage common not only to the pagan writers but also to the scholars of the church? It originated from the belief that knowledge is destined to be fertile of good works. . . . Even the three parts of the old world received the name of women." The naming-of-America passage reveals Ringmann's hand in other ways, too. In his poetry and prose Ringmann regularly amused himself by making up words, by punning in different languages, and by investing his writing with hidden meanings for his literary friends to find and savor. The passage is rich in just this sort of wordplay, much of which requires a familiarity with Greek, a language Waldseemüller didn't know.

The key to the passage, almost always ignored or overlooked, is the curious name Amerigen—a coinage that involves just the kind of multifaceted, multilingual punning that Ringmann frequently indulged in. The word combines *Amerigo* with *gen*, a form of the Greek word for "earth," creating the meaning that the author goes on to propose—"the land of Amerigo." But the word yields other meanings, too. *Gen* can also mean "born" in Greek, and the word *ameros* can mean "new," making it possible to read *Amerigen* as not only "land of Amerigo" but also "born new"—a double entendre that would have delighted Ringmann, and one that very nicely complements the idea of fertility that he associated with female names. The name may also contain a play on *meros*, a Greek word that can sometimes be translated as "place." Here *Amerigen* becomes *A-meri-gen*, or "No-place-land": not a bad way to describe a previously unnamed continent whose geography is still uncertain. To cap it all off, *Amerigo* itself is of distinctly Germanic origin, deriving from *Amalrich*, a name used by many important figures throughout German history. Immersed as he was in the study of Germania and the history of the German people, Ringmann may well have recognized this connection and slyly exploited it to give the New World a Germanic name.

Ringmann doesn't seem to have spent much time on the *Introduction to Cosmography*. The book has a hastily written feel, as though cobbled together from a few basic texts that he happened to have at hand. He often borrows directly from Sacrobosco's *Sphere*, a work on which one of Ringmann's favorite professors in Paris was an expert. The author also fills almost five full pages of the book's final chapter with a description of the known world borrowed verbatim from a popular Latin translation of a work titled the *Periegesis*—a text by the third-century Greek writer Dionysius Periegetes that was often taught in medieval schools. When it was time for Ringmann to introduce the Vespucci letter in the book, too, Ringmann didn't even bother to compose a new poem for the occasion, as was his custom. Instead he inserted a slightly modified version of the poem he had already printed in *Concerning the Southern Shore*.

None of this means that Ringmann must have written the *Introduction to Cosmography* by himself. Some of the more technical parts of the book—a few of the charts and degree tables, a warning about the placement of the equator on marine charts, a brief explanation of how to use an astrolabe, and the famous paragraph setting out the book's purpose and listing the emblems used on the map—have the feel of late additions, written in another voice. These would seem to be the work of Waldseemüller, the detail-oriented cartographer, who perhaps revised and added to the book after Ringmann had produced a working first draft. This kind of arrangement would have suited Waldseemüller just fine. He hadn't come to Saint-Dié, after all, just to push words around. He had a map to make.

*   *   *

RIGHT FROM THE START, Waldseemüller recognized that he would have to patch his world together from a number of different sources. One choice was easy to make: he would depict the known world exactly as Ptolemy had mapped it more than a thousand years earlier. This seems like a strange choice today; Waldseemüller knew perfectly well how imperfect and incomplete Ptolemy's maps were, and had access to modern marine charts that portrayed the known world much more accurately. But geographical accuracy wasn't what Waldseemüller was after when it came to the Old World. He was designing his map for humanists and scholars, not navigators and explorers, and they didn't need a modern picture of Europe, North Africa, the Mediterranean, and western Asia. What use would *that* be, after all, in studying the travels of Odysseus and Aeneas, the conquests of Alexander and Caesar, or the history of Germania? What they needed, and what

Waldseemüller decided to give them at the very center of his map, was a reconstruction of the ancient world as Ptolemy had described it.

For this part of his map Waldseemüller relied primarily on the widely respected Ulm *Geography* of 1482 (*Plate 7*). He would borrow a great deal directly from this edition of the work, including its tentative addition of northern European lands not known to Ptolemy. But the Ulm *Geography* was a Latin text, and as such Waldseemüller didn't trust it completely. Copyists and printers and translators had all introduced errors into the text during the fifteenth century, he knew. So, "with the help of others," as he put it, he also relied on at least one other version of the text—an old Greek manuscript that he assumed was closer to Ptolemy's original. Which manuscript this was isn't clear.

Having chosen his Ptolemaic sources, Waldseemüller now had to decide how to expand his world to include the many new lands beyond the limits of the *Geography* that had been explored by modern Europeans. For the Far East, he had an obvious place to turn: maps like those of Martin Behaim and Henricus Martellus, produced in the late 1480s and early 1490s. He seems to have drawn particular inspiration from a map much like the large Martellus map now owned by Yale University (*Figure 56*). Here, already merged neatly with the Ptolemaic known world, was Marco Polo's Far East: the island of Cipangu at the very far edge of the map; Cathay and Mangi at the eastern limits of mainland Asia; and the long Dragon's Tail, descending south into the Indian Ocean, where it met the Spice Islands. The map's depiction of Africa had been rendered obsolete by the recent voyages of da Gama and Cabral to India, but Waldseemüller did decide to borrow its symbolic aspect; like Martellus, he would ram the southern tip of the continent through the southern frame of his map.

Waldseemüller knew exactly where to turn for a much more accurate picture of southern Africa, and for an up-to-date picture of the New World: his copy of the Caverio chart. It was a source he knew he would have to handle with care, because, like all marine charts, it didn't account for the curvature of the earth. This meant that any geographical information he wanted to borrow from the map would first have to be converted into a set of coordinates that could be integrated with mathematical consistency into the rest of his projection. Only then, once he had fused the Old World and the New, would he be able to offer viewers the kind of world map that Ptolemy had insisted all cartographers should strive to create: one depicting the world as "a single and continuous entity."

Ptolemy at the center, Martellus in the east, Caverio in the west—Waldseemüller had his main cartographic sources lined up. But he still had lots of decisions to make about the general design of his map. There was the question of its size, for example. Ptolemy had nothing to say on this subject. But Strabo—the *other* great ancient Greek geographer whose work Europe's humanists had translated into Latin and admired—had proposed some specific guidelines. "To hold the suitable parts of the known world with clarity," Strabo had written in his *Geographical Sketches,* "and to give an appropriate display to the spectators . . . one ought to draw [the map] on a planar surface of at least seven feet."

Waldseemüller and his colleagues in Saint-Dié liked the idea of a big map. They knew they had a unique and fleeting opportunity to create a brand-new vision of the world, and they wanted to make the most of it. The bigger their map, the more detail it could provide, the more people it could reach, and, of course, the more effectively it would advertise the Gymnasium and its future geographical publications. A big map had another advantage: it could be used for teaching. Hung on a wall in a classroom, such a map would provide the perfect backdrop for a humanist professor or schoolteacher who wanted a single visual prop, easily viewed from a distance, to help him guide his students through the worlds of ancient history, classical literature, and modern geography—or to help him inspire those students with an image that made clear how the Holy Roman Empire of the German Nation had grown out of and was now being rebuilt upon the ancient empires of Greece and Rome. Didactic functions of this kind may well have been what Waldseemüller had in mind when, in his dedication to Maximilian, he explained whom he understood his map's audience to be. "I have prepared a map of the whole world for the general use of scholars," he wrote, "like an introduction, so to speak."

* * *

THE MAP WOULD be four and a half feet tall by eight feet wide, Waldseemüller decided—bigger than the Caverio and Cantino charts, bigger than the Martellus wall map, bigger, in fact, than any map ever printed. Previously when maps of this scale had been made they had been drawn or painted by hand, and were luxury items destined for the wealthy and the powerful. But Waldseemüller and his colleagues had in mind something different. They would print a thousand identical copies of their map and sell it far and wide, at an affordable price.

The printing itself would require considerable advance planning and coordination. No press could print a map of this size on a single piece of paper, Waldseemüller knew, so he decided to design his map in twelve separate but contiguous segments. First he would prepare a master sketch of the whole map on these twelve segments, and an artist would fill in the margins and the blank spaces with portraits and ornamental details. Waldseemüller would then turn over the finished master sheets to a skilled woodcutter, or a team of woodcutters, who would transfer the outlines of the twelve drawings onto twelve corresponding woodblocks—in reverse, of course, so that the final print would come out right. This was generally handled by placing the worksheets face down on the woodblocks and then cutting through the back of the drawing, following the outlines of the drawing, directly into the block. With the outlines transferred in reverse, the woodcutter would then begin focusing on the details. Anything destined to be white space—the interior of continents and islands, the margins of the map, the spaces allotted for legends—would be chipped and gouged and chiseled out, to create areas on the block that wouldn't come into contact with the paper during printing. Everything else—mountain ranges, rivers, place-names, legends, illustrations—would be left at the surface, to be shaped and sanded and oiled and varnished until the block was ready. Some of the map's lettering would be carved into the wood at the surface, and some of it would be inserted into white spaces as metal type. When at last the whole block was finished it would be clamped into a press and coated with thick, gooey ink; a single sheet of paper would be pressed down on top of it; and the result would be a single printed copy of that segment of the map, which would be hung on a line to dry.

Repeat a thousand times for each of the twelve blocks.

Waldseemüller and his colleagues knew that this was too big and complex a job for their new little press. They would have their hands full in

Saint-Dié just printing the *Introduction to Cosmography*, and they had the matter of timing to take into account, too; they wanted to release their book, their map, and their globe together. So they agreed, it seems, to divide and conquer. The *Introduction to Cosmography* they decided to print in Saint-Dié, as the Gymnasium's very first publication. Where they printed the globe isn't known, but its design was a simple one—a drastically reduced variation on the big map, cut into a single small woodblock—that may have been printed in Saint-Dié as well. But for the big map they looked to Strassburg for help.

It was the obvious choice. Waldseemüller, Ringmann, and Lud all had good contacts in the publishing industry there. The town's major printers could handle jobs of this size and complexity, and had access to thriving communities of woodcutters, artists, typesetters, proofreaders, and other specialists—not to mention big supplies of paper and ink. The printer they settled on, the evidence suggests, was Johannes Grüninger, the owner of one of Strassburg's biggest and most important presses.

Having settled on a printer, Waldseemüller returned to the design of his map. One of the most important early decisions he had to make concerned the projection he would use: that is, the mathematical system he would employ to make the round world flat. He liked the projection Martellus had used for his big map: a modified version of the second projection described by Ptolemy in the *Geography*. Unlike Ptolemy and Martellus, however, Waldseemüller had to include the New World, and to make room for it he decided to do something unprecedented: he would stretch his map a full 360 degrees of longitude. Where his map ended in the east, in other words, it would begin in the west, with nothing left out in between.

Waldseemüller chose his projection for practical reasons. But he had aesthetic ones, too. He realized, for one thing, that it would allow him to reproduce Ptolemy's known world at the center of his map with a minimum of distortion, which meant that it would come across as instantly recognizable to anybody familiar with the *Geography*. The cost of this choice was that his world would appear warped at its edges, where he planned to map the regions unknown to Ptolemy—but he realized that this in fact would create a pleasing symmetrical quality that he could deploy to symbolic effect. He could make his map's two sides—the New World on the left, the Far East on the right—appear to spread out majestically from the Old World at its center, almost like the wings of a bird about to take flight—almost, one might say, like the outstretched wings of the great double-headed eagle that Maximilian I used on his imperial standard (*Figures 68 and 69*).

Waldseemüller knew the double-headed eagle well. Visual shorthand for the Holy Roman Empire, it was an emblem dear to the hearts of the early German humanists, and one that Waldseemüller himself would use on his map to mark off Europe's political boundaries. The Holy Roman Emperors had been using the emblem for some three centuries, in large part because of its association with Rome's imperial past. In late antiquity

Figures 68 and 69. *Left:* Holy Roman Emperor Maximilian I on horseback, next to his imperial standard, the double eagle (1508). *Right:* Maximilian's standard transformed into a kind of map, with shields representing the electors and members of the Holy Roman Empire (1510).

the Roman emperors of Byzantium had used the emblem to represent their secular and religious authority, and their dreams of ruling over both East and West.

Medieval Europe's Holy Roman Emperors took over this symbolism and added some of their own. The West, once so mighty, was now old and weak, but guided by the new Caesars it could be restored to health and made to grow strong again. This was just the kind of legend of self-

Figure 70. The double eagle as world map, dedicated
to Holy Roman Emperor Maximilian II (1574).

rejuvenation that the Bible and Christians had long associated with the
eagle. According to one legend, as an eagle grew old its beak curved in on
itself, rendering the eagle unable to eat. But the eagle would break its beak
off on a rock, enabling it to eat again, and with time its youthful energy
would be restored. "The world is passing away," Saint Augustine had written
more than a thousand years earlier, "the world is losing its grip, the world is
short of breath. Do not fear, thy youth shall be renewed as an eagle."

This was precisely the kind of rebirth that Ringmann and other humanists associated with the discovery of the New World. Old Europe had closed
in on itself and was withering, but now, at last, it had bumped up against new
lands that had opened it up to the possibility of revival. Waldseemüller may
well have had such in ideas in mind, consciously or unconsciously, when he
sketched out his map. New lands had been discovered, the Old World was
better understood, and the whole world could be remade—perhaps even in
the form of Maximilian's imperial eagle itself. Although only a conjecture,
this would help explain the most memorable design feature that Waldseemüller would choose for his map: these two hemispheres placed so prominently atop the frame. One, he decided, would show the Old World and be
accompanied by a portrait of Ptolemy, and the other would portray the New

World and the Far East and be accompanied by a portrait of Vespucci. At one level, this was an obvious allusion to the images of Christ that hovered atop the medieval *mappaemundi* (*Plate 1*), and its message was clear: previously only God had been able to see the entire globe, but now, thanks to the learning of the ancients and the achievements of the moderns, epitomized by the achievements of Ptolemy and Vespucci, it was possible for *everybody* to take the whole world in at a glance. *Understand that you are god.* But those two portraits, affixed atop a great wing-shaped map, one looking east and the other looking west, also call to mind something else: the double-headed eagle. Waldseemüller planned to dedicate his map to Maximilian I, soon to be the Holy Roman Emperor of the German Nation, and what better way to honor him than by giving him the world in the form of his imperial standard? The idea would catch on: later in the century the German cartographer Georg Braun and others would dedicate maps to the Holy Roman Emperor that explicitly played with the double-eagle motif (*Figure 70*).

*   *   *

WITH THE SIZE and general look of his map determined, Waldseemüller moved on to the preparation of a master copy. He began by gathering and interpolating data from his various sources, and then preparing a comprehensive list of geographical coordinates, a daunting logistical and mathematical challenge. He then laid out his grid and started to plot those coordinates. Dot by dot, the contours of his world took shape: its continents and oceans, its islands and seas, its regions and cities and towns. Europe, Africa, and Asia all emerged in familiar form—but not the New World. Breaking with the practice of every other mapmaker of his time, Waldseemüller neither made the newly discovered lands a part of Asia nor left their continental status ambiguous. Instead he decided to surround it with water. Years before Balboa and Magellan would supposedly become the first Europeans to confirm the existence of the Pacific Ocean, Waldseemüller put it on his map.

It was a brash move, a leap of the imagination unjustified by his known sources. The Caverio chart and other early depictions of the New World had made no attempt to define its full nature or extent, after all; they had simply delineated the portions of its east coast that had been recently explored. In his letters, both familiar and public, Vespucci himself had always referred to the New World only as a previously unknown part of Asia. But Waldsee-müller had a different idea. "The earth is now known to be divided into four parts," he and Ringmann would write in the *Introduction to Cosmography*—and the fourth part, they would continue, "is found to be surrounded on all sides by the ocean."

What made Waldseemüller and Ringmann so certain defies easy under-standing. Perhaps they had access to now-lost information about an early secret Portuguese voyage through the Strait of Magellan. Perhaps Vespucci himself made such a voyage. The top of the strait lies just below 52 degrees south, some observers have pointed out, a latitude that corresponds almost exactly to the southernmost point that Vespucci claimed to have reached on his great southern voyage. There are other possibilities. Perhaps Waldsee-müller mistook what he saw on the Caverio chart—a land undefined on its western side—for a land surrounded by water. Perhaps he and Ringmann, intoxicated by the writings of Plato, Virgil, Plutarch, Seneca, and others, assumed that the ocean had indeed unloosed its chains and revealed a great new world on the other side of the earth. Or perhaps the explanation is sim-pler. Having decided to lay down a full 360 degrees of longitude on his map, perhaps Waldseemüller realized that he couldn't blur this new land at the western margins of his world, as previous mapmakers had done, but instead had to give it definition—and so he simply made an educated guess. He had plenty of raw data to work with. He had a reasonable idea of the size of the earth; he knew that Ptolemy had wrapped the known world halfway around the globe; he knew how far Martellus and Behaim had expanded Asia to the east beyond Ptolemy's limit; he knew roughly how far Colum-bus and Vespucci had sailed from Europe to the west; and he saw the new coastlines that they had discovered and charted there.

Ringmann himself, as a poet and a dreamer, may have tipped the bal-ance. Addressing Ptolemy in the version of the little geographical poem that he had printed in *Concerning the Southern Shore*, he had described the land discovered by Vespucci as "a world not known in your pictures . . . a land far under the Antarctic Pole." This was generally how scholarly geographers reacted when they first read the *Mundus novus* letter. Vespucci had clearly discovered *something* in the south, but it wasn't yet possible to say what it was. The natural thing for a cartographer to do was to reflect that uncertainty. But in the revised version of his geographical poem that he would include in the *Introduction to Cosmography*, Ringmann left all doubt behind. Most of the poem remained exactly the same—with one significant exception. Instead of describing Vespucci's new world as "a land far under the Antarctic Pole," Ringmann now described it as "a land encircled by the vast ocean."

*   *   *

HAVING SUMMONED UP the general contours of the world's four parts, Waldseemüller now could fill in the details: the geographical regions, the political divisions, the mountains and rivers, the place-names, the descriptive legends, the decorative illustrations.

Europe was a vision of the ancient past. Waldseemüller drew it as the Romans had known it, broken up into such provinces as Anglia, Hispania, Gallia, Italia, and, of course, Germania, which he portrayed the most prominently of all. Moving east, he continued to follow Ptolemy, but also began to introduce the parts of Asia that Europeans had begun to visit and explore in the thirteenth century. Beyond Eastern Europe were Tartaria and the desert wastes of Central Asia: Mongol territory. These were the lands pondered from a distance by Matthew Paris and first described for medieval Europeans by early Christian missionaries—places where, Waldseemüller noted, one could find the eaters of horse and human flesh (*hippophagi* and *anthropophagi*); tiny communities of Nestorian Christians; and the land of Gog and Magog, which Waldseemüller identified as the home of the *iudei clausi*, or "enclosed Jews."

Continuing east and traveling forward in time, Waldseemüller left Ptolemy's world behind. On the eastern side of Central Asia, in a region located vaguely between what he called Upper and Middle India, he drew a large cross signifying the kingdom of Prester John, and to its south borrowed language from the apocryphal *Letter to Prester John* to describe the great king's gem-rich lands. Then came the Far East as Marco Polo, Sir John Mandeville, and other early travel writers had described it, as Martellus and Behaim had mapped it, and as Columbus and Vespucci and other New World explorers had imagined it—although Waldseemüller didn't necessarily spell his place-names in the way that they had. First came

Cathay (which Waldseemüller called Chatay) and Manzi (Mangi), dominated by a picture of the Great Khan's magnificent City of Heaven, Kinsai
(Quinsaij); then came the eastward-jutting peninsula that Columbus had
wrongly associated with Cuba; then the great island of Cipangu (Zipangri), drawn out in the ocean at the very edge of the map, exactly as Martellus had portrayed it; and finally, in the south, the great Spice Islands.

Waldseemüller had now reached the bottom right of his map. Turning west, he passed under the Dragon's Tail and reentered Ptolemy's world.
As he made his way west across the Indian Ocean he filled in many of the
many ancient landmarks that had so obsessed fifteenth-century Europeans: Cattigara, the Aurea Chersonese, the Sinus Magnus, Taprobane, the
Prassum Promontorium. Here he relied primarily on what he found in the
Geography, as he did for Arabia Felix, the Red Sea, and much of Africa's
east coast. But when he turned south along that coast, once again he had to
leave Ptolemy behind. Now he turned to his copy of the Caverio chart, and
as he moved south he copied what he saw: a coastline crammed full of Portuguese place-names; a series of tall Portuguese flags topped with crosses,
representing the *padrões* planted along the coast by the Portuguese; and, in
the interior, a large portrait of an elephant, below which he—or his illustra-
tor—added a huddled group of naked Africans.

Now at the bottom middle of his map, Waldseemüller broke the southern tip of Africa right through his map's frame, just as Martellus had done
before him. Still using his copy of the Caverio chart as his main guide, he
then rounded the Cape of Good Hope and made his way up Africa's west
coast, first traveling north, past more Portuguese *padrões*; then turning west
and making his way along the underside of the West African bulge, past the
fort at Elmina; and then heading north again along the coast of West Africa.

For West Africa, where the new lands discovered by the Portuguese met
the old ones mapped by Ptolemy, Waldseemüller blended elements of the
Caverio chart and the *Geography*. He drew the equator running well above
the West Africa bulge, as Ptolemy had done, even though he knew from
the chart that it actually was located farther south—a confusing decision
he and Ringmann felt compelled to mention in the *Introduction to Cosmography*. He put the recently discovered Cape Verde Islands on his map, and
credited Prince Henry with their discovery—but placed them to the west

of the ancient Roman province of Libya. He referred to the Canary Islands by their classical name, the Fortunate Isles, but mapped them in the marine-chart style. He chose to include the dreaded Cape Bojador of medieval legend on his map, too, but situated directly alongside the Roman province of Getulia. But this mixing of sources didn't last for long. Once Waldseemüller reached North Africa and turned east into the Mediterranean, Ptolemy alone once again became his guide.

At last Waldseemüller turned his attention to the New World. Here he carefully copied what he had found on the Caverio chart, planting a series of flags across the region that divided it into Portuguese and Spanish territory. The wooded island in the north he gave to Portugal, labeling it simply "unknown shore." The islands of the Caribbean he gave to the Spanish and directly credited Columbus with their discovery. The mainland to the east he gave to the Spanish, too, fleshing it out exactly as Caverio had done. In the east he reproduced the names and coastal features, in the west he doodled in a generic mountain range; and beyond it all, at the very western edge of his map, he made room for the ocean—the very same body of water that, on the other side of his map, washed up on the shores of Cipangu and Cathay.

Waldseemüller stuck closely to the chart as he began to sketch the great new southern land explored so recently by Columbus, Vespucci, and Cabral. At the eastern and western ends of the continent's northern shore he placed two Spanish flags, framing an area of the coast, that included the Gulf of Paria. Copying a legend directly from the chart, he described the whole northern portion of the continent as having been discovered for the king of Castile. Moving east, crossing an invisible line of demarcation, he planted two Portuguese flags on the continent's long east coast: the first atop a ship near where Cabral had made his accidental landing, and the second at the most southerly point reached by Vespucci—which, for reasons unknown, he chose to locate at exactly 42 degrees south, drawn into his map's southern frame. In the west he placed more mountains, with more ocean beyond, and in between the two coasts, no doubt spurred on by Ringmann, he gave this new southern continent a name: America.

His world was just about complete. All that remained were the border illustrations, which he would have to ask an artist to do: the large portraits of Ptolemy and Vespucci at the top, and smaller portraits, in the standard Ptolemaic style, of the twelve puffy-cheeked classical winds. At the four corners of the map's frame, however, he left himself room for four long legends—and here, back in the realm of words, he seems once again to have turned to Ringmann for help. Ringmann's influence comes across most obviously in the legend above the New World, which contains strong echoes of what Ringmann had written in *Concerning the Southern Shore* to his friend Jacob Braun. "Many have regarded as an invention the worlds of a famous poet [Virgil]," the legend begins, "that 'a land lies beyond the stars, beyond the paths of the year and the sun, where Atlas the heaven bearer turns on his shoulder the firmament studded with blazing stars.' But now, finally, it proves clearly to be true."

At last a draft version of the map was ready, and it was astonishing to behold. By merging previously independent cartographic traditions, by bringing the Old World together with the New, by revealing and naming a new continent, and by mapping both time and space, the map exalted the spirit of learning and inquiry that had finally made it possible to imagine laying out a vision of the world as a whole. Not everybody, Waldseemüller and Ringmann knew, would be ready to accept this vision, so in the legend that appeared at the bottom right of the map they begged the indulgence of their viewers. "This one request we have to make," they wrote: "that those who are inexperienced and unacquainted with cosmography shall not condemn all this before they have learned what will surely be clearer to them later on, when they have come to understand it."

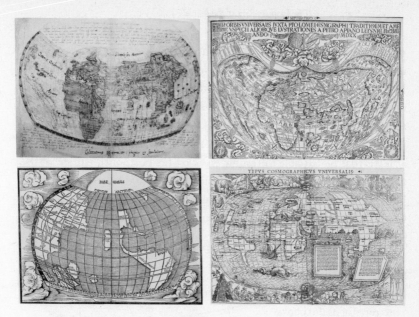

+ Clockwise from top left, maps by Glareanus, Apian, Müster, and Stobnicza +

CHAPTER NINETEEN

AFTERWORLD

We have lately composed, drawn, and printed a map of the whole world,
as a globe and as a flat plane, which is making its way through the world,
not without glory and praise.

—Martin Waldseemüller to Matthias Ringmann (1508)

THE WINTER OF 1506–1507 was a busy one for Waldseemüller. Even
after he had finished the master sheets of his big map he still had much to
do. He repeatedly had to ride down out of the mountains to talk shop with
Johannes Grüninger in Strassburg, and to consult with the artists there
who would be illustrating his map—artists who, it has been suggested, may
have had a connection to the printmaker Albrecht Dürer. He had to oversee
the design and production of his twelve woodblocks; he had to work on the

design and printing of his globe gores; he had to coordinate with Ringmann on the writing and production of the *Introduction to Cosmography*; and he had to meet with Walter Lud and his other Gymnasium colleagues to work out the practical and financial logistics of the printing jobs they were about to undertake.

But by early April most of the work was behind him—enough of it, at least, that he felt able to resume work on his postponed edition of the *Geography*. To that end, at the beginning of the month he sent a letter to Johann Amerbach, an important humanist printer based in Basel since the late 1470s. Waldseemüller had lived and worked in the town as a young man and had evidently come to know Amerbach while he was there, and now he had a favor to ask.

To the distinguished man, Master Johann Amerbach, most diligent restorer of useful books.

Greetings. I do not believe it has escaped your notice that we will be printing Ptolemy's Cosmography in the town of Saint-Dié, after having added certain new maps. And because our exemplars don't agree, I pray that you will gratify me, not as much for my sake as for that of my masters, Walter and Nicholas Lud. I believe you will do this all the more gladly because it is a matter that will benefit our common literary endeavors, for which you labor day and night.

In the library of the Dominicans near you there is a manuscript of Ptolemy written in Greek letters that I judge to be as free from errors as the original. I therefore request that you do whatever needs to be done so that we may borrow it for the period of a month, either in your name or ours. If either a security deposit or a receipt would be helpful, we will see to it that you get whatever you need at once. I would have troubled others with this had I not believed that you would gladly take on this task, and indeed would be able to succeed in it.

The globe that we have prepared for the general [world map] of Ptolemy is not yet printed, but it will be within the space of a month. And if that [Greek] Ptolemy manuscript does come to us, I will see to it that the globe is sent back to you with that same Ptolemy, along with certain other items that might be useful to your sons.

Farewell, and please ensure that we have not troubled you and sought out your help in vain. From the town of Saint-Dié, April 5, 1507.

Martin Waldseemüller, alias Ilacomylus, your most humble servant.

This letter—the only document to have survived in Waldseemüller's own hand—provides a fascinating glimpse of Waldseemüller at work: making the most of his contacts in the publishing world, dropping names to get things done, diligently tracking down an obscure edition of the *Geography*, demonstrating a trust in the accuracy of early Greek editions of the text, and defining his own geographical enterprise as nothing but a humble part of the overall humanist effort to revive and build on the learning of the ancients. His reference to Amerbach's sons is revealing, too. Two of them were university students at the time, members of the very generation of young humanists that Waldseemüller and Ringmann hoped to reach with their maps and their message.

The letter also offers clues about when the Gymnasium's map and globe were printed. Because Waldseemüller refers to the map as though it already exists, it must have been almost complete or even already in print by the time the letter was written, at the beginning of April. His claim that the globe would be finished within a month provides even-more concrete information, suggesting that the globe was probably printed near the end of April—which would correspond nicely with the date of the *Introduction to Cosmography*'s first printing: April 25.

By the end of April all of the hard work was over. Everything was in print: the giant map, the little globe, and the *Introduction to Cosmography*. The members of the Gymnasium presented an honorary copy of the whole package to a delighted Duke René, and a few years later, in 1511, Waldseemüller would recall his reaction with pride. "We have not forgotten," he wrote in dedicating his map of Europe to René's son and successor, Duke Antoine, "with what indulgent attention, with what an agreeable countenance, and with what gracious disposition [your father] accepted a general map of the world, and some other examples of our literary work that we presented to him."

With everything printed, it was finally time to get down to business. Up in Saint-Dié, freshly printed unbound sheets of the *Introduction to*

Cosmography were gathered together (at the time books were sold unbound, with covers made by the buyer or a bookbinder), loaded into barrels, and then trundled down out of the mountains by donkey and cart to Strassburg, perhaps along with copies of the globe gores. There, presumably in Grüninger's printing house, copies of the book joined copies of the big map and were sold as a package—in Grüninger's shop, in markets, in taverns, and in the bustling plaza in front of the town cathedral, where booksellers hawked their wares by stringing up loose copies of title pages around their stalls to advertise their offerings. Business was brisk. The Gymnasium would print more copies of the *Introduction to Cosmography* in Saint-Dié in August, and Grüninger himself would print another edition in Strassburg in 1509.

But Strassburg was only the beginning. The best places to market books in the early sixteenth century were regional book fairs: popular events, held once or twice a year in important European cities, where printers would get together to buy and sell their books, promote forthcoming offerings, dream up new projects, and build their literary and commercial networks. Strassburg had its own fair in 1507, and the members of the Gymnasium surely sold their work there, but they also set their sights on other fairs, including the most important of them all: the Frankfurt Book Fair, still in operation today. "The time of the Frankfurt Fair draws nigh," the great Flemish mapmaker Gerardus Mercator would tell a friend later in the century, when he had a new Ptolemy atlas to sell, "in which, if it can be done, I would hawk about freely this work of Ptolemy." Waldseemüller and his colleagues no doubt had a very similar plan. And so it was, in the spring and summer of 1507, that their map began to make its way across Germany, and beyond.

* * *

SOMETIME IN THE summer of 1507, the Benedictine scholar Johannes Trithemius received a letter from his friend William de Velde. A learned German cleric and author who moved in humanist circles, Trithemius had recently taken over as the abbot of the Würzburg monastery, some seventy miles east of Frankfurt. Aware that Trithemius had an interest in geography, de Velde reported in his letter that he had recently come across something that might be of interest to his friend. His letter is lost—but Trithemius's response is not. "You write that there is a beautifully drawn globe of the lands, seas, and islands for sale in Worms that I might wish to be able to obtain," Trithemius wrote on August 12, 1507. "But nobody will easily convince me to spend forty florins on this. A few days ago, however, I did acquire a handsome small globe for a modest sum, recently printed in Strassburg, and at the same time a large map of the world, expanded on the plane, showing the islands and regions recently discovered by the Spaniard Americus Vespucius in the Western Ocean toward the south, almost to the tenth parallel [50 degrees], with some other matters pertaining to this same exploration."

Trithemius's letter is the earliest surviving independent reference to the Waldseemüller map, and it provides the only surviving description of how the map and its supporting materials were sold. They were marketed as a package, it seems, just as the title page of the *Introduction to Cosmography* suggested they should be—and they were cheap. Globes and large wall maps were no longer just for the wealthy and powerful.

Nothing in the *Introduction to Cosmography* provided Trithemius with any information about how to assemble and mount his new map. But whoever sold it to him surely gave him instructions, and they must have been almost identical to the instructions that Johannes Grüninger would print some two decades later to accompany a different twelve-part wall map of the world: the 1525 Carta Marina of Lorenz Fries. Each edition of this map—itself based on Waldseemüller's 1516 map of the same name—came with a diagram (*Figure 71*) and the following set of instructions.

Figure 71. Diagram for the assembly of a twelve-sheet wall map.
Lorenz Fries (1525).

If you want to mount the map yourself and glue it, take a linen cloth or piece of a clean old linen sheet, put a broad board on a table or chest, and stretch the sheet out firmly, with nails hammered in all around. After that, cut the pages along the left side, so that they fit each other. The middle sheets, *aa* etc., must also be cut off lengthwise. You should try their fit before you start to glue. After that, put some glue, but not too strong, into a little pan. Warm it up, but not too hot. Then take a brush, not too small and with soft bristles. Put the pan on the board on which you have stretched the linen, and then take the first sheet, labeled *a*. Turn it around, brush the back with glue, and put it on the upper left hand, the way one writes *a* and then *b*. Have somebody around to hand you the pages, so that you can put them on fast, and so that they fit immediately. Then put a clean piece of paper over it and rub it with a piece of cloth, so that it is smooth.

Presumably mounted in this manner for classroom use, copies of the Waldseemüller map began to turn up in German universities in the decade after 1507. This may be how the young Swiss humanist Henricus Glareanus first encountered the map. In 1507, at the age of nineteen, Glareanus matriculated at the University of Cologne, where for three years he pursued a wide-ranging course of study. When he left the university, in 1510, he took with him not only the *Introduction to Cosmography* but also several sketches that he had made of the map while at the university (*Figures 2* and *3*)— the very sketches that would surface almost four hundred years later and prove that the Waldseemüller map had indeed once existed.

Untold numbers of other young Germans also came across the map at universities at about the same time. Sometime between 1514 and 1518 at the University of Tübingen, for example, a young professor of Hebrew named Sebastian Münster attended lectures on geography given by one of his colleagues, and sketched copies of more than forty maps in his lecture notebook. Two of these sketches derive from the Waldseemüller map, including this obvious copy of its New World hemisphere (*Figure 72*).

The map turned up elsewhere, too. In 1520, a young mathematician named Peter Apian, a student at the University of Vienna who had recently finished a course of study at the University of Leipzig, made an obvious copy of the map, and published it as his own work (*Figure 73*).

Strassburg, Würzburg, Cologne, Tübingen, Leipzig, Vienna: the map was getting around—as was the *Introduction to Cosmography.* Copies of the

Figure 72. Copy of Waldseemüller's western hemisphere, by Sebastian Münster (circa 1514–18). China and India are on the left, Japan and North America are in the middle; South America is on the bottom right.

book were turning up not only in Germany but all across Europe, to such an extent that in 1516, in his *Utopia,* the English humanist Sir Thomas More was able to refer offhandedly to Vespucci's "four voyages, accounts of which are now common reading everywhere"—almost certainly a reference

Figure 73. Copy of the Waldseemüller map of 1507, by Peter Apian (1520).

to the Vespucci letter that had appeared in the *Introduction to Cosmography.* Intriguingly, the *Introduction to Cosmography* may have helped More come up with the idea for his *Utopia.* More's book, after all, introduces its readers to a communitarian New World paradise far off in the ocean, supposedly discovered by one of the sailors left behind at the end of Vespucci's fourth voyage. The name Utopia itself—a word coined by More from Greek roots to mean "no-place" (*ou-topía*), "good place" (*eu-topía*), or both—bears an uncanny likeness to the name America. Both are four-syllable coinages derived from Greek, both identify a previously unknown region across the ocean and suggest it may be a kind of Nowhere Land (*A-meri-gen*, or "no-place-land," and *ou-topía*, or "no-place"), and both involve a private play on words intended for a small audience of highly educated humanists. More may have been one of the few people who actually got Matthias Ringmann's pun.

What about Vespucci himself? Did he ever come across the Waldsee-müller map or the *Introduction to Cosmography*? Did he ever learn that the New World had been named in his honor? Probably not. Neither the book nor the name is known to have made it to the Iberian Peninsula before his death, in Seville, in 1512. But they did make it there soon afterward: the name America would appear in a book printed in Spain in 1520, and some-time before 1539 Columbus's son Ferdinand, who lived in Spain, is known to have acquired a copy of the *Introduction to Cosmography*. At about this time, too, Bartolomé de Las Casas also read the book and found himself so distressed by its celebration of Vespucci that he launched his famous campaign to set the record straight. "It surprises me," he wrote in his *History of the Indies*, "that the Admiral's son Hernando, who is such a wise man, did not notice how Americo Vespucci usurped the glory of his father, especially since he had documentary proof of it as I know he has." That proof can only have been the *Introduction to Cosmography*.

Inspired by Las Casas, the Spanish would refuse to use the name America for two more centuries, but it was a lost cause from the start. The name, such a poetic and even mythical-sounding complement to the names Asia, Africa, and Europa, had appeared in exactly the right place at the right time, and there was no going back, especially once the German humanists began to spread the word. Influenced heavily by the Waldseemüller map and the *Introduction to Cosmography*, Henricus Glareanus, Sebastian Münster, and Peter Apian would all become geographical authorities in their own right, and in all of their writing and on all of their maps they would identify the New World matter-of-factly as America. Probably no work did more to ensure the acceptance and survival of the name than Münster's 1544 *Cosmography*. Reprinted at least thirty-five times and translated into at least five languages, the book was to become one of the bestselling books of the entire sixteenth century. Against that kind of competition, Las Casas and the Spanish didn't stand a chance.

Many others helped ensure the name's acceptance—most famously, the young Gerardus Mercator, destined to become the century's most influential and widely copied cartographer. At the age of twenty-six, in 1538, when designing his first map of the world, Mercator decided that the whole of the New World, not just its southern part, should be called America. The names he decided to use are the ones that have appeared on most maps ever since: North America and South America.

* * *

WALDSEEMÜLLER AND RINGMANN continued to work together in the years after 1507. Ringmann traveled to Italy again in 1508, this time returning with a Greek Ptolemy manuscript, and he and Waldseemüller appear to have completed work on their new edition of the *Geography* soon after. But their plan to print it in Saint-Dié came to nothing. Out hunting wolves in December of 1508, Duke René had a stroke and died the same day. His successor, Duke Antoine, seems to have cared little about the Gymnasium or its work.

Ringmann himself didn't have long to live. By 1509 he was suffering from chest pains—probably tuberculosis—and was exhausted. Despite his poor health he maintained a punishing work and travel schedule, but by the fall of 1511 he was at death's door. The letter he wrote that fall praising Waldseemüller's new map of Europe may have been the last time Waldseemüller ever heard from him. "Considering the haste of my composition and the gravity of my illness," Ringmann told his old friend, "you will not refuse to excuse me, dear Martin, for having written without elegance and Latin grace." Just a few weeks later, before he had even reached the age of thirty, Ringmann was dead.

Waldseemüller would live on for another eight years. But something peculiar happened after Ringmann's death: Waldseemüller stopped using the name America. In 1513, when his new edition of the *Geography* was finally printed, in Strassburg, neither of the two maps in the atlas that

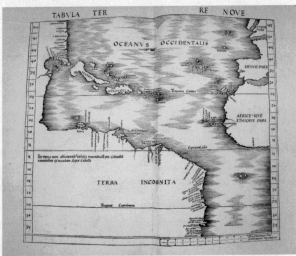

Figures 74 and 75. Top: the modern world, from Martin Waldseemüller and Matthias Ringmann's Strassburg *Geography* (1513), with the name America conspicuously absent. *Bottom:* the New World map from the same edition of the *Geography.* Here South America is called *Terra Incognita,* and only Columbus is identified as its discoverer.

portrayed the New World called it America or showed it surrounded by water (*Figures 74 and 75*).

Waldseemüller depicted the New World in very much the same way three years later, in 1516, when he produced his Carta Marina: a stunning, elaborately illustrated, and heavily annotated twelve-sheet printed map of the world, based closely on the Caverio chart. Here Waldseemüller labeled North America THE LAND OF CUBA—PART OF ASIA and called South America simply THE NEW LAND, although he did go on to identify the northern region discovered by Columbus as THE LAND OF PARIA, and the southern region discovered by Vespucci only as BRASILIA, OR THE LAND OF THE PARROTS (*Figure 76*).

As he explained in a long and curious address to the reader at the bottom left of the map, Waldseemüller designed his Carta Marina as a follow-up to the 1507 map, and as a retraction of sorts. The address—which centuries later would help Joseph Fischer identify Waldseemüller as the author of the two world maps he had just discovered at Wolfegg Castle—is both revealing and cryptic, and at times hauntingly plaintive.

> Ilacomilus, Martin Waldseemüller, wishes the reader good fortune.
>
> We will seem to you, reader, previously to have diligently presented and shown a representation of the world that was filled with error, wonder, and confusion. We do believe that the reader disagrees with us in this representation, in that we have represented irregular forms in our previous description of the land and sea (and these we certainly described with no deceiving rhetoric). As we have lately come to understand, our previous representation pleased very few people. Therefore, since true seekers of knowledge rarely color their words in confusing rhetoric, and do not embellish facts with charm but instead with a venerable abundance of simplicity, we must say that we cover our heads with a humble hood.
>
> In the past we published an image of the whole world in a thousand copies that was completed in a few years, not without hard work, and that was based on the tradition of Ptolemy, whose works are known to few because of his excessive antiquity. . . . Although it is well known that the machinery of the world has not varied since the time of Ptolemy, it is

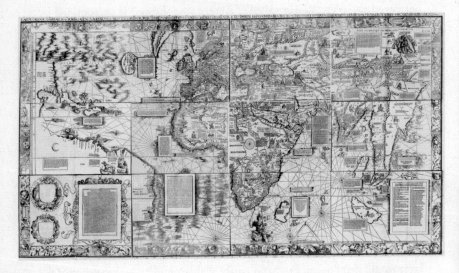

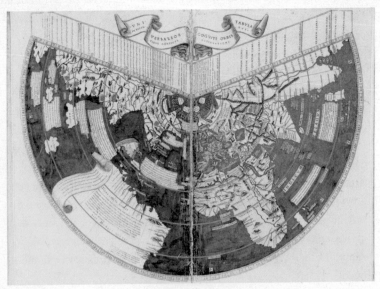

Figures 76 and 77. Top: the Carta Marina of Martin Waldseemüller (1516).
Note the close correspondence to the Caverio chart (Figure 67). *Bottom:*
the world map of Johannes Ruysch (1508). The New World is on the left.
Its northern shores are identified as a part of Asia, but the giant nature
of the land in the south is left undefined.

indeed a fact that the passage of time inverts and changes things, so that it is difficult to find one city or one region in twenty that since his time has kept its ancient name or has not been newly developed. Because of this, and because nothing in these matters is clear in hindsight, difficulties may arise in our understanding of very distant regions and cities. Where are now located Augusta, Rauricum, Elcebu, Berbetomagus, or, among the foreign maritime powers, Byzantium, Aphrodisium, Carthage, Nineveh, whose names and locations have been transferred to us with great accuracy by Ptolemy? This is, of course, a difficult question. Are they close by, next to the River Rhine, or far away and concealed? Who has knowledge of, who can tell apart, and who can make known to us the Sequani people, the Hedui, the Helvetians, the Leuci, the Vangioni, the Hagoni, the Mediomatrices, all of whom were at one time so well known? I acknowledge that it is possible that no one could now know the manners of the ancients and could come upon knowledge of Celtic Gaul and Belgium, Austrasia, Norica, Sarmatia, Scythia, Thaurica, and the Golden Chersonese, the Bay of Caticolphi, the Bay of Ganges, and the very well-known island of Taprobane. Time is expansive; it renews and brings change into the affairs of men. . . .

Moved by these considerations, I have prepared this second image of the whole world for the benefit of the learned, so that, just as the representation of the whole of the land and sea by the ancient authors stood together [in the 1507 map], not only would the new and present image of the world shine through but also the natural change that has taken place in the intervening time would be so evident that you would have a unique view of what sort of things become perishable. These things, whatever they may have been in the past and whatever they may become in the future, are presented so that this change may in no way be doubted as time goes on. It has pleased us, therefore, to create an image and description of the whole world as a marine chart, after the manner of modern cartographers. . . . We took great care in making sure that not a single word of our description be embellished in some sweet style or adorned with some kind of festive arrangement, for it is always better to speak in a humble and truthful style. For this reason we ask you to look upon us with a benevolent spirit.

Why the need for this extended apology? One possibility is that not long after publishing his map, Waldseemüller came across the map of Johannes Ruysch and decided that its picture of the New World was better (*Figure 77*). Ruysch's map, which appeared in a 1508 reprint of the Rome *Geography*, depicted North America as a part of Asia, and South America as a giant new land of a still-undetermined size and nature. Ruysch placed a scroll over the west coast of South America, which read, in part, "Inasmuch as they have not wholly explored it [South America] . . . it must remain thus imperfectly delineated until it is known in what direction it extends." Waldseemüller clearly knew Ruysch's map; to describe South America on his Carta Marina he borrowed several sentences verbatim from the legend Ruysch had placed in the middle of the continent.

There are other possibilities. Perhaps Ringmann, in the thrall of his Virgil prophecy and his language games, had convinced Waldseemüller to depict the New World more imaginatively and boldly in 1507 than Waldseemüller had ended up feeling comfortable with, and with Ringmann now gone Waldseemüller felt able to back away from his earlier work. Perhaps Waldseemüller came to feel that by combining so many different sources and traditions, his 1507 map had come to represent exactly the sort of "monstrous chaos" that his new edition of Ptolemy was supposed to dispel. Perhaps the map's symbolism and fawning dedication to Maximilian offended his French masters in Saint-Dié, who had made him a canon in the town in 1514. Or perhaps, as he read more about the discovery and exploration of the New World in the years after 1507, he began to realize that Vespucci wasn't everything his public letters had made him out to be.

Waldseemüller would produce no more maps after the Carta Marina. Four years later, on March 16, 1520, he passed away—"dead without a will," a clerk would later write when recording the sale of his house in Saint-Dié, "the late Martin Valdesmiles, formerly a canon in these parts."

Copies of the Waldseemüller map began to disappear during the decades that followed. No other fate was possible for maps slathered with glue, pasted onto old linen sheets, hammered full of tacks, hung up on walls, rolled up and unrolled time after time, scribbled on with corrections, poked and pawed at by admirers, exposed to heat and cold, to sun and damp, to smoke and soot—and, above all, rendered obsolete by the ongoing pace of geographical discovery. By the tens, scores, and hundreds, copies wore out or were discarded in favor of newer, more up-to-date, and better-printed maps. By 1570 the map had all but vanished from memory. When the mapmaker Abraham Ortelius that year published a remarkably comprehensive list of his cartographical predecessors and their maps, he mentioned Waldseemüller—and made no reference to the great 1507 map.

But one copy did survive. Sometime between 1515 and 1517, while gathering materials to help him design and print globe gores of his own, the Nuremberg mathematician Johannes Schöner acquired a copy of the map—not an original but a reprint, recent studies have concluded, because of the worn and cracked state of the woodblocks, and the kind of watermark used on the paper. Schöner didn't need the map for display or for teaching purposes. He needed it as a reference work, so he didn't assemble or mount it. Instead he bound it—along with a copy of Waldseemüller's 1516 Carta Marina, a star chart engraved by Albrecht Dürer in 1515, and celestial globe gores that he himself produced in 1517—into an oversized, wood-covered folio that he could keep in his personal library. In the years between 1515 and 1520 Schöner studied the map carefully. He drew a grid of red lines across parts of its two central sheets, presumably to help him transfer the coordinates of places he saw there to his globes—of which he produced several during those years, all of which rely heavily on the Waldseemüller map. But as the decades wore on, as newer maps became available, and as his own interests shifted from geography to astronomy, Schöner consulted the folio less and less. By the time he died, in 1545, he probably hadn't opened it in years. The Waldseemüller map was now about to begin its long sleep. But by then the map had already reached Poland and helped convince a German-speaking humanist Nicholas Copernicus, that it was time to propose a fundamentally new model of the cosmos.

* * *

COPERNICUS PUBLISHED *On the Revolutions of the Heavenly Spheres*, the astronomical treatise that laid out his new theory of the cosmos in full, in 1543, the year of his death. He had actually developed the idea much earlier, but for decades he had resisted the idea of publishing it, for fear of how it would be received. "I can readily imagine, Holy Father," he told Pope

Paul III in the opening line of his preface to *On the Revolutions*, "that as soon as some people hear that in this volume, which I have written about the revolutions of the spheres of the universe, I ascribe certain motions to the terrestrial globe, they will shout that I must be immediately repudiated together with this belief." Copernicus did discuss his ideas privately with friends during much of his lifetime, however, and even sent some of them a brief manuscript summarizing his heretical theory of the cosmos, a work today known as the *Commentariolus,* or *Little Commentary*. Many of those who admired Copernicus's ideas pressed him to go public with them, and eventually, as he wrote in his preface, they wore him down. "The scorn that I had reason to fear on account of the novelty and unconventionality of my opinion," he wrote, "almost induced me to abandon completely the work which I had undertaken. But while I hesitated for a long time and even resisted, my friends drew me back. . . . [They] repeatedly encouraged me and, sometimes adding reproaches, urgently requested me to publish this volume and finally permit it to appear after having buried among my papers and lying concealed not merely until the ninth year but by now the fourth period of nine years."

The fourth period of nine years: the phrase is convoluted, but the math is simple. Four periods of nine years is thirty-six years—and thirty-six years before 1543 is 1507.

Despite what his preface implies, Copernicus didn't write *On the Revolutions* in its entirety in 1507 and then keep it hidden for decades. The last part of the manuscript, which still survives, reveals itself to have been composed in haste, and it contains references to celestial observations made not long before Copernicus's death. There's no doubt, however, that Copernicus had developed a preliminary version of his theory and put it into writing early in the century. By 1514, for example, he seems already to have written and begun circulating the *Little Commentary*; that year, Matthew of Miechów, a physician and a professor of geography at the University of Kraków, under whom Copernicus had studied geography a decade earlier, prepared an inventory of his library and noted that in a beechwood drawer he kept something remarkable—"a manuscript of six leaves expounding the theory that the earth moves while the sun stands still!"

In the preface to *On the Revolutions*, Copernicus made sure to point out that he was not the first to suggest that the earth was in motion. Several

Greek astronomers had made the claim in antiquity, he wrote, and "it was they," he explained, "who first opened the road to the investigation of these very questions." The idea had resurfaced in modern times, too, although Copernicus didn't mention this in his preface. In 1440 the influential cardinal and humanist Nicholas of Cusa—a noted astronomer and cartographer who also was a close friend of Toscanelli's—had written that "the earth, which cannot be the center [of the cosmos], must in some way be in motion." At the very time Copernicus was developing his theory, Leonardo da Vinci was thinking along very similar lines. "The earth is not the center of the sun's orbit nor at the center of the universe," he wrote in 1511, in one of his private notebooks, "but in the center of its companion elements, and united with them."

Like so many other German-speaking humanists, Copernicus spent time studying in Italy in the late 1490s and early 1500s, and it was probably during this period that he first came across the theory that the earth was in motion. The theory clearly intrigued him, and after he returned to Poland in 1503 he began looking for ways to prove it. Sometime not long after 1507 he seems to have come across the Waldseemüller map—and what he saw on it made him realize that the newly named America was where he should start.

Both the map and the *Introduction to Cosmography* are known to have reached Poland—specifically, the University of Kraków—no later than 1512. The university was a center of humanist learning in Eastern Europe at the time; Conrad Celtis had spent time there from 1489 to 1491, and Copernicus himself had studied there between 1491 and 1495. On the faculty at the university were a number of professors who taught geography, and in 1512 one of them, a Johannes de Stobnicza, published a small geographical compendium that betrays obvious debts to the Gymnasium's geographical package. Titled *Introduction to the Cosmography of Ptolemy, with the Latitudes and Longitudes of Famous Regions and Cities*, the work declared its intention to describe the lands "that have come to our knowledge through the travels of Amerigo Vespucci and others"; it repeatedly referred to the New World as America; it identified America as a newly discovered fourth part of the world, and at one point reproduced the naming-of-America passage almost verbatim—all sure signs that Stobnicza was cribbing from *Introduction to Cosmography*. Stobnicza also printed two little hemispheric maps in his book, one of the Old World and another of the New World (*Figure 78*), and these leave no doubt whatsoever that Waldseemüller was his source.

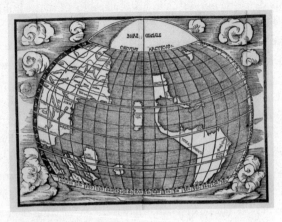

Figure 78. Copy of Waldseemüller's western hemisphere, by Johannes de Stobnicza (1512).

In the years after his return from Italy, Copernicus maintained ties with friends and former professors at the University of Kraków, several of whom actively followed reports of the New World discoveries and had strong ties to the world of German humanism and publishing. Copernicus himself had a particular interest in geography during that period, and in fact would later go on to make maps of his own. If, as seems undeniable, the *Introduction to Cosmography* and the Waldseemüller map made it to Poland in the years immediately after 1507 the odds are that Copernicus came across them. But the best evidence for such a connection comes from *On the Revolutions* itself, a work Copernicus himself claimed to have begun at that very time.

*   *   *

ON THE REVOLUTIONS is primarily a work of astronomy and geometry. But what allowed Copernicus to begin writing the book was a geographical revelation.

A fundamental question nagged at Copernicus as he considered the traditional model of the cosmos. If the cosmos was a set of interlocking spheres with the earth sitting immobile at its center, why wasn't the earth completely surrounded by water? In the preceding centuries, Copernicus knew, scholars had come up with an explanation. God had displaced the earth from the center of the cosmos, their theory went, making one of its sides bob to the surface of the watery sphere so that land and air would meet and make terrestrial life possible. This meant that the earth could no longer occupy the geometrical center of the watery sphere, or, by extension, the cosmos, but that didn't matter; what was important was the gravitational center of the cosmos, where everything finally came to rest. Because land exposed to air was lighter than land saturated with water, according to this theory, the earth's gravitational center and the center of the cosmos were still one and the same.

Copernicus knew this argument well from his student days. But he didn't buy it. Mathematically, geometrically, logically, it just didn't make sense—as he explained at the very beginning of *On the Revolutions*, in a little section titled "How Earth Forms a Single Sphere with Water."

We should not heed the peripatetics who declared that . . . the earth bulges out to some extent as it does because it is not of equal weight everywhere, on account of its cavities, its center of gravity being different from its center of magnitude. But they err through ignorance of the art of geometry. For they do not realize that the water cannot be even seven times greater and still leave any part of the land dry, unless the earth as a whole vacated the center of gravity. . . .

Moreover, there is no difference between the earth's centers of gravity and magnitude. This can be established by the fact that from the ocean inward the curvature of the land does not mount steadily in a continuous rise. If it did, it would keep the sea water out completely and in no way permit the inland seas and such vast gulfs to intrude. Furthermore, the depth of the abyss would never stop increasing from the shore of the ocean outward, so that no island or reef or any form of land would be encountered by sailors on the longer voyages.

But, Copernicus noted, inland seas and gulfs *did* exist; the Mediterranean and the Red Sea were proof enough of that. He then went on to announce that modern explorers had recently brought home even more compelling evidence—empirical evidence, not philosophical or theological evidence—against the idea of an earth that bulged out from one side of the watery sphere. "In his *Geography*," he wrote,

> Ptolemy extended the habitable area halfway around the world. Beyond that meridian, where he left unknown land, the moderns have added Cathay and territory as vast as sixty degrees of longitude, so that now the earth is inhabited over a greater stretch of longitude than is left for the ocean. To these regions, moreover, should be added the islands discovered in our time under the rulers of Spain and Portugal, and especially America, named after the ship's captain who found it. On account of its still undisclosed size it is thought to be a second group of inhabited countries. There are also many other islands heretofore unknown. So little reason have we to marvel at the existence of antipodes. . . . Indeed, geometrical reasoning about the location of America compels us to believe that it is diametrically opposite the Ganges district of India.

Only two sources in the years immediately after 1507 could have provided Copernicus with this dense cluster of very specific geographical information: the Waldseemüller map and the *Introduction to Cosmography*. The map, for example, singles out "the land of Cathay and all of southern India" as a newly discovered region that lies "beyond 180 degrees of longitude," and then goes on to place that region's easternmost tip at a point exactly 60 degrees beyond Ptolemy's limit. Everything Copernicus goes on to say, too, can be easily found in the *Introduction to Cosmography* and even more easily observed on the Waldseemüller map—the references to the kings of Spain and Portugal as the prime movers of western discovery; the particular attention paid to America and its "discoverer," Amerigo Vespucci; the remarks on the existence of previously unknown islands and the Antipodes. But most telling of all is Copernicus's final line, in which he declares America to be "diametrically opposite the Ganges district of India." Of all the maps available to Copernicus in the early sixteenth century, only the Waldseemüller map presented a vision of the world that could have prompted him to make this particular observation. Waldseemüller placed the word *Ganges* at 145 degrees of longitude on his map, just below the Tropic of Cancer, and placed the word *America* at 325 degrees of longitude, just above the Tropic of Capricorn—points on the globe that are diametrically opposite each other.

Gazing at the map, or reflecting on it later, Copernicus seems to have had an epiphany. If the earth really did bulge out of the watery sphere on one side—as the diagrams in Sacrobosco's *Sphere* (*Figure 26*) and Pierre d'Ailly's *Image of the World* (*Figure 43*) showed it to do—then by definition the ocean had to get deeper and deeper the farther away one sailed from the shores of the known world. In this model of the cosmos the earth simply could not protrude from the sphere of the water in two diametrically opposed places—but that's exactly what Copernicus saw happening on the Waldseemüller map. In front of him, based on the empirical observations of modern explorers and laid out on a majestic scale by humanist cosmographers whose learning he could trust, he had a dramatic new picture of the entire world that showed a giant new southern land lying diametrically opposite the known world in the middle of the ocean. Only one explanation seemed possible: there simply could not be a watery sphere. "From all these facts, finally," Copernicus concluded, "I think it is clear that land and water together press upon a single center of gravity; that the earth has no other center of magnitude; that, since earth is heavier, its gaps are filled with water; and that consequently there is little water in comparison with land, even though more water perhaps appears on the surface."

Quite suddenly, at its very core, the traditional model of the cosmos had fallen apart, and Copernicus at last felt himself free to begin working out what a sun-centered cosmos might look like. There were only two possible ways to explain the motion of the heavens, after all: either the heavens or the earth had to be moving. If the earth-centered model produced discrepancies that could not be explained away, and that, in fact, were directly at odds with observable geographical reality, then why not consider the other possibility? Aristotle, Copernicus noted, had argued that the natural motion of all heavenly spheres was circular. If this was the case, then why not assume that the earth moved in this very same way? "We regard it as a certainty," he explained,

that the earth, enclosed between poles, is bounded by a spherical surface. Why then do we still hesitate to grant it the motion appropriate by nature to its form [and instead] attribute a movement to the entire universe, whose limit is unknown and unknowable? Why should we not admit with regard to the daily rotation that the appearance [of motion] is in the heavens, and the reality in the earth? . . . For when a ship is floating calmly along, the sailors see its motion mirrored in everything outside, while on the other hand they suppose that they are stationary, together with everything on board. In the same way, the motion of the earth can unquestionably produce the impression that the entire universe is rotating.

* * *

COPERNICUS WOULDN'T ALLOW *On the Revolutions* to be published until 1543, at the very end of his life. But his ideas got around well before that. By the 1530s his new theory of the cosmos had attracted admirers, some of whom began to allude to it in print. Fittingly, the earliest of these allusions emerged in a work not of astronomy but of geography, in the form of a remarkable little map that—slyly, for those in the know—brought together the new worlds of both Waldseemüller and Copernicus (*Figure 79*).

Generally attributed to Sebastian Münster, the map appeared in a little volume of travel writing titled *The New World of Regions and Islands Unknown to the Ancients*, a work edited by Simon Grynäus and published in 1532. As is often noted, the map betrays many debts to the Waldseemüller map. Among other similarities, Münster's map presents viewers with an unusual blend of Ptolemaic and non-Ptolemaic sources; it identifies the southern portion of the New World as America; it locates many land formations exactly where they appear on the Waldseemüller map, latitudinally and longitudinally; and it uses many of the very same place-names, including Zipangri for Cipangu, Spagnolla for Hispaniola, and Isabella for Cuba. Münster also relied on other sources to make his map (it reflects an awareness of Magellan's recent circumnavigation of the globe, for example), but its overall vision is one inspired by the Waldseemüller map—which, of course, Münster had sketched in his notebook some fifteen years earlier, during the geography classes he had attended as a young university professor. But Münster on his map also made a nod to Copernicus that's rarely noticed. At the top and the bottom of his map, as part of the ornamentation of its border, two little cherubs appear. One hovers above the North Pole, and the other below the South Pole, and both turn the world around on a crank (*Figure 80*).

The message, one that would have dumbfounded Waldseemüller, is clear. At last mapped as a whole, the world has four parts—and together they have begun to turn.

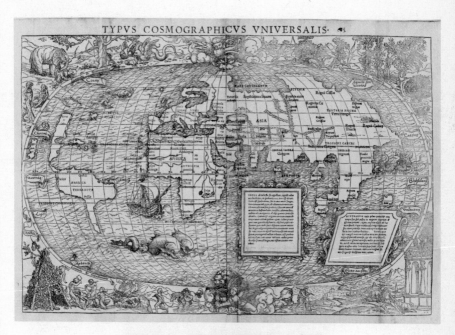

Figures 79 and 80. Top: world map generally attributed to Sebastian Münster (1532), demonstrating a clear debt to the Waldseemüller map of 1507. *Bottom:* detail from the Münster map, showing a cherub turning the world on a crank eleven years before Copernicus published his theory that the earth revolved around the sun.

+ Alexander the Great +

Epilogue

THE WAY OF THE WORLD

*Wonder is the movement of the man who does not know on his way
to finding out. . . . Such is the origin of philosophy.*

—Albertus Magnus (circa 1250)

NEAR THE BEGINNING of his *Chronicle*, Matthew Paris, the Benedictine monk with whom this book began, drew a portrait of Alexander the Great. At first glance the portrait appears to be just one of Matthew's many doodles. But it deserves a closer look, because with a wonderfully light touch it captures the new spirit of inquiry that was emerging in Matthew's day and would animate the many different voyages of geographical and intellectual discovery that made the Waldseemüller map possible.

Depicted in the garb of a Christian king and symbolizing the imperial and scholarly heritage of antiquity, Alexander holds a T-O globe in his hand. It's a classic pose, drawn countless times during the Middle Ages to suggest the power and wisdom of a ruler. But Matthew subverts the form. His Alexander isn't static. He doesn't sit stiffly on his throne and stare blankly out from the page. He's an active, inquisitive figure. With his head cocked to one side, one hand on his hip, and one eye closed, he holds the globe in his hand and takes its measure. What he sees are the three parts of the known world, but you can tell he's wondering about something else—a fourth part, perhaps somewhere across the ocean, on the back side of the globe. He's trying to catch a glimpse of the unknown.

When medieval authorities discussed the fourth part of the world, they had an actual place in mind, of course. But they also understood the expression to have a metaphysical significance. It represented a basic truth about the human condition: that the unknown will always be with us. This was dogma in Christian Europe for centuries. The coordinates of all creation had been projected onto a cosmic map of space and time, with the earth and human history at its center, and, as the *mappaemundi* taught, only God could see and understand it as a whole. But that dogma had begun to erode when Matthew drew his portrait of Alexander. Emboldened by a newly empirical and humanistic spirit, growing numbers of Europeans cocked their heads, wondered about the world, and set out to take its measure for themselves. Their motives differed, and they moved in fits and starts. But gradually, guided to a surprising degree by a Church increasingly global in its outlook and imperial in its ambitions, their efforts fed into a collective quest for knowledge, power, and wealth the likes of which had never before been seen. That quest was at once mystical, rapacious, evangelical, self-centered, grand, inspiring, and often delusional—and nothing charts its full course better than the Waldseemüller map of 1507.

The map reveals world upon world. It shows the world as Plato imagined it, with the inhabitants of the Mediterranean Basin bustling and scurrying around at the water's edge like ants and frogs around a marsh. It shows the world as Ptolemy mapped it in the second century, with Rome at the height of its geographical and imperial reach. Irish monks bob up and down in tiny boats in the North Atlantic, hoping for a glimpse of paradise. Mongol hordes sweep across Central Asia and into Europe, and in response Friar John and Friar William, clad only in their robes and sandals, plod eastward in search of the Great Khan. Marco Polo treks all the way to the Sea of Cathay, and Sir John Mandeville goes farther still, taking his readers beyond the realm of the Earthly Paradise, out into the ocean, past the great island of Cipangu, and right back around the globe to where his journey had begun. Moving in the other direction, the Vivaldi brothers sail out beyond the Pillars of Hercules and into the Atlantic, hoping to find a new route to India. European sailors chart Mediterranean and Atlantic coastlines, and snap the known world into a new kind of focus. Petrarch, Boccaccio, and the early humanists revive the study of ancient geography; Manuel Chrysoloras carries Ptolemy's *Geography* to Florence; and the learned men, at their gatherings in Constance and Florence, try to synthesize the geographical accounts of the ancients and the moderns. Guided by Prince Henry, by the Catalan charts, by Ptolemy and Fra Mauro, and by the stars, the Portuguese sail south past Cape Bojador, east under West Africa, and south again across the equator, following a coast that refuses to end. The aged Paolo Toscanelli tells them to sail west, not east, to reach the Indies, and Bartolomeu Dias makes landfall near the Cape of Good Hope, on the underside of the world.

Columbus is there too. Four times he sails across the Atlantic, and after each voyage he reports momentous news: he has reached the Indies; he has located Ophir and Tarshish; he has approached the Earthly Paradise; he has found the Aurea Chersonese. Cabot, the Corte-Reals, and Cabral crisscross those waters—and then there's Vespucci, maddening, unknowable Vespucci, who sails deep into the southern hemisphere, and whose reports of a vast new land lie somewhere between truth and fiction. New coastlines rise up in isolated stretches on the early New World marine charts; one of those charts makes its way to the little mountain town of Saint-Dié, in Lorraine; and there, in the minds of Martin Waldseemüller and Matthias Ringmann, those coasts resolve themselves at last into a new, fourth part of the world.

The Waldseemüller map contains all of this, and more. It's the world viewed as a whole from above, a godlike vision of the earth suddenly accessible to all. It's a supplement to Ptolemy, and an introduction to cosmography. It's a record of the past, a commentary on the present, and a dream of the future; a world at once ancient and medieval and modern. It's the Holy Roman Empire transformed into the double eagle, it's the fulfillment of ancient prophecy, and it's the backdrop for something new: a modern epic of Western discovery and manifest destiny in which European explorers, like Odysseus and Alexander and Aeneas before them, wander the known world, roam the high seas, and arrive at unknown shores. It's a birth certificate for the world that came into being in 1492—and it's a death warrant for the one that was there before. It's a cosmic revelation: a globe wrenched out of its age-old place at the center of the cosmos and set free, at last, to spin and wobble its way around the sun.

Geography as we now understand it, as an independent branch of learning devoted to the description of the earth and its features, is something very new. For millennia the discipline was something less than that—and something more. It was a tool used by philosophers and theologians to probe the mysteries of existence and to trace the course of human history. It laid out the boundaries of the known. By placing the fourth part of the world inside those boundaries and then extending them all the way around the world for the first time, the Waldseemüller map helped usher in the modern geographical era, an achievement for which it deserves an important place in the history of ideas. But the map also represents the culmination of that much older tradition in which geography is philosophy—and in which the appearance on a map of the fourth part of the world is a humbling reminder of all that still remains unknown. What the map ultimately charts, in other words, is nothing less than the contours of the human experience itself: the never-ending attempt to imagine a place for ourselves in the world.

• The modern world, by Martin Waldseemüller (date uncertain) •

Appendix

THE STEVENS-BROWN MAP

I hope to find time to write you shortly about a map I have discovered which bears the name America earlier than anything known, and somehow the idea has got into my mind that it is Waldseemüller's long lost St. Dié map.

—Henry Newton Stevens Jr. to John Nicholas Brown (1896)

IN DECEMBER OF 1893, at a book auction in London, the American rare-book and map dealer Henry Newton Stevens Jr. bought an incomplete copy of the 1513 Strassburg *Geography*. The copy consisted of a fragment of Ptolemy's text and just one map: Martin Waldseemüller's well-known 1513 depiction of the modern world (*Figure 74*). The offering attracted no other bidders, and Stevens was able to obtain it for the bargain price of two pounds, four shillings.

Stevens gave little thought to what he had bought. As one of the part-
ners at Henry Stevens, Son & Stiles, an antiquarian bookseller located just
across the street from the British Museum, he routinely scooped up cheap,
imperfect copies of rare texts and then stored them away in his shop. If later
he acquired a better but nevertheless incomplete copy of one of these texts,
he could often extract pages from his fragmentary copies and use them to
"restore" his better texts, so that he could sell them as complete.

Stevens didn't touch his Ptolemy fragment for a few years, until, in
the mid-1890s, he had reason to dig it out of storage—"with a view," he
would later recall, "to improving another imperfect copy which had just
been acquired." But, to his astonishment, when he looked over the map this
time he noticed something that he hadn't seen before: the word AMERICA,
printed in block letters across the New World.

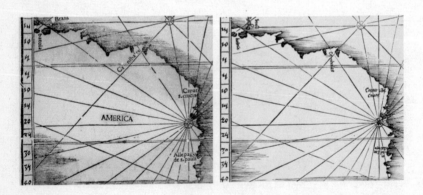

Figures 81 and 82. *Left*: the New World bearing the name America, as it
appears on the Stevens-Brown map (opening illustration of this Appendix).
Right: the New World bearing no name, as it appears on the standard 1513
Waldseemüller map of the modern world (*Figure 74*).

This was most strange. Many copies of Waldseemüller's 1513 world
map had survived into the nineteenth century, but all of them—printed
as they were after Waldseemüller's apparent retraction of the word
America—left the New World nameless. In size, title, and general appearance, Stevens's map was almost identical to the 1513 map, but as he
inspected it closely Stevens realized that it had been printed from an entirely

different woodblock, and that it contained two important differences: it used inserted roman type for place-names, instead of carved Gothic letters, and it included a few place-names that didn't appear on the 1513 map—most notably, America (*Figures 81 and 82*).

The discovery thrilled Stevens. At the time, historians of cartography considered Peter Apian's 1520 copy of Waldseemüller's big 1507 map (*Figure 73*) to be the earliest surviving printed map containing the name America. Stevens realized that his map might be several years earlier than Apian's, and that perhaps what he had in his possession was a prototype of the 1513 map, engraved and printed sometime before April of 1507—sometime, that is, before the Gymnasium had postponed work on its new edition of Ptolemy. Stevens even began to imagine that his map might be the long-lost 1507 map itself.

Stevens promptly sent news of his find to one of his most important customers, John Nicholas Brown, an avid American collector of books and maps relating to the discovery of the New World. Brown was the scion of the family that in the previous century had helped found Brown University, in Providence, Rhode Island, and his collection was (and still is) housed in the university's rare-book library, the John Carter Brown Library. "Please regard this as confidential at present," Stevens told Brown after suggesting to him that his map might be the Waldseemüller map of 1507. "It is not for sale yet, but you shall have first offer when it is."

Needless to say, Brown responded with interest. But he also asked for proof of the authenticity and primacy of Stevens's map. Stevens threw himself into the task, and in the years that followed was gradually devoured by an obsession with his map. With the help of friends at the British Museum he investigated every aspect of the map he could think of: its place in cartographical history, its relationship to the other parts of the Gymnasium's geographical output, its correspondences to and tiny differences from the 1513 map, the kind of paper it was printed on, the provenance of its watermark, the style of lettering used, and so on. By 1900, based on what he discovered in his research, he felt he had been able to prove that the map had been engraved and printed in Nuremberg, sometime after 1505 and before April of 1507, and he decided, moreover, that his map was not only the prototype of the 1513 map but also the first printed map ever to use the word

America or show any part of the New World. John Nicholas Brown died in 1900, but Stevens pressed ahead with his investigations nevertheless, and began sending updates to George Parker Winship, who ran the John Carter Brown Library and who had been deputized as its buying agent by Brown's mother, Sophia Augusta Brown.

In December of 1900, Stevens finally made his official case to Winship. "You can make up your mind at once that you are going to be convinced," he wrote in a cover letter accompanying a lengthy report. "The Museum people are quite satisfied I have proved my case. . . . As the ginger beer merchant said when asked to guarantee his beer . . . 'No pop, no pay, but you had better get your money ready all the same.'"

Stevens's argument, in brief, was that his map had been designed in Saint-Dié but engraved and printed in Nuremberg sometime before April of 1507, probably in either 1505 or 1506. No printing press had yet been established in Saint-Dié at that point, Stevens contended, so the members of the Gymnasium had had their map produced in Nuremberg, perhaps as a kind of test run for their edition of the *Geography*. But then Waldseemüller and Ringmann set aside the *Geography* in order to produce their big 1507 map, their globe, and the *Introduction to Cosmography*. By 1513, when his edition of the *Geography* was finally printed in Strassburg, Waldseemüller had backed away from the name America, so naturally the name was dropped from the new version of the map that was engraved and printed that year.

Surprisingly, almost no scholars have examined the map since Stevens laid out his argument. Among the few who have, the informal consensus is that Stevens—whose vested interest in building up the historical importance of his map as he was trying to sell it made him anything but a neutral observer—overstated his case. It would make sense that the map was produced at the same time as the 1507 wall map and globe gores, as part of the overall package of "America" materials being produced by the Gymnasium in those years; but it would also make sense for the map to be kind of a reissue, produced sometime after the name America had really begun to catch on. Whatever its origins, the map is still undeniably a treasure, with mysterious history that cries out for further study, but it doesn't seem to be what Stevens so desperately sought to prove that it was.

But George Parker Winship didn't know that. When he received Stevens's report on the map, at the end of 1900, he was convinced by its claims—and on February 28, 1901, Sophia Augusta Brown authorized him to buy it. "After all that you tell me of Mr. Stevens's researches regarding *the* Map and your convictions as to its genuineness," she told Winship in a letter, "I am ready to buy it of Mr. Stevens for £1000, feeling sure that my dear Son would have done the same had he lived."

A sense of triumph reigned on both sides of the Atlantic. George Parker Winship and Sophia Augusta Brown believed they had acquired a map for the ages—one that would become the jewel of their collection and bring the John Carter Brown Library international fame and renown. Stevens, for his part, had not only made himself a spectacular profit but also earned himself the honor of discovering a landmark map in the history of cartography. That spring and summer Winship and Stevens exchanged a series of letters in which, already dreaming of the map on display at the library, they discussed the kind of lavish case that should house it, and the exact kind of gold lettering that should identify it.

The moment of triumph didn't last long. In July, Joseph Fischer made his famous visit to Wolfegg Castle, and word of what he had discovered reached Winship in October. Reeling from the news, Winship composed a letter to Stevens on October 18 and posed an anxious question that must have pierced Stevens directly in the heart.

"What is this new Waldseemüller 1507 MSS map," Winship wrote, "that has turned up at Wolfegg Castle?"

Acknowledgments

I HAVE RELIED ON the kindness of both strangers and friends in researching and writing this book, and one of the happy results is that many of
those strangers are now friends. I owe a great deal to my agent, Rafe
Sagalyn, who helped me understand this as a broad rather than a narrow
book; to many of my colleagues at *The Atlantic Monthly* in Boston, who
provided me with much wise counsel as I embarked on this project; to my
parents, Jim and Valerie, and my sister, Alison, experienced book writers
all, who have helped and guided me in more ways than I can express; and
to my aunt, Jane Lester, who regularly provided me with a most welcome
pied-à-terre in Washington during my visits to the Library of Congress.
I'm indebted to the entire wonderful staff at Free Press, especially Ellen
Sasahara, the book's designer, and my two editors: Bruce Nichols, with
whom I began the book, and Hilary Redmon, who has seen it through to
completion with a rare mix of enthusiasm, insight, and sensitivity. Special
thanks also go to three other Free Press staffers: Carol de Onís for her
admirably thorough and thoughtful copyediting, Sydney Tanigawa for all
of her help in shepherding the book through production, and to Jill Siegel,
for guiding me so ably through the unfamiliar waters of book promotion.
Many thanks also to Daniel Crewe, the editorial director of my English
publisher, Profile Books, whose careful reading of my initial manuscript
helped me strengthen this book greatly.

Among those I accosted for help and expert guidance as I researched
and wrote this book are Peter Barber, Danny Barenholz, Theodore
Cachey, Tony Campbell, Yelitza Claypoole, Alan Cooper, Susan Danforth,
Carol Delaney, Anthony Grafton, Ernesto Guerra, Felipe Fernández-
Armesto, Ronald Grim, Kathy Hayner, Brian Jay Jones, Robert
Karrow, Corby Kummer, Hans-Jörg Künast, Jill Lepore, R. Jay Magill,
Berndt Mayer, An Mertens, Peter Meurer, Thomas Nadler, Benjamin
Olshin, Kieran O'Mahony, Monique Peletier, John Pike, Richard Ring,

Albert Ronsin, Sara Schecter, Felicitas Schmeider, Benjamin Schwarz, Zur Shalev, Rodney Shirley, Larry Silver, Geoffrey Symcox, Nathaniel Tayler, Kim Veltman, Benjamin Weiss, Colin Wells, Ashley West, Scott Westrem, and Margarita Zamora. For research help I'm deeply grateful to Maud Streep, David Thoreson, Dustin Heestand, and Preston Copeland. I studied Latin privately with Paul Anders and Tom Burke, of the Ancient Studies Institute, in Cambridge, Massachusetts, during much of the time I was writing—sessions that aided me immeasurably in appreciating the sources I was working with. Robb Menzi, Bill Pistner, and Chris Stone provided valuable moments of sub-zero perspective.

Several people deserve special mention for making more of their time and expertise available to me than I had any right to expect, and in many cases for reading some or all of this book in manuscript form. They are Evelyn Edson, W. Ralph Eubanks, Elizabeth Fisher, Alfred Hiatt, Peter Jackson, Christine Johnson, the late John Larner, Cullen Murphy, and Kirsten Seaver. I'm very grateful to Ted Widmer, the head of the wonderful John Carter Brown Library, at Brown University, for making me an Invited Research Scholar at the library, and for inviting me to give two talks about this book there, in 2007 and 2008—talks that helped me clarify my sense of what I wanted the book to accomplish. I owe a particularly great debt to two members of the Geography and Map Division at the Library of Congress: John Hebert, the head of the division, who introduced me to the Waldseemüller map when I knew very little about it, and who, with heartening enthusiasm, embraced the idea of this book from the outset; and John Hessler, one of the division's senior cartographic librarians, who surely knows more about the Waldseemüller map than anybody else alive. Without John Hessler's unfailingly generous guidance and support, this book wouldn't have been possible. I owe my greatest debt, however, to Catherine Claypoole, my wife and best reader, who has done so very much, at work and at home, to make it possible for me to write this book—and who, in ways tangible and intangible, has contributed to it far more than she knows. This is her book too.

Needless to say, all errors of fact or judgment in this book are entirely my own.

Notes

OPENING QUOTE

xiii "The earth is placed": Isidore, *Etymologies* 14, in Barney et al., *The "Etymologies" of Isidore*, 285, 293. Translation slightly modified.

PROLOGUE: AWAKENING

1 "But where is this Waldseemüller map?": Thacher, *The Continent of America*, 151, 157.
2 "INTRODUCTION TO COSMOGRAPHY": Fischer and von Wieser, *The "Cosmographiae Introductio"*: Latin original on i; translation mine.
2 "seven years after": Fischer and von Wieser, *The "Cosmographiae Introductio."* Latin original on ciii; translation mine.
3 "These parts": Fischer and von Wieser, *The "Cosmographiae Introductio."* Latin original on ciii; translation mine.
4 "It is manifest": Vespucci, *Letters*, 126–27.
5 "It is well here to consider": Vespucci, *Letters*, 126–27.
5 "Everybody knows the crafty wiles": Irving, *A History of New York*, 70.
5 "broad America": Emerson, *English Traits*, 154–55.
5 "The universe, astonished": Lester and Foster, *The Life and Voyages*, 306–7.
6 "His name was given": Irving, *A History of the Life*, 276.
7 "This extremely rare book": Humboldt, *Examen Critique*, IV. French original on 100; translation mine.
7 "I have had the pleasure": Humboldt, *Examen Critique*, IV. French original on 33; translation mine.
7 "Whoever he was": Sumner, handwritten note, inside front cover, Waldseemüller and Ringmann, *Cosmographiae introductio*, Houghton Library, Harvard University (Houghton *AC85 Su662 Zz507w).
8 "The purpose of this little book": Fischer and von Wieser, *The "Cosmographiae Introductio."* Latin original on the pullout page between xxviii and xxxix; English translation on the pullout page between 66 and 67. Translation somewhat modified.
8 "To the best of my ability": Fischer and von Wieser, *The "Cosmographiae Introductio."* Latin original on iii–iv; English translation on 34. Translation somewhat modified.
9 "sheets": Fischer and von Wieser, *The "Cosmographiae Introductio."* Latin original on xxxvii; English translation on 78.
9 "is found to be surrounded": Fischer and von Wieser, *The "Cosmographiae Introductio."* Latin original on xxx; English translation on 70.
9 "endless Asian land": Vespucci, *Letters*, 15.
10 "Ever since Humboldt": Soulsby, "The First Map," 202.
10 "Much of what is written": Nordenskiöld, *Facsimile-Atlas*, 69, note 2.

10 "of special interest": *Geographical Journal*, 329.

10 "the Deodatian": Fischer and von Wieser, *The Oldest Map*. Latin original on 9; translation mine.

10 "I have depicted it": Fischer and von Wieser, *The Oldest Map*. Latin original on 9; translation mine.

12 "The mystery of the map": Thacher, *The Continent of America*, 157.

13 "a most friendly welcome": Fischer, *The Oldest Map*, 269.

13 "Posterity, Schöner gives this to you": Fischer, *The Oldest Map*. Latin original on 2; translation mine.

15 "A general description": Waldseemüller map of 1507, legend at bottom left. The corresponding text in the *Introduction to Cosmography* can be found in Fischer and von Wieser, *The "Cosmographiae Introductio."* Latin original on xlv; English translation on 88. Translation modified slightly.

16 "A greeting from Waldseemüller": Fischer, *The Oldest Map*, 270.

16 "In the past": Waldseemüller, Carta Marina, legend at bottom left of map. Unpublished translation by John Hessler, senior cartographic librarian, Library of Congress. Translation slightly modified.

16 "Geographical students": Soulsby, "The First Map," 203.

17 "There has lately been made": "LONG SOUGHT MAP DISCOVERED," *New York Times*, March 2, 1902, SM5.

17 "before the so-called Reformation": Seaver, *Maps, Myths, and Men*, 326.

18 "What more suitable": Stevens, *Rare Americana*, viii.

Chapter One: Matthew's Maps

25 "It is the vocation": Bernard, *The Letters*, 503. Translation slightly modified.

26 three five-hundred-page volumes: See Paris, *English History*.

26 a whole flock of sheep: Leclerq, *The Love of Learning*, 155.

27 "I desire and wish": Lewis, *The Art of Matthew Paris*, 49.

27 "Turning the pages": Lewis, *The Art of Matthew Paris*, 45.

27 "The earth is placed": Isidore, *Etymologies* 14.1.1, in Barney et al., *The "Etymologies" of Isidore*, 285. Translation slightly modified.

28 "If the earth were flat": Thorndike, *The Sphere of Sacrobosco*, 121.

29 "and there is no other explanation": Thorndike, *The Sphere of Sacrobosco*, 121.

29 the library at St. Albans: This and the other references to texts in the library at St. Albans come from Ker, *Medieval Libraries of Great Britain*, 164–68.

29 "You will observe": *The Dream of Scipio* 20, in *On the Good Life*, 349–50.

31 "It is divided": Isidore of Seville, *Etymologies* 14.2.1, in Barney et al., *The "Etymologies" of Isidore*, 285.

32 "eastward in Eden": Genesis 2:8 (King James translation).

32 "The glory of the God": Ezekiel 43:2 (King James translation).

32 "established two parts": Lactantius, *The Divine Institutions* 2.10, in Cassidy, *The Sea Around Them*, 46.

33 "a global empire": Cosgrove, *Apollo's Eye*, 16 note 40.

34 "enthroned above the circle": Isaiah 40:22 (King James translation).

34 symbolized the Cross: Lanman, "The Religious Symbolism," 21.

34 "Most appropriate is this division": Maur, Raban, *De universo*, 12.2, in Kimble, *Geography in the Middle Ages*, 32.

34 "Paradise is a place": Isidore of Seville, *Etymologies* 14.3.2, in Barney et al., *The "Etymologies" of Isidore*, 285.

34 "This is Jerusalem": Ezekiel 5:5 (King James translation).

36 "Know that there were three reasons": Baxandall, *Painting & Experience*, 41.

39 "This gospel of the kingdom": Matthew 24:14 (King James translation).

40 "What great learning there was": J. R. S. Phillips, "The Outer World of the European Middle Ages," in Schwartz, *Implicit Understandings*, 45.

41 "The Welshman left his hunting": Phillips, *The Medieval Expansion*, 46.

42 "Going to Jerusalem": Constable, *Religious Life*, 134.

42 "This is an image of Paradise": Leclerq, *The Love of Learning*, 165.

42 "Lord, restore me": Gribble, *The Early Mountaineers*, 4.

44 "not of peace": Constable, "Religious Life," 133.

44 "Who will give us wings": Connolly, "Imagined Pilgrimage," 598.

Chapter Two: Scourge of God

45 "In this year, that human joys": Paris, *English History*, I, 312–13.

45 "on behalf of the whole": Paris, *English History*, II, 131–32. All quotes about this episode in the following paragraphs are drawn from the above two pages.

46 "Let us leave these dogs": Paris, *English History*, II, 132.

46 "Where have such people": Paris, *English History*, II, 348.

46 a Hungarian friar named Julian: Jackson, *The Mongols and the West*, 60–61, and Dienes, "Eastern Missions," 226–40.

47 "devastate all the lands": Dienes, "Eastern Missions," 239.

47 *Tartarus*: Rachewiltz, *Papal Envoys to the Great Khans*, 73.

47 "Through the power of God": Phillips, *The Medieval Expansion of Europe*, 60.

47 slaughter of whole cities: Rachewiltz, "Some Remarks," 25.

47 Fierce Ruler: Jackson, *The Mongols and the West*, 55.

48 "The men are inhuman": Paris, *English History*, I, 312–13.

49 "Unknown tribes came": Dawson, *Mission to Asia*, xii.

49 "We know not where they hid": Dawson, *Mission to Asia*, xii.

49 "The Tartars entered our land": Jackson, *The Mongols and the West*, 49, note 101. Translation mine.

49 "We took no precautions against them": Jackson, *The Mongols and the West*, 49.

49 massacred on the spot: Jackson, *The Mongols and the West*, 49.

50 "He related that not many years before": Slessarev, *Prester John*, 27–28. Translation slightly modified.

51 "I, Prester John": Silverberg, *The Realm of Prester John*, 42.

51 "When we ride forth": Silverberg, *The Realm of Prester John*, 43.

52 "Our magnificence": Silverberg, *The Realm of Prester John*, 42.

52 "I believe that there are more Christians": Beckingham and Hamilton, *Prester John*, 242.

52 "A new and mighty protector": Silverberg, *The Realm of Prester John*, 71.

52 "commonly called Prester John": Silverberg, *The Realm of Prester John*, 71.

52 Some rumors suggested" Silverberg, *The Realm of Prester John*, 146.

53 "scourge of God": Dienes, "Eastern Missions," 235.

53 "I am aware": Jackson, *The Mongols and the West*, 60.

54 "If, which God forbid, they invade": Paris, *English History*, I, 346.

54 "The dangers foretold": Paris, *English History*, I, 339.

54 "nobody to piss against a wall": Jackson, *The Mongols and the West*, 65.

54 "a remedy against the Tartars": Jackson, *The Mission of Friar William*, 28.

55 "We feared that we might be killed": Dawson, *Mission to Asia*, 3.

55 "King of the Tartars": Rachewiltz, *Papal Envoys*, 87.

56 "It is not without cause": Dawson, *Mission to Asia*, 75–76.
56 "desperately ill": Dawson, *Mission to Asia*, 52.
56 "When we were journeying": Dawson, *Mission to Asia*, 29–30.
56 "barbarian nations": Dawson, *Mission to Asia*, 52.
56 "On the first Friday": Dawson, *Mission to Asia*, 53.
57 sent by the pope: Dawson, *Mission to Asia*, 53.
57 carrying a message: Jackson, *The Mongols and the West*, 46 and 90.
57 "Together with the two Tartars": Dawson, *Mission to Asia*, 57.
57 "We came across many skulls": Dawson, *Mission to Asia*, 58.
57 "innumerable ruined cities": Dawson, *Mission to Asia*, 59.
57 a journey of almost three thousand miles: Rachewiltz, *Papal Envoys*, 96.
57 "more wretched": Dawson, *Mission to Asia*, 6.
57 "We were so weak": Dawson, *Mission to Asia*, 57–58.
58 "So many gifts": Dawson, *Mission to Asia*, 64.
58 "had been among the Tartars": Dawson, *Mission to Asia*, 66.
58 "Men who knew": Dawson, *Mission to Asia*, 65.
58 "Having taken counsel for making peace": Dawson, *Mission to Asia*, 83–84.
59 "We traveled throughout the winter": Dawson, *Mission to Asia*, 69.
59 "He wrote a great book": Coulton, *From St. Francis to Dante*, 135. Translation modified somewhat, based on the version that appears in Biller, *The Measure of Multitude*, 231.
60 "told us that they firmly": Dawson, *Mission to Asia*, 68.
60 and all of his people: Jackson, *The Mongols and the West*, 98.
61 the Mongols had not only converted: Jackson, *The Mission of William of Rubruck*, 36.
62 "Whether Sartaq believes": Jackson, *The Mission of Friar William*, 126.
62 "a slight smile": Jackson, *The Mission of Friar William*, 133.
62 "The nurse begins": Jackson, *The Mission of Friar William*, 282.
62 "I am to take you to Mangku Chan": Jackson, *The Mission of Friar William*, 136.
63 "A man named David": Jackson, *The Mission of Friar William*, 249.
63 "what Isidore says": Jackson, *The Mission of Friar William*, 129.
63 "I enquired about the monsters": Jackson, *The Mission of Friar William*, 201.
63 "The Nestorians called him King John": Jackson, *The Mission of Friar William*, 122.
63 "We baptized there": Jackson, *The Mission of Friar William*, 253.
64 "in the hope of a profitable venture": Polo, *The Travels*, 34.

Chapter Three: The Description of the World

65 "From the time when": Polo, *The Travels*, 33. Translation slightly modified.
65 "the Great Khan of all the Tartars": Polo, *The Travels*, 36.
65 "ends of the earth": Polo, *The Travels*, 35.
65 "argue and demonstrate plainly": Polo, *The Travels*, 36.
66 "On the day when we see this": Polo, *The Travels*, 120.
66 "plenary authority": Polo, *The Travels*, 39.
66 "all of the privileges and letters": Polo, *The Travels*, 39.
66 "snow and rain and flooded rivers": Polo, *The Travels*, 39.
66 "showed him such favor": Polo, *The Travels*, 41.
66 "strange seas": Polo, *The Travels*, 42.
67 a number of local languages: Larner, *Marco Polo*, 186.
67 "undecipherable enigma": Heers, "De Marco Polo à Christophe Colombe," 125. Translation mine.
68 "These people came and presented garments": Franke, "Sino-Western contacts," 54–55.
69 the Loppy Ears: This and the following examples come from the glossary titled "Notes on

Chinese Names and Terms" in Birrell, *The Classic of Mountains and Seas.* See also Schiffeler, *The Legendary Creatures.*

70 "all the great wonders": Polo, *The Travels,* 33.
70 "in due order": Polo, *The Travels,* 33.
70 "Now let us leave these regions": Polo, *The Travels,* 335.
70 "all the things he had seen": Polo, *The Travels,* 33.
70 "to afford entertainment": Polo, *The Travels,* 33.
71 "Let me begin with Armenia": Polo, *The Travels,* 46.
71 "the maps and writings": Polo, *The Travels,* 303.
71 prepared by Chinese or Persian cosmographers: This and other possible sources are cited in Larner, *Marco Polo,* 90.
72 "The first of these": Polo, *The Travels,* 206–7.
72 "truest and best": Polo, *The Travels,* 119.
73 "If men had really been sent": Polo, *The Travels,* 120.
73 "the finest and most splendid city": Polo, *The Travels,* 213.
73 "City of Heaven": Polo, *The Travels,* 213.
73 "like Venice": Polo, *The Travels,* 214.
73 "shops in which every craft": Polo, *The Travels,* 215.
73 "the port for all the ships": Polo, *The Travels,* 237.
74 "here and there": Polo, *The Travels,* 236.
74 "Jin-pön-kuo": Larner, *Marco Polo,* 94.
74 "the choicest of all foods": Polo, *The Travels,* 248.
74 "They have gold": Polo, *The Travels,* 244.
75 "islands of the Indies": Polo, *The Travels,* 303.
75 "surprise everyone . . . The truth is": Polo, *The Travels,* 252–53.
77 "the whole earth": Genesis 9:19 (King James translation).
77 "all the earth": Romans 10:18 (King James translation).
77 "But in regard to the story": Saint Augustine, *City of God* 16.9, 49–51.
78 "Apart from these three parts": Isidore of Seville, *Etymologies* 14.5.17, in Barney et al., *The "Etymologies" of Isidore,* 293.
79 "past all credence": John Larner, *Marco Polo,* 115.
79 "the most extensive traveler": Pelliot, *Notes on Marco Polo,* I, 602.
79 "kindly exalter": Jackson, *The Mongols and the West,* 174.
79 several Mongol envoys were baptized: Schmieder, "The Mongols as non-believing apocalyptic friends," 10. Also see Critchley, *Marco Polo's Book,* 149.
80 "Know ye that many of our fathers": Dawson, *The Mongol Mission,* xxix.
80 "confirming his patriarchal authority": Dawson, *The Mongol Mission,* xxx.
80 "When we have taken Jerusalem": Dawson, *The Mongol Mission,* xxx.
80 "had set up at least thirty-four monasteries": Schmieder, "Clash of Civilizations," 17.
81 "I am of the opinion": Larner, *Marco Polo,* 113.
81 "according to what the merchants say": Yule, *Cathay and the Way Thither,* II, 292.
81 in Mongol garb: Phillips, "The Outer World of the European Middle Ages," in Schwartz, *Implicit Understandings,* 54.
81 Asian and European rats: McNeill, *Plagues and Peoples,* 134, 142, 146.

Chapter Four: Through the Ocean Sea

83 "Let us sail westward": Barron and Burgess, *The Voyage of Saint Brendan,* 26.
83 "When I say": Polo, *The Travels,* 248.
84 "A man could go around the world": Woodward, "Medieval *Mappaemundi,*" 321.
86 "the Earth begins": Mandeville, *The Travels,* 183.

86 "You should realize": Mandeville, *The Travels*, 185.

87 "those countries girdling": Mandeville, *The Travels*, 129.

87 "I have often thought": Mandeville, *The Travels*, 129.

88 "children of the king of India": Cassidy, *The Sea Around Them*, 87.

88 "A fog so thick": Barron and Burgess, *The Voyage of Saint Brendan*, 27.

89 "pillar in the sea": Barron and Burgess, *The Voyage of Saint Brendan*, 53.

89 "There is in the ocean": Honorius Augustodunensis, *Imago Mundi* 1.35 ("Sardinia"), Latin original at http://12koerbe.de/arche/imago.htm. Translation mine.

89 "The Lost Island": For a high-quality digital facsimile of the Ebstorf Map, the original of which was destroyed by Allied bombing during World War II, see http://weblab. unilueneburg.de/kulturinformatik/projekte/ebskart/content/start.html

90 The author refers to the map: Nangis, "Gesta," 444.

90 "Before the harbor": Schoff, *The Periplus of the Erythraean Sea*, 22–23.

92 the shape of an oak leaf: Pliny, *Natural History* 3.5.43, 35.

92 Almost two hundred marine charts: Campbell, "Portolan Charts," 373.

93 "the south-pointing needle": Gurney, *Compass*, 37.

93 "Mariners at sea": Brown, *The Story of Maps*, 127–28.

93 "Divide the cover": Brown, *The Story of Maps*, 129.

94 "an infernal spirit": Westropp, "Brasil and the Legendary Islands," 241.

95 Portugal, Spain, England: Phillips, *The Medieval Expansion of Europe*, 147.

95 "through the Ocean Sea": Rogers, "The Vivaldi Expedition." Latin original on 37; translation mine.

Chapter Five: Seeing Is Believing

96 "Every point of the earth": Bacon, *The Opus majus*, I, 272.

97 "Whosoever exercises": Edson, *The World Map*, 62.

99 "writings and remedies": Clegg, *The First Scientist*, 96.

100 "All the sciences are connected": Bridges, *The Life and Work of Roger Bacon*, 139.

100 "those who have in great measure traveled": Howe and Woodward, "Roger Bacon."

100 "Men without number": Howe and Woodward, "Roger Bacon."

100 "He who has gained": Bacon, *The Opus majus*, I, 204.

100 "The whole truth of things": Bacon, *The Opus majus*, I, 234.

102 "neither place nor void nor time": Aristotle, *On the Heavens* 1.9, in Leggatt, *Aristotle*, 91.

103 "And God said": Genesis 1:9 (King James translation).

103 "This quiet and peaceful heaven": Baumer, *Main Currents*, 68.

104 "surround the Earth": Thorndike, *The Sphere of Sacrobosco*, 119.

104 "front face of the Earth": Goldstein, "The Renaissance Concept of the Earth," 33.

105 "Aristotle suggests": Howe and Woodward, "Roger Bacon."

105 "Esdras tells us in his fourth book": Howe and Woodward, "Roger Bacon."

106 "southeast of India": Pliny, *Natural History* 6.24.88, 405.

106 "Some men from this place": Howe and Woodward, "Roger Bacon."

107 "Since these zones": Howe and Woodward, "Roger Bacon."

108 "This is the first axiom": Howe and Woodward, "Roger Bacon."

108 "We sorely need": Howe and Woodward, "Roger Bacon."

109 "his book about the form of the world": Gautier Dalché, "Le souvenir de la *Géographie*," 95. Translation mine.

109 "collected the latitudes": Gautier Dalché, "Le souvenir de la *Géographie*," 105. Translation mine.

109 "What is still missing in the preliminaries": Berggren and Jones, *Ptolemy's "Geography*," 19.

Chapter Six: Rediscovery

113 "Who can doubt": Petrarch, *Letters on Familiar Matters*, I, 293.
113 the sole surviving account: All of the quotations from this account come from Franzini and Bouloux, *Îles du Moyen Âge*, 9–13. Translation mine.
113 traces of the Vivaldi brothers' expedition: Diffie and Winius, *Foundations*, 25.
115 Plutarch reported: Sertorius 8.2, in Plutarch, *Lives* 8, 3.
115 "About the Fortunate Isles": Pliny, *Natural History* 6.37.203–5, 489–90.
116 "The four men": This portion of Boccaccio's *De Canaria et insulis* derives from the partial English translation by Peter Hulme in Schwartz, *Implicit Understandings*, 181.
118 "Each famous author": Robinson, *Petrarch*, 25–26.
118 "darkness and dense gloom": Petrarch, *Opera omnia*, 187.
118 "There will follow a better age": Petrarch, *Africa* 9.451–57, cited in Findlen, *The Italian Renaissance*, 234.
120 "About to write of the exploits": Bouloux, *Culture et Savoirs*, 201. Translation mine.
121 the copy Petrarch had made: See Bouloux, *Culture et Savoirs*, 160, and Gormley, Rouse, and Rouse, "The Medieval Circulation," 302 ff.
122 "The harbor had been formed": Virgil, *Aeneid* 3.530–36, 91.
122 "Many things cause errors": Bouloux, *Culture et Savoirs*, 195–96. Translation mine.
124 "Powerful on the sea": Cachey, *Petrarch's Guide to the Holy Land*, 30.
124 "I decided not to travel": *Seniles* 9.2, cited in Cachey, *Petrarch's Guide to the Holy Land*, 22.
124 "Once India": Petrarch, *Letters on Familiar Matters*, II, 36.
125 "I have been around": Bouloux, *Culture et Savoirs*, 201. Translation mine.
125 "Within the memory of our fathers": Petrarch, *The Life of Solitude*, 267.
126 "prince of Fortune": Bouloux, *Culture et Savoirs*, 252. Translation mine.
126 "Canaria, Ningaria, Pluviaria": Bouloux, *Culture et Savoirs*, 252. Translation mine.
126 "On Mountains": Bouloux, *Culture et Savoirs*, 309. The Latin title is *De montibus, syluis, fontibus, lacubus, fluminibus, stagnis, seu paludibus, de nominibus maris*.
127 "Which one of these opinions": Bouloux, *Culture et Savoirs*, 212. Translation mine.
128 "The more they advanced": Franzini and Bouloux, *Îles du Moyen Âge*, 11. Translation mine.

Chapter Seven: Ptolemy the Wise

129 "World cartography": Berggren and Jones, *Ptolemy's "Geography,"* 58.
129 "lamenting for her bridegroom": Bisaha, "Petrarch's Vision," 288.
129 "How the Catholic Faith was of old diffused": Petrarch, *The Life of Solitude*, II, 243–44. All of the quotes in this paragraph are taken from these two pages.
131 "Read through geographical writings": Cassiodorus, *Institutions* 1.25, 157–58.
132 "to enable the Roman maiden": Pliny, *Natural History* 6.20.54, 379.
132 "lies beyond the stars": Virgil, *Aeneid* 6.795–97, 171.
132 "chit chat about places": Larner, "The Church and the Quattrocentro Renaissance," 35.
133 "The goal of regional cartography": Berggren and Jones, *Ptolemy's "Geography,"* 57.
133 "harbors, towns, districts": Berggren and Jones, *Ptolemy's "Geography,"* 57.
133 "the mathematical method": Berggren and Jones, *Ptolemy's "Geography,"* 58.
133 "the proportionality of distances": Berggren and Jones, *Ptolemy's "Geography,"* 58.
138 "Concerning the land journey": Berggren and Jones, *Ptolemy's "Geography,"* 67.
138 "Concerning the sail": Berggren and Jones, *Ptolemy's "Geography,"* 68.
139 "It is agreed": Berggren and Jones, *Ptolemy's "Geography,"* 64.
140 "Our present object": Berggren and Jones, *Ptolemy's "Geography,"* 59.

140 "does not conveniently allow": Berggren and Jones, *Ptolemy's "Geography,"* 82.
141 "a semblance of the spherical surface": *Ptolemy's "Geography,"* 83.
141 "Many parts of our *oikoumene*": Berggren and Jones, *Ptolemy's "Geography,"* 63.
144 "I desire to see": Bisaha, "Petrarch's Vision," 308.
144 "indolent and deceitful Greeklings": Bisaha, "Petrarch's Vision," 308.
146 "Heroic verses": Berggren and Jones, *Ptolemy's "Geography,"* 49.
146 "A mighty marvel": Dilke and Dilke, "The Adjustment of Ptolemaic Atlases," 118.
146 traced back to Planudes' original: Diller, "The Oldest Manuscripts," 67.

Chapter Eight: The Florentine Perspective

147 "This great work": Cosgrove, *Apollo's Eye*, 109.
148 "the fifth element": Staley, *The Guilds of Florence*, 562.
148 "have spread their wings": Edson, *The World Map*, 133.
149 "Consider a little": Witt, *Hercules at the Crossroads*, 129–130.
150 "To you, men of Florence": Kohl, *The Earthly Republic*, 150.
151 "the home of the study": Witt, *Hercules at the Crossroads*, 68 note 54. Translation mine.
151 "If anything happens": Cassiodorus, *Institutions* 1.25, 156.
154 he wrote to Salutati in 1396: Holmes, *The Florentine Enlightenment*, 9.
154 "Tomorrow I shall reach": Setton, "The Byzantine Background," 57.
155 "Oh, how much patience": Witt, *Hercules at the Crossroads*, 287.
155 "I leave to my sons": Diller, "The Greek Codices," 316. Translation mine.
156 "Does not this present age": This and the other Angeli quotes in this paragraph come from Hankins, "Ptolemy's *Geography*," 120.
157 "Let him wander": Satire 3, 55–66, in Ariosto, *The Satires*, 61.
160 "picture of the world": Grafton, *Leon Battista Alberti*, 244.
161 "Nothing more convenient": Alberti, *On Painting*, 65–66.
162 "raises us above the limits": Cosgrove, *Apollo's Eye*, 109.
162 "Understand that you are god": *The Dream of Scipio* 24, in Cicero, *On the Good Life*, 353.
162 "The deity that the painter has": Veltman, "Leonardo da Vinci," 17, note 50.
163 "By the ancients man was termed": Kemp, *Leonardo da Vinci*, 11.
164 "In fifteen entire figures": Veltman, "Leonardo da Vinci," 2.
165 Consider the visual similarities: For an intriguing but controversial look at the similarities between Leonardo's Vitruvian Man and symbolic Christian *mappaemundi*, and on the possible influence of Ptolemy's *Geography* on the advent of linear perspective, see Edgerton, "From Mental Matrix," in Woodward, *Art and Cartography*, 10–50.
165 "You have a god's capacity": *The Dream of Scipio* 24, in Cicero, *On the Good Life*, 353.

Chapter Nine: Terrae Incognitae

166 "Our *oikoumene* is bounded": Berggren and Jones, *Ptolemy's "Geography,"* 108.
167 Estimates of the size of the Council: Kremple, "Cultural Aspects," 6–13.
167 an escort of five hundred people: d'Ailly, *Ymago Mundi*, I, 84; translation mine.
167 some thirty thousand horses: d'Ailly, *Ymago Mundi*, I, 84; translation mine.
168 "doing nothing in Constance": Gordan, *Two Renaissance Book Hunters*, 195.
168 "Salvation itself": Holmes, *The Florentine Enlightenment*, 64.
169 "Few things have been accomplished": Kremple, "Cultural Aspects," 57.
169 "This book was purchased": Kremple, "Cultural Aspects," 71.
169 "Purchased by Paul Vladimir": Kremple, "Cultural Aspects," 72, note 84.
169 "This book, which I sought": Gautier Dalché, "L'oeuvre géographique," 326, note 35. Translation mine.

172 "Let me untangle": Mela, *De chorographia* 1.2, in Romer, *Pomponius Mela's Description*, 33.

172 "The uplifted earth": Mela, *De chorographia* 1.2, in Romer, *Pomponius Mela's Description*, 34.

173 "From the beginning of the book": Gautier Dalché, "L'oeuvre géographique," 332. Translation mine.

173 "Water does not surround the whole earth": Cosgrove, *Apollo's Eye*, 77.

174 "It seems necessary to affirm": Gautier Dalché, "L'oeuvre géographique," 337. Translation mine.

175 "It seems that the image": d'Ailly, *Ymago Mundi*, I, 152; translation mine.

175 statements that he found: See Morison, *Admiral*, 93–94.

175 "It is apparent": d'Ailly, *Ymago Mundi*, I, 212. Translation mine.

176 "Toward the equator": Hiatt, "Blank Spaces," 238.

177 no earlier map: For this claim see, for example, Hiatt, "Blank Spaces," 238.

177 "more complete": Gautier Dalché, "L'oeuvre géographique," 374. Translation mine.

178 "Ptolemy was perhaps unaware": Gautier Dalché, "L'oeuvre géographique," 375. Translation mine.

179 "the port and key": Ropes, *Council of Constance*, 520.

Chapter Ten: Into African Climes

180 "It is found": Azurara, *The Chronicle*, II, 236.

181 "Not one hundredth part": Silverberg, *The Realm of Prester John*, 138.

181 the region's Christian population: Russell, *Prince Henry*, 122.

181 "more potent than any man": Jordanus, *Mirabilia Descripta*, 45–46.

182 "the Christians of Nubia": Relano, *The Shaping of Africa*, 56.

183 "the end of Africa": Chet van Duzer, unpublished manuscript.

184 "to carry out very great deeds": Azurara, *The Chronicle*, I, 27.

184 "shouting out 'St. James'": Azurara, *The Chronicle*, I, 66.

184 "If there chanced to be": All of the quotations in this paragraph come from Azurara, *The Chronicle*, I, 28–29.

185 "No one knows": Phillips, *The Medieval Expansion*, 172.

185 "Up to his time": Azurara, *The Chronicle*, I, 27.

186 "For, said the mariners": Azurara, *The Chronicle*, I, 31.

186 "yet there was not one": Azurara, *The Chronicle*, I, 30.

186 "touched by the self-same terror": Azurara, *The Chronicle*, I, 33–34.

188 "pilots who knew the harbors": This and the other *Canarian* quotations in this paragraph come from Bontier and Le Verrier, *The Canarian*, 92.

188 "a book by a mendicant friar": Bontier and Le Verrier, *The Canarian*, 96.

189 "I traveled along the coast": Marino, *El Libro*, 49.

189 "Nubia and Etiopia": Marino, *El Libro*, 65–67.

189 "as black as pitch": Marino, *El Libro*, 61.

189 "a Genoese man": Marino, *El Libro*, 62.

189 "some very high mountains": Marino, *El Libro*, 63.

190 "governs many great lands . . . Prester John, the Patriarch": Marino, *El Libro*, 61–63.

191 fifteen days: Marino, *El Libro*, 59.

191 "great slaughter": Azurara, *The Chronicle*, I, 37.

192 "having done such small service": Azurara, *The Chronicle*, I, 40.

193 "salvation for the lost souls": Azurara, *The Chronicle*, I, 51.

193 "the houses . . . turn their blame": Azurara, *The Chronicle*, I, 61.

193 "delight": Azurara, *The Chronicle*, I, 50.

194 "He reflected with great pleasure": Azurara, *The Chronicle*, I, 83.

195 "never lowered sail": Azurara, *The Chronicle*, I, 99.

195 "the first to be taken by Christians": Azurara, *The Chronicle*, I, 100.

195 "Let us press on": Azurara, *The Chronicle*, II, 174.

196 "After this year": Azurara, *The Chronicle*, II, 289.

196 Formally titled *Inter caetera*: All quotations from this bull in this paragraph come from Davenport, *European Treaties*, 31–32.

197 Poggio had already drafted: Larner, "The Church and the Quattrocento Renaissance," 37.

197 "When I asked my many Portuguese friends": Hammond, *Travelers in Disguise*, vii. Latin original appears in Bracciolini, *Opera Omnia*, 379.

Chapter Eleven: The Learned Men

198 "We know very well": Gautier Dalché, "The Reception of Ptolemy's Geography," 310.

198 "Heretics who are called Nestorians": Hammond, *Travelers in Disguise*, 10.

199 "towards the extreme confines": Hammond, *Travelers in Disguise*, 18.

199 "The natives of India": Hammond, *Travelers in Disguise*, 31.

199 "excels . . . the men . . . the merchants": Hammond, *Travelers in Disguise*, 25.

199 "the houses and palaces": Hammond, *Travelers in Disguise*, 18.

200 The Greek delegation to the Council: The details that follow come from Gill, *The Council of Florence*, 141.

200 "that the Catholic church scattered": Gill, *The Council of Florence*, 322.

202 "the meetings of learned men": Hammond, *Travelers in Disguise*, 7.

202 "I was seized with desire": Hammond, *Travelers in Disguise*, 40.

203 "So far as our records inform us": Hammond, *Travelers in Disguise*, 7. I have changed Sumatra to Taprobane, based on what appears in the original Latin and in Major, *India in the Fifteenth Century*, 4.

204 "another Plato": Kremple, "Cultural Aspects," 177.

204 "statesmen and commanders": Strabo, *Geography* 1.1, 3.

205 "Homer declares": Strabo, *Geography* 1.3, 5.

205 "I, Jacomo Barbarigo": Larner, *Marco Polo*, 138–39.

206 "The Ethiopians": Strabo, *Geography* 1.6, 9.

206 "As for Libya": Herodotus, *The Histories* 4.44, 285.

206 "the circuit of Africa": Pliny, *Natural History* 5.1.8, 223.

207 "gently hones itself": Romer, *Pomponius Mela's Description*, 39–40.

208 "a map of the world": Galvão, *Discoveries of the World*, 67.

208 "Don Henry, the King's third son": Kimble, "The Laurentian Map," 30.

209 Pedro may have played: See, for example, Nowell, "Henry the Navigator," 62–67.

209 "Sent by Prester John himself: Gill, *The Council of Florence*, 324, note 4.

209 Zare'a Ya'qob, and his regal title, Constantine: Gill, *The Council of Florence*, 324, note 4.

210 Adam and Eve: Boxer, *The Portuguese Seaborne Empire*, 27.

210 "South from Greenland": Williamson, *The Cabot Voyages*, 11.

211 "In the year 734": Ravenstein, *Martin Behaim*, 77.

211 "The island was larger": Plato, *Timaeus*, 41.

213 "I believe that the Earth": Plato, *The Works*, 179.

213 in the Atlantic: Strabo, *Geography* 1.2.18, 95.

213 "two inhabited worlds": Strabo, *Geography*, 1.4.6, 243.

213 "far out at sea": Plutarch, "The Face on the Moon," *Moralia*, 181–83.

213 "beyond the stars": Virgil, *Aeneid* 6.795–96, 171.

213 "An age will come": Seneca, *Medea*, 375–79. Translation from Morison, *Admiral*, 54.

214 "Better the turban of the Turk": For a discussion of this remark, which is cited in almost

every discussion of the fall of Constantinople, see Sevcenko, "Intellectual Repercussions," 315, note 47.

Chapter Twelve: Cape of Storms

215   "Let none doubt": Ravenstein, *Martin Behaim*, 71.
215   entertained an offer to write: Larner, "The Church and the Quattrocento Renaissance," 37.
216   "A large *mappamundi*": Gautier Dalché, "The Reception of Ptolemy's *Geography*," 334.
217   "Some authorities write": Edson, *The World Map*, 154.
220   "Paolo, physician, to Fernão, greetings": This translation of the Toscanelli letter comes from Griffin, *Las Casas on Columbus*, 34–36. The spelling of a few place-names has been changed to conform to those used elsewhere in this book. For details on the various translations about the Toscanelli correspondence, and a summary of the controversy that surrounds it, see N. Sumien, *La Correspondance du Savant Florentin Paolo dal Pozzo Toscanelli avec Christophe Colombe*.
223   "all the trade, lands, and barter": Davenport, *European Treaties*, 44.
223   "past that southern shore": Davenport, *European Treaties*, 31.
224   "Last year our men": Parry, *The Discovery of the Sea*, 134.
226   "a King the most powerful": Werner, "Prester John," 164.
226   "wherefore the king": Crone, *The Voyages*, 127.
226   "ships by sea": Crone, *The Voyages*, 127.
227   "cinnamon and pepper": Alvares, *The Prester John of the Indies*, 374.
227   "could easily": Alvares, *The Prester John of the Indies*, 374.
227   "Prester John is out there": Rogers, *The Quest for Eastern Christians*, 122.
228   "giving great hope": Randles, "Bartolomeu Dias," 23.
229   "Each day we strive": Randles, "The Evaluation," 57.
229   "that great and notable Cape": Randles, "Bartolomeu Dias," 26.
229   "Dias had been chosen like Joshua": Randles, "Bartolomeu Dias," 28.
229   "a new myth of the dangers": Randles, "The Evaluation," 57.
234   "Be it known that on this Apple": Ravenstein, *Martin Behaim*, 71.
234   "To the Most Serene and Invincible João": Morison, *Journals*, 15–17.
236   blue eyes, a fair, ruddy complexion: See Griffin, *Las Casas*, 25.

Chapter Thirteen: Colombo

236   "Just as one thing leads": Colón, *The History of the Life and Deeds*, 39.
236   "a big talker and boastful": Morison, *Admiral*, 71.
238   "I have sailed in every part": Colón, *The History of the Life and Deeds*, 35.
239   "the Island of Brasylle": Williamson, *The Cabot Voyages*, 188.
240   "Thomas Croft of Bristol": Williamson, *The Cabot Voyages*, 188–89.
240   "Consequently he wrote down": Colón, *The History of the Life and Deeds*, 39.
241   "Noting your magnificent and great desire": Griffin, *Las Casas on Columbus*, 34.
241   "The said voyage is not only possible": Morison, *Journals*, 14.
242   "that [Columbus] confer with D. Diogo Ortiz": Morison, *Admiral*, 71.
242   "They all considered": Morison, *Admiral*, 71.
243   "When the traveler leaves Kan-chau": Polo, *The Travels*, 103.
244   "By Jordan, if a man lies flat": Taylor, *The Haven-Finding Art*, 129.
244   "very low down over the sea": Crone, *The Voyages of Cadamosto*, 61.
244   "At dawn we made sail": Crone, *The Voyages of Cadamosto*, 55.
245   "I had a quadrant": Crone, *The Voyages of Cadamosto*, 101.

247 "he was a foreigner": Taviani, *Christopher Columbus*, 172.
248 "a man of great intelligence": Morison, *Admiral*, 35.
249 "the hearts of some members": Larner, *Marco Polo*, 113.
249 "the German inventor": Grafton, *Leon Battista Alberti*, 331.
250 "Every poor scholar": Anderson, *The Annals*, I, lii.
250 "matching modern with ancient": Clough, "The New World," 297.
250 "The opposite has been proved": Buron, *Ymago Mundi*, III, 742. Translation mine.
251 "Julius [Solinus] teaches": Buron, *Ymago Mundi*, III, 742. Translation mine.
251 "Note the voyage": Rogers, *The Quest for Eastern Christians*, 67.
251 "Men from Cathay": Buron, *Ymago Mundi*, III, 743. Translation mine.
251 "Biggest city in the world . . . Quinsay is 25 miles": Buron, *Ymago Mundi*, III, 740. Translation mine.
251 "15 thousand ships": Taviani, *Christopher Columbus*, 453.
253 "The length of the land . . . The sea cannot cover": Buron, *Ymago Mundi*, III, 660. Translation mine.
253 "Note that any degree": Buron, *Ymago Mundi*, I, 224. Translation mine.
253 "is not uninhabitable": Buron, *Ymago Mundi*, I, 196. Translation mine.
253 "He described and depicted": Buron, *Ymago Mundi*, I, 208. Translation mine.
254 "completely navigable": Buron, *Ymago Mundi*, I, 208. Translation mine.
254 Columbus claimed: The distances cited in this paragraph all derive from Morison, *Admiral*, 68.
254 "The earth lies": Randles, "The Evaluation," 53.
254 "The only way to explore": Griffin, *Las Casas on Columbus*, 50–51.
255 "no prospect": Griffin, *Las Casas on Columbus*, 59.
255 "By our command": Morison, *Admiral*, 105.

Chapter Fourteen: The Admiral

256 "He saw so many islands": Morison, *Journals*, 95.
257 "Your Highnesses, as Catholic Christians": Morison, *Journals*, 48.
258 "I intend to make": Morison, *Journals*, 49.
258 "making or drawing": Taviani, *Christopher Columbus*, 445.
258 "very skilled in the art": Morison, *Admiral*, 35.
258 "The Admiral made a sphere": Symcox and Formisano, *Italian Reports*, 102.
258 "on which it seems the Admiral": Morison, *Journals*, 57.
258 "left astern the islands": Morison, *Journals*, 60.
259 "so that if the voyage": Morison, *Journals*, 52.
259 Extracts from Las Casas's abstract: All of the entries that follow come from Morison, *Journals*, 53–62.
260 "Presently they saw": Morison, *Journals*, 64.
261 "I believe that they would easily": Morison, *Journals*, 65.
261 "I . . . saw so many islands": Morison, *Journals*, 68.
262 "by signs": Morison, *Journals*, 67.
262 "There was a king there": Morison, *Journals*, 67.
262 "To lose no time": Morison, *Journals*, 67.
262 "I do nothing else": Morison, *Journals*, 76.
262 "a mine of gold": Morison, *Journals*, 72.
262 "in the manner of Moorish tents": Morison, *Journals*, 74.
262 "I shall . . . depart": Morison, *Journals*, 78.
263 "This night at midnight": Morison, *Journals*, 80.

263 "a dog that didn't bark": Morison, *Journals*, 82.
263 "mines of gold and pearls": Morison, *Journals*, 83.
264 "that this Cuba was a city": Morison, *Journals*, 84.
264 "It is certain": Morison, *Journals*, 86.
265 "Hebrew and Aramaic": Morison, *Journals*, 86.
265 "nothing that resembled a city": Morison, *Journals*, 90.
265 "men with one eye": Morison, *Journals*, 88.
265 so closely echo passages: See, for example, the twentieth chapter of *Mandeville's Travels*.
265 Columbus had studied the *Travels*: Taviani, *Christopher Columbus*, 448, 450.
266 "He would enter some of the wealthiest: For a fascinating study of the importance of the south and the equatorial regions in Columbus's thought, see Wey Gómez, *The Tropics*.
266 "Seeing the grandeur": Morison, *Journals*, 116 note 3.
266 "every day we understand": Morison, *Journals*, 117.
266 "a belt that, in place of a purse": Morison, *Journals*, 131.
266 "Cipangu, which they called Cybao": Morison, *Journals*, 133.
266 "secure from shoals and rocks": Morison, *Journals*, 134.
267 "Observe the humanity": Morison, *Journals*, 136 note 8.
267 "So many good things": Morison, *Journals*, 138.
268 "He says that he hopes": Morison, *Journals*, 139.
268 "And there I found": Morison, *Journals*, 182.
269 "In this Española": Morison, *Journals*, 185.
269 "a thousand other things of value": Morison, *Journals*, 186.
269 "Seven years from today": Zamora, *Reading Columbus*, 194–95.
269 "He believed that this discovery": All of the quotes in this paragraph come from Morison, *Admiral*, 344.
270 "As you can see": Morison, *Admiral*, 355.
271 "one called Columbus": All quotes from the Hanibal Ianuarius letter come from Symcox and Rabitti, *Italian Reports*, 27–28.
271 "the Islands newly discovered": Morison, *Admiral*, 375.
271 "The king of Spain discovered": Symcox and Formisano, *Italian Reports*, 27.
271 "titled it *The History of the Discovery*": Cachey, "The literary response," 66.
271 "A certain Colonus": Morison, *Admiral*, 383.
273 "by drawing and establishing a line": Davenport, *European Treaties*, 77.
273 "one hundred leagues toward the west": Davenport, *European Treaties*, 77.
273 "This sea belongs to us": Vander Linden, "Alexander VI and the Demarcation," 16.
273 "the line which you said": Davies, "Columbus Divides the World," 339.
273 "When I sailed from Spain": Morison, *Journals*, 285.
273 "in great and disturbing numbers": Morison, *Admiral*, 372.
274 "grant him 300 leagues": Davies, "Columbus Divides the World," 340.
274 "If the sea chart you undertook": Griffin, *Las Casas on Columbus*, 89.
274 rough map: Morison, *Admiral*, 345.
275 "Your Highnesses will see the land": Zamora, *Reading Columbus*, 111.
276 "to discover the mainland of the Indies": Jane, *The Four Voyages*, 114.
276 "he would be able to return": Jane, *The Four Voyages*, 118.

Chapter Fifteen: Christ-Bearer

278 "It pleased our Redeemer": Vigneras, "Saint Thomas," 86.
278 "Solomon overlaid the house": 1 Kings 6:21–22 (King James translation).
279 "the uttermost parts": Matthew 12:42 (King James translation).

279 "gold, and silver, ivory, and apes": 1 Kings 10:22 (King James translation).

279 "many navigators": Columbus, *The "Libro de las Profecías,"* 88.

279 "The kingdom of Tarshish": Buron, *Ymago Mundi,* 304–6. Translation from Morison, *Journals,* 22.

280 "corporeal roads": Bacon, *The Opus majus,* I, 204.

280 "Amber is certainly found": Flint, *The Imaginative Landscape,* 70–71, note 71.

281 "Before we made the big island": Morison, *Journals,* 227.

281 "The Admiral sent": Morison, *Journals,* 244.

281 "This island is Tarshish": Varela, *Textos,* 311. Translation mine.

282 "a boundary or straight line": Davenport, *European Treaties,* 95.

282 "Nobody wants to live": Morison, *Journals,* 217.

284 "In regard to what you say": Williamson, *The Cabot Voyages,* 203.

284 "the eastern, western, and northern sea": Williamson, *The Cabot Voyages,* 204.

284 "That Venetian of ours": Williamson, *The Cabot Voyages,* 207–8.

286 "His Majesty here has gained": Williamson, *The Cabot Voyages,* 209–10.

286 "toward the region": Williamson, *The Cabot Voyages,* 233.

286 "for the stopping of the English": Williamson, *The Cabot Voyages,* 234.

286 "He is believed": Williamson, *The Cabot Voyages,* 225.

287 "Cathay is not so far to the north": Buron, *Ymago Mundi,* III, 747.

288 "to make discoveries": Ravenstein, *A Journal,* 1.

288 "to test the theory": Griffin, *Las Casas on Columbus,* 185.

288 "On Friday, 13 July": Morison, *Journals,* 263–64.

289 "suffered an attack": Morison, *Journals,* 264.

289 "No lands in the world": Morison, *Journals,* 273.

289 "cut in the manner of Castile": Morison, *Journals,* 268.

289 "a vast gulf": Morison, *Journals,* 274.

289 "Neither the Ganges": Morison, *Journals,* 278.

290 "Greatly did the Admiral": Morison, *Journals,* 275.

290 "I have come to believe": Morison, *Journals,* 279.

290 "The Earthly Paradise is a pleasant spot": Delumeau, *History of Paradise,* 53–54.

291 "The learned conclude": Delumeau, *History of Paradise,* 53.

291 One authoritative table: Scafi, *Mapping Paradise,* 231.

292 "I have always read that the world": Morison, *Journals,* 285–26.

293 "The order of space": Scafi, *Mapping Paradise,* 126–7.

294 "An enterprise has been carried out": Arciniegas, *Why America?,* 183.

295 "The truth is that he was": Columbus, *The "Libro de las Profecías,"* 72.

296 "Here begins the book": Columbus, *The "Libro de las Profecías,"* 101.

296 "Surely the isles": Isaiah 60:9 (King James translation).

296 "Joachim the Abbot": Columbus, *The "Libro de las Profecías,"* 239.

297 "From the creation of the world": Columbus, *The "Libro de las Profecías,"* 109.

298 "huge moving masses of solid snow": Symcox and Rabitti, *Italian Reports,* 55.

299 "They also believe that it joins": Symcox and Rabitti, *Italian Reports,* 56–57.

Chapter Sixteen: Amerigo

302 "I trust Your Magnificence": Vespucci, *Letters,* 17.

303 "Amerigo Vespucci to Lorenzo": Vespucci, *Letters,* 45. All *Mundus Novus* quotes in this section come from Vespucci, *Letters,* 45–55.

307 "There is a barrel": Baxandall, *Painting & Experience,* 86.

308 "Going back and forth": Arciniegas, *Why America?,* 85.

309   "Inform yourself there": Pohl, *Amerigo Vespucci*, 31.
310   "Amerigo Vespucci, my agent": Fernandez-Armesto, *Amerigo*, 55.
310   "Morigo Vespucci": Diffie and Winius, *Foundations*, 460.
310   "The present letter": Vespucci, *Letters*, 3.
311   "I departed with two caravels": Vespucci, *Letters*, 3.
313   "It was my intention": Vespucci, *Letters*, 4.
313   "Given the little progress": Vespucci, *Letters*, 5.
314   "Almost the majority": Vespucci, *Letters*, 9.
314   "Sailing along the coast": Vespucci, *Letters*, 11.
315   "houses built with great skill": Vespucci, *Letters*, 13.
315   "They all go about naked": Vespucci, *Letters*, 11.
315   "did not want our friendship": Vespucci, *Letters*, 11.
315   "We would put them to rout": Vespucci, *Letters*, 11.
315   "there was a conjunction": Vespucci, *Letters*, 8.
316   "We were thirteen months": Vespucci, *Letters*, 15.
316   "continental land": Vespucci, *Letters*, 11.
317   "I have resolved": Vespucci, *Letters*, 17.
317   "You will have heard": Vespucci, *Letters*, 19.
318   "has discovered a new world": Diffie and Winius, *Foundations*, 194.
318   "infinite cinnamon": Vespucci, *Letters*, 26.
318   "I hope in this voyage": Vespucci, *Letters*, 24.
319   "This voyage I embark upon": Vespucci, *Letters*, 27.
319   "When last I wrote": Vespucci, *Letters*, 29.
319   "We passed by that land": Vespucci, *Letters*, 29–30.
319   "The caravels that were sent": Arciniegas, *Why America?*, 352–53.
320   "The men tell of many miracles": Vespucci, *Letters*, 35.
320   "The most remarkable of all": Vespucci, *Letters*, 30.
320   "Amerigo Vespucci will be arriving": Arciniegas, *Why America?*, 351–52.
321   "My Dear Son": Vespucci, *Letters*, 101.
322   "to find out or learn secretly": Arciniegas, *Why America?*, 418.
322   "It is impossible to procure": Greenlee, *The Voyage*, 123.
322   "two mariner's charts": Arciniegas, *Why America?*, 418.

Chapter Seventeen: Gymnasium

327   "[Concerning] the unknown world": Lud, *Speculum orbis*, cited in Thacher, *The Continent of America*, 151.
330   "Emulate, noble men": Forster, *Selections*, 43–47.
331   led to culture by the Druids: Spitz, "Conrad Celtis," 365.
331   "Our little Thomas": Davis, *Society and Culture*, 196.
332   "A number of provinces": Hiatt, "Mutation and Nation," in Shalev and Burnett, *Ptolemy's Geography,* **forthcoming.**
333   "a land beyond": Virgil, *Aeneid* 6.795–96, 171.
333   "Master Ringmann Philesius": Ringmann, *De ora antarctica,* cited in Thacher, *The Continent of America*, 126–27.
334   "a world not known": Ringmann, *De ora antarctica*. Latin original in Thacher, *The Continent of America*, 162. Translation mine.
335   "the exactly repeatable pictorial statement"; Ivins, *Prints and Visual Communication*, 2.
336   "There are some Greek writers": Pliny, *Natural History* 25.4–5, cited in Ivins, *Prints and Visual Communication*, 14.

336 The first printed edition: For the dates and editions listed in this paragraph, see Campbell, *The Earliest Printed Maps*, 122–141.

339 "The islands recently discovered": For this and the Cantino legends that follow, see Nebenzahl, *Atlas of Columbus*, 34.

341 "This continent extends": Anghiera, *De orbe novo* 10, 271.

341 "It is claimed": Anghiera, *De orbe novo* 10, 271.

341 "henceforth he will not again": Fernandez-Armesto, *Amerigo*, 177.

344 "In many places": Skelton, "Bibliographical Note," in Ptolemy, *Geographia*, v.

Chapter Eighteen: World Without End

346 "It has seemed best": Waldseemüller map, top right legend.

346 "The principal cause": Vespucci, *Letters*, 57–58.

347 "a description of the chief lands": Fischer and von Wieser, *The "Cosmographiae Introductio."* Latin original on xlv; English on 88.

348 "a cultural and political operation": Vespucci, *Letters*, xxxii.

349 "Since we had already been ten months": Vespucci, *Letters*, 90–91.

351 "If my companions": Vespucci, *Letters*, 47.

354 "Looking at this same map": Waldseemüller and Ringmann, *Instructio manuductionem*, cited in Thacher, *The Continent of America*, 155.

356 "our poet Gallinarius": Fischer and von Wieser, *The "Cosmographiae Introductio."* Latin original on xxvii; English on 66.

356 "I humbly beg of you": Waldseemüller and Ringmann, *Instructio manuductionem*, cited in Thacher, *The Continent of America*, 153. Translation slightly modified.

356 "These parts have in fact": Fischer and von Wieser, *The "Cosmographiae Introductio."* Latin original on xxx; translation mine.

357 "Why are all the virtues": From the introductory essay to Lilio Giraldi, *Syntagma of the Muses*, cited in Charles, *The Romance*, 81.

357 but also "born new": Private correspondence with Elizabeth Fisher, Professor of Classics, George Washington University.

357 translated as "place": Kadir, *Columbus*, 60.

358 borrows directly from Sacrobosco's Sphere: Hessler, *The Naming of America*, 60.

359 "with the help of others": Fischer and von Wieser, *The "Cosmographiae Introductio."* Latin original on iii; translation mine.

360 "a single and continuous entity": Berggren and Jones, *Ptolemy's "Geography,"* 57.

360 "To hold the suitable parts": Strabo, *Geography* 2.5.10, translation from Berggren and Jones, *Ptolemy's "Geography,"* 32.

361 "I have prepared a map": Fischer and von Wieser, *The "Cosmographiae Introductio."* Latin original on iii; English on 34.

361 The printing itself: For details on the early printing of maps from woodblocks, see Campbell, "The Woodcut Map"; Grenacher, "The Woodcut Map"; and Woodward, *Five Centuries of Map Printing*, especially Robinson, "Mapmaking and map printing," and Woodward, "The woodcut technique," which are chapters 1 and 2, respectively.

364 "The world is passing away": Augustine, Sermon 81. 8, cited in Brown, *Augustine of Hippo*, 296.

366 "The earth is now known": Fischer and von Wieser, *The "Cosmographiae Introductio."* Latin original on xxx; English on 70.

367 "a land encircled by the vast ocean": Fischer and von Wieser, *The "Cosmographiae Introductio."* Latin original on xl; English on 82.

Chapter Nineteen: Afterworld

371 "We have lately composed, drawn, and printed": Waldseemüller, "Architecturae et perspectivae rudimenta," in Reisch, *Margarita philosophica*, cited in Thacher, *The Continent of America*, 133.

371 The printmaker Albrecht Dürer: Fischer and von Wieser, *The Oldest Map*, 19. This connection has yet to be substantiated but it is at least plausible; Dürer knew of Waldseemüller and his maps, and worked with several of the Strassburg humanists with whom Waldseemüller associated. See Johnson, *Carta Marina*, 27. Suggestively, a miniature version of Dürer's famous woodcut illustration of a rhinoceros, first printed in 1515, appears on Africa on Waldseemüller's 1516 Carta Marina.

372 "To the distinguished man": Original Latin in Thacher, *The Continent of America*, 128. Translation substantially modified. See also the translation in Amerbach, *The Correspondence*, 97–98.

373 "We have not forgotten": Waldseemüller and Ringmann, *Instructio manuductionem*, cited in Thacher, *The Continent of America*, 153.

374 "The time of the Frankfurt Fair": Thacher, *The Continent of America*, 192.

375 "You write that there is a beautifully drawn globe": Cited in Thacher, *The Continent of America*, 151. Translation somewhat modified.

376 "If you want to mount the map yourself": Johnson, *Carta Marina*, 52.

378 "four voyages, accounts of which": More, *Utopia*, I, 10.

378 "It surprises me": Schwartz, *Putting "America" on the Map*, 129.

379 "Considering the haste of my composition": Waldseemüller and Ringmann, *Instructio manuductionem*, cited in Schmidt, *Histoire littéraire*, 127. Translation mine.

381 "Ilacomilus, Martin Waldseemüller, wishes the reader": Previously unpublished translation by John Hessler, senior cartographic librarian, Library of Congress.

384 "Inasmuch as they have not": Nebenzahl, *Atlas of Columbus*, 50.

384 "dead without a will": Ronsin, *Le nom de l'Amerique*, 119, and 265, note 51. Translation mine.

385 recent studies have shown: See, for example, Harris, "The Waldseemüller World Map."

386 "I can readily imagine, Holy Father": Copernicus, *On the Revolutions*, 3.

386 "The scorn that I had reason to fear": Copernicus, *On the Revolutions*, 3.

387 "a manuscript of six leaves": Rose, untitled review, 138. See also Biskup, *Regesta*, no. 91, 63.

387 "it was they who first opened": Copernicus, *On the Revolutions*, 8.

387 "the Earth, which cannot be the center": Goldstein, "The Renaissance Concept," 35, citing Cusanus, Nicolaus, *Of Learned Ignorance*, trans. G. Heron (London: Routledge & Kegan Paul, 1954).

387 "The earth is not the center": Goldstein, "The Renaissance Concept," 35, citing Richter, *The Notebooks*, II. 137, no. 858.

388 "Titled *Introduction to the Cosmography*": Uminski, *Poland Discovers America*, 17. The title in Latin begins *Introductio in Ptholome: Cosmographiam*.

388 reproduced the naming-of-America passage: See Uminski, *Poland Discovers America*, 18. For the original Latin, see Bartlett, *Bibliotecha Americana*, I, 54.

389 "We should not heed the peripatetics": Copernicus, *On the Revolutions*, 9.

390 "Ptolemy extended the habitable area": Copernicus, *On the Revolutions*, 10.

391 "From all these facts, finally": Copernicus, *On the Revolutions*, 10.

392 "We regard it as a certainty": Copernicus, *On the Revolutions*, 16.

Epilogue: The Way of the World

397 "Wonder is the movement": Albertus Magnus, *Opera*, 6:30, as translated in Cunningham, *Woe or Wonder*, 80.

398 "a death warrant": I'm indebted to Richard Ring, the special-collections librarian at the Providence Public Library, for this observation.

Appendix: The Stevens-Brown Map

399 "I hope to find time": Henry N. Stevens Jr. to John Nicholas Brown, private correspondence, April 1, 1896. John Carter Brown Library at Brown University.

400 "with a view to improving": Stevens, *The First Delineation*, ix.

401 "Please regard this as confidential": Henry N. Stevens Jr. to John Nicholas Brown, private correspondence, April 1, 1896. John Carter Brown Library at Brown University.

402 "You can make up your mind": Henry N. Stevens Jr. to George Parker Winship, private correspondence, December 5, 1900. John Carter Brown Library at Brown University.

403 "After all that you tell me": Sophia Augusta Brown to George Parker Winship, private correspondence, February 28, 1901. John Carter Brown Library at Brown University.

403 "What is this new Waldseemüller": Remark quoted in letter from Henry N. Stevens to George Parker Winship, October 26, 1901. John Carter Brown Library at Brown University.

Works Cited

Ailly, Pierre d'. *Ymago Mundi de Pierre d'Ailly*. Edited by Edmond Buron. Paris: Maisonneuve Frères, 1930.

Alberti, Leon Battista. *On Painting* and *On Sculpture*. Edited and translated by Cecil Grayson. London: Phaidon, 1972.

Alvares, Francisco. *The Prester John of the Indies* . . . Edited by Charles Fraser Beckingham and George Wynn Huntingford. Translated by Lord Stanley of Alderly. Cambridge, U.K.: Hakluyt Society, 1961.

Amerbach, Johannes. The Correspondence of Johann Amerbach: Early Printing in its Social Context. Edited and translated by Barbara C. Halporn. Ann Arbor: University of Michigan Press, 2000.

Anderson, Christopher. *The Annals of the English Bible*. Edited by Hugh Anderson. London: Jackson, Walford, & Hodder, 1862.

Anghiera, Pietro Martire d'. *De orbe novo*. Translated by Francis Augustus MacNutt. New York: G. P. Putnam's Sons, 1912.

Arciniegas, Germán. *Why America? 500 Years of a Name: The Life and Times of Amerigo Vespucci*. Translated by Harriet de Onís. Bogotá: Villegas Editores, 2002.

Ariosto, Ludovico. *The Satires of Ludovico Ariosto: A Renaissance Autobiography*. Translated by Peter DeSa Wiggins. Athens: Ohio University Press, 1976.

Aristotle. *On the Heavens I and II*. Edited and translated by Stuart Leggatt. Warminster, U.K.: Aris & Phillips, 1995.

Augustine, Saint, Bishop of Hippo. *The City of God Against the Pagans*. Vol. 5 (Books 16–18, chs. 1–35). Translated by Eva Matthews Sanford and William McAllen Green. Loeb Classical Library. Cambridge, Mass.: Harvard University Press, 1957–1972.

Azurara, Gomes Eannes de. *The Chronicle of the Discovery and Conquest of Guinea*. Translated by Charles Raymond Beazley and Edgar Prestage. 2 vols. London: Hakluyt Society, 1896–99.

Bacon, Roger. *The Opus Majus of Roger Bacon*. 2 vols. Translated by Robert Belle Burke. Philadelphia: University of Pennsylvania Press, 1928.

Barron, W. R. J., and Glyn S. Burgess, eds. *The Voyage of Saint Brendan: Representative Versions of the Legend in English Translation*. Exeter, U.K.: University of Exeter Press, 2002.

Bartlett, John Russell. *Bibliotecha Americana: A Catalogue of Books Relating to North and South America* . . . Cambridge, Mass.: Riverside Press, 1875.

Baumer, Franklin L., ed. Main Currents of Western Thought. New Haven: Yale University Press, 1978.

Baxandall, Michael. *Painting & Experience in Fifteenth-Century Italy: A Primer in the Social History of Pictorial Style*. 2nd ed. Oxford: Oxford University Press, 1988.

Beckingham, Charles Fraser. *Between Islam and Christendom: Travellers, Facts, and Legends in the Middle Ages and the Renaissance*. London: Variorum Reprints, 1983.

Beckingham, Charles Fraser, and Bernard Hamilton, eds. *Prester John, the Mongols, and the Ten Lost Tribes*. Aldershot, U.K.: Variorum, 1996.

Berggren, J. Lennart, and Alexander Jones, eds. and trans. *Ptolemy's "Geography": An Annotated Translation of the Theoretical Chapters*. Princeton: Princeton University Press, 2000.

Bernard of Clairvaux. *The Letters of St. Bernard of Clairvaux.* Translated by Bruno Scott James. Stroud, U.K.: Sutton, 1998.

Biller, Peter. *The Measure of Multitude: Population in Medieval Thought.* Oxford: Oxford University Press, 2000.

Birrell, Anne, trans. *The Classic of Mountains and Seas.* New York: Penguin Putnam, 1999.

Bisaha, Nancy. "Petrarch's Vision of the Muslim and Byzantine East." *Speculum* 76, no. 2 (April 2001): 284–314.

Biskup, Marian, ed. *Regesta Copernicana (Calendar of Copernicus' Papers).* Warsaw: Ossolineum, Polish Academy of Science Press, 1973.

Boccaccio, Giovanni. "De Canaria." In *Tutte le Opere di Giovanni Boccaccio,* 970–86. Originally edited by Manlio Pastore Stocchi, 1964. Reprint edited by Vittore Branca. Milan: Mondadori, 1992.

Bontier, Pierre, and Jean Le Verrier. *The Canarian, or Book of the Conquest and Conversion of the Canarians in the Year 1402, by Messire Jean de Bethencourt.* Edited and translated by Richard Henry Major. London: Hakluyt Society, 1872.

Bouloux, Nathalie. *Culture et Savoirs Géographiques en Italie au XIVe Siècle.* Turnhout, Belgium: Brepols, 2002.

Boxer, Charles Ralph. *The Portuguese Seaborne Empire: 1415–1825.* New York: Knopf, 1969.

Brewer, John Sherren, ed. *Monumenta Franciscana.* Vol. 1. London: Longman, 1858.

Bridges, John Henry. *The Life and Work of Roger Bacon: An Introduction to the "Opus Majus."* Edited by H. Gordon Jones. London: Williams & Norgate, 1914.

Brown, Lloyd. *The Story of Maps.* New York: Dover, 1979. Originally published in 1949 by Little, Brown & Company.

Brown, Peter Robert Lamont. *Augustine of Hippo: A Biography.* Berkeley: University of California Press, 1967.

Bruni, Leonardo. "Panegyric to the City of Florence." In *The Earthly Republic: Italian Humanists on Government and Society,* 121–75. Edited and translated by Benjamin G. Kohl and Ronald G. Witt, with Elizabeth B. Welles. Philadelphia: University of Pennsylvania Press, 1978.

Cachey, Theodore J. "The literary response of Renaissance Italy to the New World encounter." *Claudel Studies* 15, no. 2 (1988): 66–75.

——. trans. *Petrarch's Guide to the Holy Land: Itinerary to the Sepulcher of Our Lord Jesus Christ.* Notre Dame, Ind.: University of Notre Dame Press, 2002.

Campbell, Tony. *The Earliest Printed Maps, 1472–1500* (Berkeley: University of California Press, 1987)

——. "Portolan Charts from the Late Thirteenth Century to 1500." In J. B. Harley and David Woodward, eds., *Cartography in Prehistoric, Ancient, and Medieval Europe and the Mediterranean.* Vol. 1 of *The History of Cartography,* 371–463. Chicago: University of Chicago Press, 1987.

——. "The Woodcut Map Considered as a Physical Object: A New Look at Erhard Etzlaub's 'Rom Weg' Map of c. 1500." *Imago Mundi* 30 (1978): 79–91.

Cassidy, Vincent H. *The Sea Around Them: The Atlantic Ocean, A.D. 1250.* Baton Rouge: Louisiana State University Press, 1968.

Cassiodorus. *"Institutions of Divine and Secular Learning" and "On the Soul."* Translated by James W. Halporn. Liverpool: Liverpool University Press, 2003.

Charles, Heinrich. *The Romance of the Name "America."* New York: Author, 1926.

Cicero. *On the Good Life: [Selected writings of] Cicero.* Translated by Michael Grant. Harmondsworth, U.K.: Penguin, 1971.

Clegg, Brian. *The First Scientist: A Life of Roger Bacon.* London: Constable, 2003.

Clough, Cecil H. "The New World and the Italian Renaissance." In *The European Outthrust and Encounter: The First Phase (c. 1400–c. 1700)*, 291–328. Edited by Cecil H. Clough and Paul Edward Hedley Hair. Liverpool: Liverpool University Press, 1994.

Colón, Fernando. *The History of the Life and Deeds of the Don Admiral Christopher Columbus.* Edited by Ilaria Caraci Luzzana. Translated by Geoffrey Symcox and Blair Sullivan. Repertorium Columbianum, vol. 13. Turnhout, Belgium: Brepols, 2004.

Columbus, Christopher. *The Book of Prophecies.* Edited by Roberto Rusconi. Translated by Blair Sullivan. Repertorium Columbianum, vol. 3. Berkeley: University of California Press, 1997.

——. *The "Libro de las Profecías" of Christopher Columbus.* Translated by Delno C. West and August Kling. Gainesville: University of Florida Press, 1991.

Connolly, Daniel K. "Imagined Pilgrimage in the Itinerary Maps of Matthew Paris." *The Art Bulletin* 81, no. 4 (December 1999): 598–622.

Constable, Giles. *Religious Life and Thought (11th–12th Centuries)*. London: Variorum Reprints, 1979.

Copernicus, Nicolaus. *On the Revolutions.* Edited by Jerzy Dobryckzi. Translated by Edward Rosen. Baltimore: Johns Hopkins University Press, 1992.

Cosgrove, Denis. *Apollo's Eye: A Cartographic Genealogy of the Earth in the Western Imagination.* Baltimore: Johns Hopkins University Press, 2001.

Coulton, George Gordon, ed. *From St. Francis to Dante: A translation of all that is of primary interest in the chronicle of the Franciscan Salimbene . . .* London: Duckworth, 1908.

Critchley, John. *Marco Polo's Book.* Aldershot, U.K.: Variorum, 1992.

Crone, Gerald Roe, ed. and trans. *The Voyages of Cadamosto and Other Documents on Western Africa in the Second Half of the Fifteenth Century.* London: Hakluyt Society, 1937.

Cunningham, J. V. *Woe or Wonder: The Emotional Effect of Shakespearean Tragedy.* Denver: University of Denver Press, 1951.

Davenport, Frances Gardiner, ed. *European Treaties Bearing on the History of the United States and Its Dependencies to 1648.* Washington, D.C.: Carnegie Institution of Washington, 1917.

Davies, Arthur. "Columbus Divides the World." *The Geographical Journal* 133, no. 3 (September 1967): 337–44.

Davis, Natalie Zemon. *Society and Culture in Early Modern France: Eight Essays.* London: Duckworth, 1975.

Dawson, Christopher, ed. *Mission to Asia.* Toronto: University of Toronto Press, 1980. Originally published as *The Mongol Mission.* London: Sheed & Ward, 1955.

Delumeau, Jean. *History of Paradise: The Garden of Eden in Myth and Tradition.* Translated by Matthew O'Connell. New York: Continuum, 1995.

Dienes, Mary. "Eastern Missions of the Hungarian Dominicans in the First Half of the Thirteenth Century." *Isis* 27, no. 2 (August 1937): 225–41.

Diffie, Bailey W., and George D. Winius. *Foundations of the Portuguese Empire: 1415–1580.* Vol. 1 of *Europe and the World in the Age of Expansion.* Edited by Boyd C. Shafer. Minneapolis: University of Minnesota Press, 1977.

Dilke, Oswald A. W., and Margaret S. Dilke. "The Adjustment of Ptolemaic Atlases to Feature the New World." In *The Classical Tradition and the Americas*, 117–34. Edited by Wolfgang Hasse and Meyer Rheinhold. New York: de Gruyter, 1994.

Diller, Aubrey. "The Greek Codices of Palla Strozzi and Guareno Veronese." *Journal of the Warburg and Courtauld Institutes* 24, no. 3/4 (July–December 1961): 313–21.

——. "The Oldest Manuscripts of Ptolemaic Maps." *Transactions and Proceedings of the American Philological Association* 71 (1940): 62–67.

Edson, Evelyn. *The World Map, 1300–1492.* Baltimore: Johns Hopkins University Press, 2007.

Emerson, Ralph Waldo. *English Traits.* London: G. Routledge, 1857.

Fernandez-Armesto, Felipe. *Amerigo: The Man Who Gave His Name to America*. New York: Random House, 2006.

Findlen, Paula, ed. The Italian Renaissance: The Essential Readings. Malden, Mass.: Blackwell, 2002.

Fischer, Joseph, and Franz Ritter von Wieser. *The "Cosmographiae Introductio" of Martin Waldseemüller in Facsimile*. Edited by Charles George Herbermann. New York: U.S. Catholic Historical Society, 1907.

———. *The Oldest Map with the Name America of the Year 1507 [Die älteste Karte mit dem Namen Amerika aus dem Jahre 1507]* . . . London: Stevens, Son, & Stiles [Innsbruck: Wagner], 1903.

———. "The Oldest Map with the Name 'America,' and How It Was Found, Told by Its Discoverer, Rev. Joseph Fischer, S. J." *Benziger's Magazine* 4, no. 2 (April 1902): 269–70.

Flint, Valerie. *The Imaginative Landscape of Christopher Columbus*. Princeton: Princeton University Press, 1992.

Forster, Leonard, ed. and trans. *Selections from Konrad Celtis: 1459–1508*. Cambridge, U.K.: Cambridge University Press, 1948.

Franke, Herbert. "Sino-Western contacts under the Mongol empire." *Journal of the Hong Kong Branch of the Royal Asiatic Society* 6 (1966): 49–72.

Franzini, Antoine, and Nathalie Bouloux. *Îles du Moyen Âge*. Paris: Presses Universitaires de Vincennes, 2004.

Friedman, John Block. *The Monstrous Races in Medieval Art and Thought*. Syracuse: Syracuse University Press, 2000. Originally published in 1981 by Harvard University Press.

Gadol, Joan. *Leon Battista Alberti: Universal Man of the Early Renaissance*. Chicago: University of Chicago Press, 1970.

Galvão, Antonio. *The Discoveries of the World, from Their First Originall Vnto the Yeere of Our Lord 1555*. Edited by Richard Hakluyt. London: Impensis G. Bishop, 1601.

Gautier Dalché, Patrick. "L'oeuvre géographique du Cardinal Fillastre († 1428). Représentation du monde et perception de la carte à l'aube des découvertes." *Archives d'Histoire Doctrinale et Littéraire du Moyen Âge* (1992): 319–83.

———. "The Reception of Ptolemy's Geography (End of the Fourteenth to Beginning of the Sixteenth Century)." In David Woodward, ed., *Cartography in the European Renaissance*. Vol. 3, bk. 1 of *The History of Cartography*, 285–364. Chicago: University of Chicago Press, 2007.

———. "Le souvenir de la *Géographie* de Ptolémée dans le monde latin médiéval." *Euphrosyne: Revista e Filologia Clássica* 27 (Lisbon, 1999): 79–106.

The Geographical Journal 10, no. 3 (September 1897): 323–33.

Gill, Joseph. *The Council of Florence*. Cambridge, U.K.: Cambridge University Press, 1961.

Goldstein, Thomas. "The Renaissance Concept of the Earth in Its Influence upon Copernicus." *Terrae Incognitae* 4 (1972): 19–51.

Gordan, Phyllis Walter Goodhart, ed. and trans. *Two Renaissance Book Hunters: The Letters of Poggius Bracciolini to Nicolaus de Niccolis*. New York: Columbia University Press, 1974.

Gormley, Catherine M., Mary A. Rouse, and Richard H. Rouse. "The Medieval Circulation of the *De Chorographia* of Pomponius Mela." *Mediaeval Studies* (Pontifical Institute of Mediaeval Studies, Toronto) 46 (1984): 266–320.

Grafton, Anthony. *Leon Battista Alberti: Master Builder of the Italian Renaissance*. Cambridge, Mass.: Harvard University Press, 2002.

Greenlee, William Brooks, ed. and trans. *The Voyage of Pedro Alvares Cabral to Brazil and India: From Contemporary Documents and Narratives*. London: Hakluyt Society, 1938.

Grenacher, F. "The Woodcut Map: A Form-Cutter of Maps Wanders through Europe in the First Quarter of the Sixteenth Century." *Imago Mundi* 24 (1970): 31–41.

Gribble, Francis Henry. *The Early Mountaineers*. London: T. F. Unwin, 1899.

Griffin, Nigel, ed. and trans. *Las Casas on Columbus: Background and the Second and Fourth Voyages.* Repertorium Columbianum. Vol. 7. Turnhout, Belgium: Brepols, 1999.

Gurney, Alan. *Compass: A Story of Exploration and Innovation.* New York: Norton, 2004.

Hammond, Lincoln Davis, ed. *Travelers in Disguise: Narratives of Eastern Travel by Poggio Bracciolini and Ludovico de Varthema.* Translated by John Winter Jones. Cambridge, Mass.: Harvard University Press, 1963.

Hankins, James. "Ptolemy's *Geography* in the Renaissance." In *The Marks in the Fields: Essays on the Uses of Manuscripts,* 119–27. Edited by Rodney G. Dennis, with Elizabeth Falsey. Cambridge, Mass.: Harvard University Press, 1992.

Harris, Elizabeth. "The Waldseemüller World Map: A Typographical Appraisal." *Imago Mundi* 37 (1985): 30–53.

Heers, Jacques. "De Marco Polo à Christophe Colomb: comment lire le *Devisement du monde?,*" *Journal of Medieval History* 10 (1984): 125–43.

Herodotus. *The Histories.* Translated by Aubrey De Sélincourt; revised with introductory matter and notes by John Marincola. New York: Penguin, 1996.

Hessler, John, ed. and trans. *The Naming of America: Martin Waldseemüller's 1507 World Map and the "Cosmographiae Introductio."* London: Giles; Easthampton, Mass.: distributed by Antique Collectors' Club Limited, 2008.

Hiatt, Alfred. "Blank Spaces on the Earth." *Yale Journal of Criticism* 15, no. 2 (2002): 223–50.

——. "Mutation and Nation: The 1513 Strasbourg Ptolemy." In Zur Shalev and Charles Burnett, eds., *Ptolemy's Geography in the Renaissance.* (London: Warburg Institute, forthcoming.)

Holmes, George. *The Florentine Enlightenment, 1400–50.* London: Weidenfeld & Nicolson, 1969.

Howe, Herbert M., trans., and David Woodward, ed. "Roger Bacon: The Fourth Part of the *Opus Maius:* Mathematics in the Service of Theology: Sections of interest to the history of geographical thought." Unpublished working translation, 1996, www.geography.wisc.edu/histcart/bacon.html.

Humboldt, Alexander von. *Examen critique de l'histoire de la géographie du nouveau continent . . .* Paris: Gide, 1836–39.

Irving, Washington *A History of the Life and Voyages of Christopher Columbus . . .* Vol. 3. New York: G. & C. Carvill, 1828.

——. [Diedrich Knickerbocker, pseud.]. *A History of New York, from the Beginning of the World to the End of the Dutch Dynasty . . .* Vol. 1. New York: Inskeep & Bradford, 1809.

Isidore of Seville. *The "Etymologies" of Isidore of Seville.* Translated by Stephen A. Barney, J. A. Beach, Oliver Berghof, and W. J. Lewis. New York: Cambridge University Press, 2006.

Ivins, William M. *Prints and Visual Communication.* Cambridge, Mass.: MIT Press 1969. Originally published in 1953 by Harvard University Press.

Jackson, Peter, trans. *The Mission of Friar William of Rubruck: His Journey to the Court of the Great Khan Möngke.* London: Hakluyt Society, 1990.

——. *The Mongols and the West, 1221–1410.* Harlow, U.K.; New York: Pearson/Longman, 2005.

Jane, Cecil, ed. and trans. *The Four Voyages of Columbus: A History in Eight Documents, Including Five by Christopher Columbus in the Original Spanish with English Translations.* New York: Dover, 1988.

Johnson, Hildegard Binder. *Carta Marina: World Geography in Strassburg, 1525.* Minneapolis: University of Minnesota Press, 1963.

Jordanus, Catalani. *Mirabilia descripta: The Wonders of the East.* Translated by Henry Yule. London: Hakluyt Society, 1863.

Kadir, Djelal. *Columbus and the Ends of the Earth: Europe's Prophetic Rhetoric as Conquering Ideology.* Berkeley: University of California Press, 1992.

Kemp, Martin. *Leonardo da Vinci: Experience, Experiment, and Design.* Princeton: Princeton University Press, 2006.

Ker, Neil Ripley. *Medieval Libraries of Great Britain: A List of Surviving Books.* London: Royal Historical Society, 1941.

Kimble, George Herbert Tinley. *Geography in the Middle Ages.* London: Methuen, 1938.

———. "The Laurentian Map with Special Reference to Its Portrayal of Africa." *Imago Mundi* 1 (1935): 29–33.

Kremple, Frederich Awalde. "Cultural Aspects of the Councils of Constance and Basel." Ph.D. diss., University of Minnesota, 1954.

Lanman, Jonathan T. "The Religious Symbolism of the T in T-O Maps." *Cartographica* 18, no. 4 (Winter 1981): 18–22.

Larner, John. "The Church and the Quattrocentro Renaissance in Geography." *Renaissance Studies* 12, no. 1 (March 1998): 26–39.

———. *Marco Polo and the Discovery of the World.* New Haven: Yale University Press, 2001.

Leclercq, Jean. *The Love of Learning and the Desire for God: A Study of Monastic Culture.* Translated by Catharine Misrahi. New York: Fordham University Press, 1961.

Lester, C. Edwards, and Andrew Foster. *The Life and Voyages of Americus Vespucius . . .* New York: Baker & Scribner, 1846.

Lewis, Suzanne. *The Art of Matthew Paris in the "Chronica Majora."* Berkeley: University of California with Corpus Christi College (Cambridge), 1987.

Loomis, Louise Ropes, trans. *The Council of Constance: The Unification of the Church.* Edited by John Hine Mundy and Kennerly M. Woody. New York: Columbia University Press, 1961.

Ludd, Guelterus, *Speculi orbis succinctis sed neque poenitenda neque inelegans declaratio.* Strassburg: Grüninger, 1507. Boston: [n.p.], 1924.

Major, Richard Henry, ed. *India in the Fifteenth Century.* London: Hakluyt Society, 1857.

Mandeville, John, Sir. *The Travels of Sir John Mandeville.* Translated by C.W.R.D. Moseley. U.K.; New York: Penguin, 1983.

Marino, Nancy F., ed. and trans. *El Libro del Conocimiento de Todos los Reinos (The Book of Knowledge of All Kingdoms).* Tempe, Ariz.: Arizona Center for Medieval and Renaissance Studies, 1999.

McNeill, William Hardy. *Plagues and Peoples.* New York: Anchor Books Doubleday, 1976.

More, Thomas. *Utopia.* Revised ed. Edited and translated by George M. Logan and Robert M. Adams. Cambridge, U.K.: Cambridge University Press, 2002.

Morison, Samuel Eliot. *Admiral of the Ocean Sea: A Life of Christopher Columbus.* 1 vol. Boston: Little, Brown, 1942.

———. ed. and trans. *Journals and Other Documents on the Life and Voyages of Christopher Columbus.* New York: Heritage, 1963.

Nangis, Guillame de. "Gesta Sanctae Memoriae Ludovici Regis Franciae." In *Recueil des Historiens des Gaules et de la France,* 20. Edited by J. Naudet and P. C. F. Daunou. Paris: Imprimerie Royale, 1840.

Nebenzahl, Kenneth. *Atlas of Columbus and the Great Discoveries.* Chicago: Rand McNally, 1990.

Nordenskiöld, Adolf Erik. *Facsimile-Atlas to the Early History of Cartography with Reproductions of the Most Important Maps printed in the XV and XVI centuries.* Translated by Johan Adolf Ekelof and Clements R. Markham. Stockholm: P. A. Norstedt, 1889.

Paris, Matthew. *English History from the Year 1235 to 1273.* 3 vols. Translated by J. A. Giles. London: H. G. Bohn, 1852.

Parry, J. H. *The Discovery of the Sea.* New York: Dial, 1974.

Pelliot, Paul. *Notes on Marco Polo.* Vol. 1. Paris: Imprimerie Nationale, 1959.

Petrarch, Francesco. *Letters on Familiar Matters.* Translated by Aldo S. Bernardo. New York: Italica, 2005.

——. *The Life of Solitude.* Translated by Jacob Zeitlin. Westport, Conn.: Hyperion, 1924.

——. *Opera quae extant omnia.* Basel, Switzerland: Sebastianum Henricpetri [*sic*], 1581.

Phillips, J. R. S. *The Medieval Expansion of Europe.* 2nd ed. Oxford: Oxford University Press, 1998.

Phillips, Seymour. "The Outer World of the European Middle Ages." In *Implicit Understandings: Observing, Reporting, and Reflecting on the Encounters Between Europeans and Other Peoples in the Early Modern Era,* 23–63. Edited by Stuart B. Schwartz. New York: Cambridge University Press, 1994.

Phillips, William D., Jr., ed. and trans. *Testimonies from the Columbian Lawsuits.* Repertorium Columbianum. Vol. 8. Turnhout, Belgium: Brepols, 2000.

Plato. *Plato.* Vol. 9 (Timaeus, Critias, Cleitophon, Menexenus, Epistles). Translated by R. G. Bury. Loeb Classical Library. Cambridge, Mass.: Harvard University Press, 1999.

——. *The Works of Plato.* Selected and edited by Irwin Edman. New York: The Modern Library, 1928.

Pliny the Elder. *Natural History.* Vol. 2 (Books 3–7). Translated by H. Rackham. Loeb Classical Library. Cambridge, Mass.: Harvard University Press, 1942.

Plutarch. *Lives.* Vol. 8 (Sertorius and Eumenes, Phocion and Cato and Younger). Translated by Bernadotte Perrin. Loeb Classical Library. Cambridge, Mass.: Harvard University Press, 1989.

Pohl, Frederick Julius. *Amerigo Vespucci: Pilot Major.* New York: Columbia University Press, 1944.

Polo, Marco. *The Travels of Marco Polo.* Translated by Ronald Latham. New York: Penguin, 1958.

Ptolemy. *Geographia: Strassburg, 1513.* Theatrum Orbis Terrarum: A Series of Atlases in Facsimile. 2nd ser., vol. 4. Amsterdam: Theatrum Orbis Terrarum, 1966.

Rachewiltz, Igor de. *Papal Envoys to the Great Khans.* Stanford: Stanford University Press, 1971.

——. "Some Remarks on the Ideological Foundations of Chingis Khan's Empire." *Papers on Far Eastern History* 7 (March 1973): 21–36.

Randles, W. G. L. "Bartolomeu Dias and the Discovery of the South-East Passage Linking the Atlantic to the Indian Ocean (1488)." *Revista da Universidade de Coimbra* 34 (1987): 19–28.

——. "The Evaluation of Columbus's 'India' Project by Portuguese and Spanish Cosmographers in the Light of the Geographical Science of the Period." *Imago Mundi* 42 (1990): 50–64.

Ravenstein, Ernest George. *Martin Behaim, His Life and His Globe.* London: G. Philip & Son, 1908.

Ravenstein, Ernest George, ed. and trans. *A Journal of the First Voyage of Vasco da Gama.* London: Hakluyt Society, 1898.

Reisch, Gregor. *Margarita philosophica.* Strassburg: Schott, 1508.

Relano, Francesc. *The Shaping of Africa: Cosmographic Discourse and Cartographic Science in Late Medieval and Early Modern Europe.* Burlington, Vt.: Ashgate, 2002.

Richter, Jean Paul, ed. *The Notebooks of Leonardo da Vinci.* New York: Dover, 1970.

Ringmann, Matthias. *Be [i.e., de] ora antarctica per regem portugallie pridem inuenta.* Strassburg: Hupfuff, 1505.

Robinson, James Harvey, ed., with Henry Winchester Rolfe. *Petrarch: The First Modern Scholar and Man of Letters . . .* New York: Putnam, 1898.

Rogers, Francis Millet. *The Quest for Eastern Christians.* Minneapolis: University of Minnesota Press, 1962.

——. "The Vivaldi Expedition." In *Seventy-third Annual Report of the Dante Society,* 31–45. Cambridge, Mass.: Dante Society of America, 1955.

Romer, F. E., trans. *Pomponius Mela's Description of the World.* Ann Arbor: University of Michigan Press, 1998.

Ronsin, Albert. *Le nom de l'Amérique: l'invention des chanoines et savants de Saint-Dié.* Strasbourg, France: La Nuée Bleue, 2006.

Rose, William J. Untitled review. *Isis* 16, no. 1 (July 1931): 136–38.

Russell, Peter Edward. *Prince Henry "The Navigator": A Life.* New Haven: Yale University Press, 2000.

Scafi, Alessandro. *Mapping Paradise: A History of Heaven on Earth.* Chicago: University of Chicago Press, 2006.

Schiffeler, John William. *The Legendary Creatures of the Shan Hai Ching.* San Francisco: [n.p.], 1978.

Schmidt, Charles. *Histoire littéraire de l'Alsace à la fin du XVe et au commencement du XVIe siècle.* Paris: Sandoz et Fischbacher, 1879.

Schmieder, Felicitas. "Clash of Civilizations in the Medieval World: Christian Strategies for Diplomacy and Conversion among the Mongols." *Annuario della Casa Romena di Venezia* 3 (Bucharest, 2001): 10–28.

——. "The Mongols as non-believing apocalyptic friends around the year 1260." *Journal of Millennial Studies* 1, no. 1 (Spring 1998), http://www.mille.org/publications/summer98/fschmieder.pdf.

Schoff, Wilfred H., trans. *The Periplus of the Erythraean Sea; Travel and Trade in the Indian Ocean, by a Merchant of the First Century.* New York: Longmans, Green 1912.

Schwartz, Seymour. *Putting "America" on the Map.* Amherst, N.Y.: Prometheus, 2007.

Schwartz, Stuart B., ed. *Implicit Understandings: Observing, Reporting, and Reflecting on the Encounters Between Europeans and Other Peoples in the Early Modern Era.* New York: Cambridge University Press, 1994.

Seaver, Kirsten. *Maps, Myths, and Men: The Story of the Vinland Map.* Stanford: Stanford University Press, 2004.

Setton, Kenneth. "The Byzantine Background to the Italian Renaissance." *Proceedings of the American Philosophical Society* 100, no. 1: 1–76.

Sevcenko, Ihor. "Intellectual Repercussions of the Council of Florence." *Church History* 24, no. 4 (December 1955): 291–323.

Silverberg, Robert. *The Realm of Prester John.* London: Phoenix, 2001. Originally published in 1972 by Doubleday & Co.

Slessarev, Vsevolod. *Prester John: The Letter and the Legend.* Minneapolis: University of Minnesota Press, 1959.

Soulsby, Basil H. "The First Map Containing the Name America." *Geographical Journal* 19, no. 2 (February 1902): 201–9.

Spitz, Lewis William. "Conrad Celtis, Career Humanist: A Case Study in the Nature of the Northern Renaisance." Ph.D. diss., Harvard University, 1954.

Staley, Edgcumbe. *The Guilds of Florence.* London: Methuen, 1906.

Stevens, Henry Newton. *The First Delineation of the New World and the First Use of the Name America on a Printed Map* . . . London: Stevens, Son, & Stiles, 1928.

——. *Rare Americana including the Original Waldseemüller Maps of 1507 and 1516.* Auction catalogue. London: Stevens, Son, & Stiles, 1907.

Strabo. *Geography.* Vol. 1 (Books 1–2). Translated by Horace Leonard Jones. Loeb Classical Library. New York: G. P. Putnam's Sons, 1917–33.

Sumien, N. *La correspondance du savant florentin Paolo dal Pozzo Toscanelli avec Christophe Colombe.* Paris: Société d'éditions géographiques, maritimes, et coloniales, 1927.

Symcox, Geoffrey, and Giovanna Rabitti, eds. *Italian Reports on America, 1493–1522: Letters, Dispatches, and Papal Bulls.* Translated by Peter D. Diehl. Repertorium Columbianum. Vol. 10. Turnhout, Belgium: Brepols, 2001.

Symcox, Geoffrey, and Luciano Formisano, eds. *Italian Reports on America, 1493–1522: Accounts by Contemporary Observers.* Translated by Theodore J. Cachey, Jr., and John C. McLucas. Repertorium Columbianum. Vol. 12. Turnhout, Belgium: Brepols, 2002.

Tanner, Norman P., ed. *Decrees of the Ecumenical Councils.* London: Sheed & Ward, 1990.

Taviani, Paolo Emilio. *Christopher Columbus: The Grand Design.* London: Orbis, 1985.

Taylor, Eva Germaine Rimington. *The Haven-Finding Art: A History of Navigation from Odysseus to Captain Cook.* New York: Abelard-Schuman, 1957.

Thacher, John Boyd. *The Continent of America: Its Discovery and Its Baptism.* New York: W. E. Benjamin, 1896.

Thorndike, Lynn, ed. and trans. *The Sphere of Sacrobosco and Its Commentators.* Chicago: University of Chicago Press, 1949.

Uminski, Sigmund H. *Poland Discovers America.* Vol. 1 of *The Poles in the Americas.* New York: Polish Publication Society of America, 1972.

Vander Linden, H. "Alexander VI and the Demarcation of the Maritime and Colonial Domains of Spain and Portugal, 1493–1494." *The American Historical Review* 22, no. 1 (October 1916): 1–20.

Varela, Consuelo, ed. *Textos y Documentos Completos: Relaciones de Viajes, Cartas y Memoriales.* Madrid: Alianza, 1982.

Veltman, Kim H. "Leonardo da Vinci: Studies of the Human Body and Principles of Anatomy." Unpublished translation of German original in Klaus Schreiner, ed., *Gepeinigt, begehrt vergessen,* 287–308. Munich: W. Fink, 1992. Available online at http://www.sumscorp.com/leonardoindex.htm.

Vespucci, Amerigo. *Letters from a New World: Amerigo Vespucci's Discovery of America.* Edited by Luciano Formisano. Translated by David Jacobson. New York: Marsilio, 1992.

Vigneras, Louis-André. "Saint Thomas, Apostle of America." *Hispanic American Historical Review* 57, no. 1 (February 1977): 82–90.

Virgil. *The Aeneid.* Translated by W. F. Jackson Knight. Penguin Classics 151. Revised ed. Baltimore: Penguin, 1958.

Waldseemüller, Martin, and Matthias Ringmann. *Instructio manuductionem praestans in cartam itinerariam Mart. Hilacomili.* Strassburg: Grüninger, 1511.

Werner, A. "Prester John and Benin." In "Notes and Queries," *Bulletin of the School of Oriental Studies, University of London* 2, no. 1 (1921): 163–65.

Westropp, Thomas Johnson. "Brasil and the Legendary Islands of the North Atlantic." *Proceedings of the Royal Irish Academy* 30, sec. C, no. 8 (August 1912): 223–63.

Wey Gomez, Nicolás. *The Tropics of Empire: Why Columbus Sailed South to the Indies.* Cambridge, Mass.: The MIT Press, 2008.

Williamson, James Alexander. *The Cabot Voyages and Bristol Discovery Under Henry VII.* Cambridge, U.K.: Hakluyt Society, 1962.

Witt, Ronald G. *Hercules at the Crossroads: The Life, Works, and Thought of Coluccio Salutati.* Durham, N.C.: Duke University Press, 1983.

Woodward, David. "Medieval *Mappaemundi.*" In J. B. Harley, and David Woodward, eds., *Cartography in Prehistoric, Ancient, and Medieval Europe and the Mediterranean.* Vol. 1 of *The History of Cartography,* 286–370. Chicago: University of Chicago Press, 1987.

Woodward, David, ed. *Art and Cartography: Six Historical Essays.* Chicago: University of Chicago Press, 1987.

———. *Five Centuries of Map Printing.* Chicago: University of Chicago Press, 1975.

Yule, Henry, ed. and trans. *Cathay and the Way Thither: Being a Collection of Medieval Notices of China.* Vol. 2. London: Hakluyt Society, 1866.

Zamora, Margarita. *Reading Columbus.* Berkeley: University of California Press, 1993.

Further Reading

INTEREST IN EARLY world maps has boomed in the past few decades, and the study of the history of cartography itself has recently undergone something of a renaissance. The most important reference work in this rapidly expanding field is a massive, multivolume series titled *The History of Cartography*, published by the University of Chicago Press. Three volumes of the series have been published to date, two of which are of particular relevance to the period and themes covered in this book: *Cartography in Prehistoric, Ancient, and Medieval Europe and the Mediterranean* (Vol. 1, 1987) and *Cartography in the European Renaissance* (Vol. 3, 2007). Other recently published, informative, and often beautiful-to-look-at introductions to the history of cartography include Peter Barber's *The Map Book* (2005), Dennis Cosgrove's *Apollo's Eye* (2001), Evelyn Edson's *The World Map* (2007) and *Mapping Time and Space* (1997), Kenneth Nebenzahl's *Atlas of Columbus and the Great Discoveries* (1990), Vincent Virga's *Cartographia* (2007), Peter Whitfield's *The Image of the World* (1994), and John Noble Wilford's *The Mapmakers* (1981, revised in 2000). Several earlier surveys remain of interest and value, although they betray their age in places: Leo Bagrow's *History of Cartography* (1964), C. Raymond Beazely's *Dawn of Modern Geography* (1897), Lloyd Brown's *The Story of Maps* (1949), and G. R. Crone's *Maps and Their Makers* (1953).

Interest in the Waldseemüller map itself has also grown dramatically in recent years. Since 2007 the map has been on permanent display in the Great Hall of the Library of Congress's Jefferson Building, along with Waldseemüller's Carta Marina, and the two maps there form the centerpiece of a fascinating exhibit titled "Exploring the Early Americas: The Jay I. Kislak Collection." The Waldseemüller map and the Carta Marina can also be viewed side-by-side online as part of the "Exploring the Early Americas" Web site, at loc.gov/exhibits/earlyamericas/maps/html/. A high-resolution digital image of the Waldseemüller map can also be consulted and downloaded for free from the Library by going to memory.loc.gov/ammen/gmdhtml/dsxphome.html and searching for the name "Waldseemüller." For real-time updates on the ongoing study of the map at the Library, see "Warping Waldseemüller," a private blog maintained by

John Hessler, a senior research librarian in the Library's Geography and Map Division, at warpinghistory.blogspot.com. And, with the help of Free Press, I have produced my own annotated version of the map, which can be found at: Simonandschuster.com/4thpartmap.

A number of books and articles on the Waldseemüller map have also recently appeared in English. These include Peter W. Dickson's self-published *The Magellan Myth: Reflections on Columbus, Vespucci, and the Waldseemüller map of 1507* (2007); Christine Johnson's "Renaissance German Cosmographers and the Naming of America," in *Past and Present* (Vol. 1, no. 1, May 2006); John W. Hessler's *The Naming of America: Martin Waldseemüller's 1507 World Map and the "Cosmographiae introductio"* (2007) and "Warping Waldseemüller: A Phenommenological and Computational Study of the 1507 World Map," in *Cartographica* (Vol. 44, 2006); and Seymour I. Schwartz's *Putting "America" on the Map: The Story of the Most Important Graphic Document in the History of the United States* (2007).

Several earlier Waldseemüller-related works in English are also well worth seeking out. Among them are Elizabeth Harris's "The Waldseemüller World Map: A Typographic Appraisal," in *Imago Mundi* (Vol. 37, 1985); Joseph Fischer and Franz von Wieser's *The "Cosmographiae Introductio" of Martin Waldseemüller in Facsimile* (1907) and *The Oldest Map with the Name America of the Year 1507* (1903); Hildegard Binder Johnson's *Carta Marina: World Geography in Strassburg, 1525* (1963); the Waldseemüller chapter in Robert W. Karrow's *Mapmakers of the Sixteenth-Century and Their Maps* (1993); Franz Laubenberger and Steven Rowans's "The Naming of America," in *Sixteenth-Century Journal* (Vol. 13, no. 4. Winter 1982); Kirsten Seaver's *Maps, Myths, and Men* (2004), which argues convincingly that Joseph Fischer, the discoverer of the Waldseemüller map, went on to forge the famous Vinland map; the bibliographical note on the work of Waldseemüller and the Gymnasium Vosagense by R.A. Skelton in *Geographia: Strassburg, 1513*, a facsimile of Waldseemüller and Ringmann's 1513 edition of Ptolemy's *Geography* (1966); John Boyd Thacher's *The Continent of America: Its Discovery and Its Baptism* (1896); and the chapter on Waldseemüller in Hans Wolff's *America: Early Maps of the New World* (1992).

For the many other subjects and themes discussed in *The Fourth Part of the World*, see the bibliography.

Permissions and Credits

COLOR CREDITS

Plate 1: The Psalter Map (1265). Copyright © The British Library Board, all rights reserved 2009 (Add. MS 28681 fol 9).

Plate 2: Christ crucified on a T-O map. From a manuscript of Isidore of Seville, *Etymologies* (thirteenth century). Biblioteca Medicea Laurenziana, Florence. Photo courtesy of the Ministry for the Public Good and Cultural Activities. All further reproduction is prohibited. Photo © Donato Pineider.

Plate 3: Saint Brendan and companions. From a manuscript copy of *The Voyage of St. Brendan the Abbot* (1460). Universitätsbibliothek Heidelberg (Cod. Pal. germ. 60, fol. 179v).

Plate 4: Italy on a world map by Matthew Paris (circa 1255). Courtesy of the Master and Fellows of Corpus Christi College (MS 26, fol. vii v.).

Plate 5: Italy on a fourteenth-century marine chart. Courtesy of the Geography and Map Division, Library of Congress.

Plate 6: The Catalan Atlas (1375). Bibliothèque Nationale de France.

Plate 7: The world of Claudius Ptolemy, Ulm Geography (1482). Courtesy of the Rare Book and Special Collections Division, Library of Congress.

Plate 8: The world map of Simon Marmion (1460). Bibliothèque royale de Belgique/ Koninklijke Bibliotheek van België, Section des Manuscrits/Afdeling Handschriften (MS9231, fol. 281 v.).

Plate 9: The world map of Henricus Martellus (circa 1489–1490). Copyright © The British Library Board, all rights reserved (Add. MS 15.760, fols. 68v-69r).

Plate 10: Detail from the Cantino chart (1502). Biblioteca Estense Universitaria, Modena, Italy.

Plate 11: America on the Waldseemüller map of 1507. Courtesy of the Geography and Map Division, Library of Congress.

CHAPTER OPENER CREDITS

Chapters 1 to 18: Details from the Waldseemüller map. Courtesy of the Geography and Map Division, Library of Congress.

Chapter 19: Four Waldseemüller Derivatives. From top left, clockwise, see credits for Figures 3, 73, 79, 78.

Epilogue: Alexander the Great. From Matthew Paris's *Chronica maiora (circa 1255)*. Courtesy of the Master and Fellows of Corpus Christi College (MS 26, f. 12v.).

Appendix I: The Stevens-Brown map (16th century, date uncertain). Courtesy of the John Carter Brown Library, Brown University.

BLACK-AND-WHITE CREDITS

Figure 1: Title page, Cosmographiae introductio *(1507).* Courtesy of the Rare Book and Special Collections Division, Library of Congress.

Figure 2: Map of the western hemisphere, Henricus Glareanus (1510) and *Figure 3: Map of the world, Henricus Glareanus (1510).* University Library of Munich (Cim. 74).

Figure 4: Globe gores, Martin Waldseemüller (1507). Courtesy of the James Ford Bell Library, University of Minnesota.

Figure 5: The Waldseemüller world map (1507). Courtesy of the Geography and Map Division, Library of Congress.

Figure 6: The medieval cosmos, Matthew Paris. Courtesy of the Master and Fellows of Corpus Christi College (MS 385, p. 119).

Figure 7: Sailors and ship. Sacrobosco, *Sphera mundi cum tribus commentis (1499).* Courtesy of the Trustees of the Boston Public Library/Rare Books.

Figure 8: Zonal map, Matthew Paris. Courtesy of the Master and Fellows of Corpus Christi College (MS 385, Part II, p. 178).

Figure 9: T-O map, Gregorio Dati. From a fifteenth-century manuscript of Dati's *La sfera (before 1435).* Courtesy of the Trustees of the Boston Public Library/Rare Books.

Figure 10: T-O map, Matthew Paris (circa 1255). Courtesy of the Master and Fellows of Corpus Christi College (MS 385, Part II, p. 152).

Figure 11: Frederick II holding a T-O globe, Matthew Paris (circa 1255). Courtesy of the Master and Fellows of Corpus Christi College (MS 16, fol. 127 r).

Figure 12: World map, Matthew Paris (circa 1255). Courtesy of the Master and Fellows of Corpus Christi College (MS 26, fol. vii v.).

Figure 13: The Lambeth Palace Map (circa 1300). From Nennius, *Historia Britonum* (MS 371), fol. 9v. Courtesy of the Lambeth Palace Library.

Figure 14: The monstrous races. From Sebastian Münster, *Cosmographiae universalis* (1553). Courtesy of the Trustees of the Boston Public Library/Rare Books.

Figure 15: Pilgrimage itinerary, Matthew Paris (circa 1255). Courtesy of the Master and Fellows of Corpus Christi College (MS 26, fol. i v.).

Figure 16: The Mongols through Western eyes, Matthew Paris (circa 1255). Courtesy of the Master and Fellows of Corpus Christi College (MS 16, fol. 166).

Figure 17: Chinese monstrous beings. From the *Shan-hai ching t'u-shuo* (Shanghai: Hui un t'ang shu-chü, 1917), as reproduced in Schiffeler, *The Legendary Creatures of the Shan Hai Ching* (1978).

Figure 18: World map from a Marco Polo manuscript (fourteenth century). Manuscripts, Maps, and Pictures, Kungl. Biblioteket, National Library of Sweden (M. 304).

Figure 19: T-O map with fourth continent. Courtesy of Kloster Einsiedeln (Codex Eins. 263 [973], fol. 182 r.). Digital image courtesy of Hill Museum & Manuscript Library, Saint John's Abbey and University, Collegeville, Minnesota.

Figure 20: The Carte Pisane (circa 1275). Bibliothèque Nationale de France.

Figure 21: Anonymous marine chart (fourteenth century). Courtesy of the Geography and Map Division, Library of Congress.

Figure 22: Wind rose, from the marine chart of Jose Aguiar (1492). Wikimedia Commons. Original chart at the Beinecke Rare Book and Manuscript Library, Yale University.

Figure 23: World map, Petrus Vesconte (1321). Copyright © The British Library Board, all rights reserved 2009 (Add. MS 27376*, fols 187v-188r).

Figure 24: The Christian cosmos, Hartmann Schedel, Nuremburg Chronicle *(1492).* Courtesy of the Trustees of the Boston Public Library/Rare Books.

Figure 25: The Earth at the exact center of the cosmos. Detail from Figure 24. Courtesy of the Trustees of the Boston Public Library/Rare Books

Figure 26: The off-center Earth. From Sacrobosco, *Sphera mundi cum tribus commentis (1499).* Courtesy of the Trustees of the Boston Public Library/Rare Books.

Figure 27: Diagram by Roger Bacon. From Bacon's *Opus maius (circa 1267).* Copyright © The British Library Board, all rights reserved 2009 (Royal MS 7 F VII, 85r).

Figure 28: The islands of the Atlantic. From the marine chart of Angelino Dulcert (1339). Bibliothèque Nationale de France.

Figure 29: Armillary sphere. From Peter Apian, *Cosmographicus liber (1524).* Courtesy of the Trustees of the Boston Public Library/Rare Books.

Figure 30: Latitude and longitude. From Peter Apian, *Cosmographicus liber (1524).* Courtesy of the Trustees of the Boston Public Library/Rare Books.

Figure 31: Instruments used to observe the heavens. From Peter Apian, *Instrument Buch* (1533). Courtesy of the Rare Book and Special Collections Division, Library of Congress.

Figure 32: Perceived height of the Pole Star. From Sacrobosco, *Sphaera Mundi (1490).* Courtesy of the Trustees of the Boston Public Library/Rare Books.

Figures 33 and 34: Ptolemy's first and second projections. From Ptolemy's *Geography* (facsimile of a 1430 manuscript). Courtesy of the Geography and Map Division, Library of Congress.

Figure 35: Ptolemy's third projection. From Ptolemy's *Geography (1525).* Courtesy of the Rare Book and Special Collections Division, Library of Congress.

Figure 36: Plot of the mountain ranges of Asia (fifteenth century). ÖNB/Vienna, Picture Archive, Cod. 5266, fol. 92r.

Figure 37: World map, Ptolemy's Geography *(circa 1300).* Biblioteca Apostolica Vaticana, Rome (Urb. Gr. 82, ff. 60v-61r).

Figure 38: View of Florence (circa 1352–1358). Detail from the *Madonna della Misericordia,* Loggia del Begallo, Florence. Wikimedia Commons.

Figure 39: View of Florence (1492). From Hartmann Schedel's *Nuremburg Chronicle.* Wikimedia Commons.

Figure 40: Perspectograph. Detail from *Hydraulic Devices and Study of Figure in Front of a Perspectograph (circa 1478–1480).* Codex Atlanticus (CA 5r). Copyright Biblioteca Ambrosiana, Milan (Auth. No. 129/08).

Figure 41: Vitruvian Man. Detail from *The Proportions of the Human Figure, after Vitruvius (circa 1490).* Galleria dell'Accademia, Venice. Photo courtesy of the Ministry for the Public Good and Cultural Activities.

Figure 42: Cardinal Fillastre's world map (circa 1418). From *De situ orbis*, Pomponius Mela (Reims BM. MS 1321, fol. 12), Bibliothèque Municipale de Reims.

Figure 43: Off-center earth. From Pierry d'Ailly, *Imago mundi* (*early fifteenth century manuscript*). Bibliothèque royale de Belgique/Koninklijke Bibliotheek van België, Section des Manuscrits/Afdeling Handschriften (MS 21198–204, fol. 2v).

Figure 44: Zonal world map. From Pierry d'Ailly, *Imago mundi* (*early fifteenth century*). Bibliothèque royale de Belgique/Koninklijke Bibliotheek van België, Section des Manuscrits/Afdeling Handschriften (MS 21198–204, fol. 4).

Figure 45: Map of northern Europe, Claudius Clavus (1424). Bibliothèque Municipale © Ville de Nancy (MS 441 [354]).

Figure 46: Africa and the Atlantic. Detail from the Viladestes chart (1413). Bibliothèque Nationale de France.

Figure 47: The Catalan-Estense world map (circa 1450). Biblioteca Estense Universitaria, Modena, Italy.

Figure 48: The Kangnido map (fifteenth century). Ryukoku University Library, Kyoto, Japan.

Figure 49: The world map of Albertin de Virga (circa 1411–1415). From Kamal, Youssouf, *Monumenta Cartographica*, Vol. 4, fasc. iii, fol. 1377. Courtesy of the Geography and Map Division, Library of Congress. The original map disappeared in 1932 and has never been recovered.

Figure 50: Antilia and Satanezes. From the Pizzigano chart (1424). Courtesy of the James Ford Bell Library, University of Minnesota

Figure 51: The Fra Mauro world map (circa 1459). By permission of the Biblioteca Nazionale Marciana, Venice.

Figure 52: Reconstruction of Toscanelli's map. From Kretschmer, *Die Entdeckung Amerika's in ihrer Bedeutung für die Geschichte des Weltbildes*, Table 6 (1892). Courtesy of the Geography and Map Division, Library of Congress.

Figure 53: Detail from the "Ginea Portogalexe" chart (circa 1489). Copyright © The British Library Board, all rights reserved 2009 (Egerton MS 73, fol. 33A).

Figures 54 and 55: World map, Rome Geography (1478) and Prassum Promontorium. Facsimile maps from *Cosmographia. Roma, 1478* (Amsterdam: Theatrum Orbis Terrarum, 1966). Courtesy of the Geography and Map Division, Library of Congress.

Figure 56: The wall map of Henricus Martellus (circa 1489–90). Courtesy of the Beinecke Rare Book and Manuscript Library, Yale University.

Figure 57: The globe of Martin Behaim, showing the Far East as European geographers imagined it in 1492.

Figure 58: Mariner's astrolabe. Facsimile of the title page of *A Regiment for the Sea*, by William Bourne (1574). Courtesy of the Library of Congress.

Figure 59: Columbus title page. From *De Insulis nuper in mari Indico repertis* (Basel: I.G., 1494). Courtesy of the Jay I. Kislak Collection, Rare Book and Special Collections Division, Library of Congress.

Figure 60: The northern discoveries of John Cabot. Detail from the world map of Johannes Ruysch, from the 1508 reprint of the Rome *Geography* of 1507. Courtesy of the Rare Book and Special Collections Division, Library of Congress.

Figure 61: Dante's pear-shaped world. From Dante, *Commedia,* (Florence: Giunta, 1506). Reproduced from the original held by the Department of Special Collections of the University Libraries of Notre Dame.

Figure 62: Detail from the La Cosa chart (circa 1500). Courtesy of the Museo Naval, Madrid.

Figure 63: The Christ-bearer. Frontispiece from R. H. Major, ed, *Select Letters of Christopher Columbus* (London, 1870). Courtesy of the Library of Congress.

Figures 64 and 65: Ptolemy's southeast Asia (1478) and Cattigara. Facsimile detail from *Cosmographia. Roma, 1478* (Amsterdam: Theatrum Orbis Terrarum, 1966). Courtesy of the Geography and Map Division, Library of Congress.

Figure 66: The Cantino chart (1502). Biblioteca Estense Universitaria, Modena Italy.

Figure 67: The Caverio chart (circa 1504–1505). Bibliothèque Nationale de France.

Figure 68: Maximilian I on horseback. Hans Burgkmair (1508). Erlangen, Graphische Sammlung der Universität.

Figure 69: Maximilian's imperial standard. Hans Burgkmair (1510). Erlangen, Graphische Sammlung der Universität.

Figure 70: The world as double eagle. Georg Braun (1574). Herzog August Bibliothek Wolfenbüttel (K2.6).

Figure 71: Diagram for the assembly of a twelve-sheet wall map. From *Uslegung der mercarthen,* by Lorenz Fries (Strassburg: Grüninger, 1525 edition). Courtesy of the John Carter Brown Library, Brown University.

Figure 72: Copy of Waldseemüller's western hemisphere. Sebastian Münster (Brown 1514–1518). Department of Prints and Early Manuscripts, Munich Staatsbibliothek (Cod. Lat 10691).

Figure 73: Copy of the Waldseemüller map of 1507. By Peter Apian, in Solinus, *Polyhistor* (1520). Courtesy of the Rare Book and Special Collections Division, Library of Congress.

Figure 74: The modern world, Martin Waldseemüller (1513). Courtesy of the Norman B. Leventhal Map Center, Boston Public Library.

Figure 75: The New World, Martin Waldseemüller (1513). From the 1513 Strassburg edition of Ptolemy's *Geography.* Courtesy of the Geography and Map Division, Library of Congress.

Figure 76: The Carta Marina, Martin Waldseemüller (1516). Courtesy of the Jay I. Kislak Collection, Library of Congress.

Figure 77: World map, Johannes Ruysch (1508). From the 1508 reprint of the Rome *Geography* of 1507. Courtesy of the Rare Book and Special Collections Division, Library of Congress.

Figure 78: Western hemisphere, Jan de Stobnicza. From Stobnicza, *Introductio in Ptolemei Cosmographiam* (1512). Courtesy of the Rare Book and Special Collections Division, Library of Congress.

Figures 79 and 80: World map generally attributed to Sebastian Münster. From Grynäus, *Novus orbis regionum* (1532). Courtesy of the Rare Book and Special Collections Division, Library of Congress.

Figure 81: The New World bearing the name America. Detail from opening illustration of Appendix I.

Figure 82: The New World bearing no name. Detail from Figure 74.

Index

The Fourth Part of the World

The Epic Story of the Map That Named America

Toby Lester

Author Q&A

ABOUT THIS GUIDE

The following author Q&A is intended to help you find interesting and rewarding approaches to your reading of *The Fourth Part of the World*. We hope these enhance your enjoyment and appreciation of the book. For a complete listing of reading group guides from Simon and Schuster, visit BookClubReader.com.

The Library of Congress bought the Waldseemüller map in 2003 for $10 million, which is by far the most money the Library has ever paid for anything—and almost $2 million more than was paid at auction not long before for an original printed copy of the Declaration of Independence. Why was the United States willing to pay so much for a *map*?

It's the sole surviving copy of the map that literally gave America its name. That alone makes it worth the price. But it's also the first map to show the New World surrounded by water; the first map to suggest the existence of the Pacific Ocean; one of the first maps to show the full coastline of Africa; and one of the first to portray the entire globe, using a full 360 degrees of longitude. This is an astonishing set of milestones, which all together make this the first map to depict the world roughly as we know it today. It's the mother of the modern world map. In that light, $10 million seem like a bargain.

Columbus "discovered" the New World, right? So why aren't North and South America called North and South Columbia?

In 1507 the Italian merchant Amerigo Vespucci was much more of a celebrity in Europe than Columbus—who went to his grave in 1506 believing he'd reached Asia. Columbus, despite what we like to think today, had *not* revolutionized geographical thought. Instead, he assumed he'd confirmed what he'd seen on maps of his day: the age-old idea of a world that consisted of three parts, Europe, Africa, and Asia. But in the early 1500s Vespucci had reported the astonishing news that the New World extended thousands of miles below the equator, into a southern quadrant of the globe, where most geographers had assumed there was no land. When Martin Waldseemüller and his colleagues came across this news, they decided that what Vespucci had been exploring must be a part of the world new to Europeans: a fourth part of the world. Which is why, after naming the new continent in Vespucci's honor, they put the name America in the south, on what today is Brazil.

You say people in the Middle Ages didn't believe the world was flat. How do you know?

People have long known the world is round. In the second century A.D. the Greek geographer Claudius Ptolemy produced an entire book, titled *The Geography*, in which he taught readers how to think of the world as a globe, and to determine points of latitude and longitude on it, which they could then plot on grids that accounted for the curvature of the earth: map projections. One of the most popular medieval textbooks describing the earth and its place in the cosmos, too, titled simply *The Sphere*, pointed out the obvious: that a sailor atop the mast of a ship can see land approaching before the rest of the crew on deck. So the idea that people laughed at Columbus because they assumed he would fall off the edge of the earth is pure bunk. Europeans in 1492 had a good idea of the size and shape of the earth, and the court advisors in Portugal and Spain who initially rejected Columbus's plans did so because they knew full well that sailing west from Europe all the way to Asia would be impossible. Had Columbus not accidentally bumped into the New World when he did, he and his crew would almost surely have perished at sea.

You describe the map in your preface as "a constantly shifting mosaic of geography and history, people and places, stories and ideas, truth and fiction." Can you elaborate?

This whole book is an attempt to bring that idea to life. Today maps tells us where things are. But in the Middle Ages and the early Renaissance they offered people something much richer, stranger, and more complex. Medieval maps were sometimes called histories, for example, and historical texts were sometimes called maps; a world map therefore provided a picture not only of space but of *time*. It was an idiosyncratic, imaginative guide to the full drama of human history, played out against a geographical backdrop. The Waldseemüller map itself is a glorious patchwork of many different kinds of map, each of which has embedded in it all sorts of ideas and stories and points of view. It offers a kaleidoscopic way of peering back at the multifaceted history of geographical exploration, intellectual inquiry, imperial ambition, social change, cross-cultural exchange, and much more. The map was made at a hinge moment in history, when the Middle Ages were giving way to the Renaissance, and when the European world view was about to shift dramatically—and to my mind there's no more revealing way of seeing all of these forces as they collide than by looking closely at this one great map.

In the sixteenth century the astronomer Nicholas Copernicus famously proposed that the sun didn't revolve around the earth, but vice versa. You end your book by suggesting that when Copernicus saw America for the first time, as it was laid out on the Waldseemüller map, the sight prompted him to rethink the nature of the cosmos. Can you explain?

To understand why the sight of America on a map would help Copernicus rethink the cosmos, you have start with the idea of the earth as a globe at its exact center, surrounded by a set of concentric spheres: of water, air, and fire, and then of the various celestial objects. This was an ancient model, and in the Middle Ages scholars realized it had a fundamental problem: it suggested that the earth should be completely submerged in water. Which it obviously wasn't. Europe, Asia, and Africa—the contiguous landmass that made up the known world—were exposed to the air. By the late Middle Ages scholars had managed to explain this inconsistency away; to render life possible, they argued, God had displaced the earth from the exact center of the watery sphere, making it bob to the surface at one side. This meant, by definition, that the part of the globe that lay opposite the exposed part *had* to be submerged. But when Copernicus saw Waldseemüller's America—a giant new continent rising up out of the ocean on the other side of the globe from the known world—he realized that the traditional model of the cosmos could no longer hold, and that he was free to propose something new.